Global Environmental Studies

Editor-in-Chief

Ken-ichi Abe, Research Institute for Humanity and Nature, Kyoto, Japan

Series Editors

Daniel Niles, Research Institute for Humanity and Nature, Kyoto, Japan

Hein Mallee, Kyoto Prefectural University, Kyoto, Japan

This series introduces the research undertaken at, or in association with, the Research Institute for Humanity and Nature. Located in Kyoto, Japan, RIHN is a national institute conducting fixed-term, multidisciplinary and international research projects on pressing areas of environmental concern. RIHN seeks to transcend the common divisions between the humanities and the social and natural sciences, and to develop synthetic and transformative description of humanity in the midst of a dynamic, changeable nature. The works published here will reflect the full breadth of RIHN scholarship in this transdisciplinary field of global environmental studies.

Masaaki Okamoto · Takamasa Osawa ·
Wahyu Prasetyawan · Akhwan Binawan
Editors

Local Governance of Peatland Restoration in Riau, Indonesia

A Transdisciplinary Analysis

 Springer

Editors
Masaaki Okamoto
Center for Southeast Asian Studies
Kyoto University
Kyoto, Japan

Takamasa Osawa
Institute of Liberal Arts and Science
Kanazawa University
Kanazawa, Ishikawa, Japan

Wahyu Prasetyawan
Syarif Hidayatullah Islamic State
University
Banten, Indonesia

Akhwan Binawan
Perkumpulan Ara Sati Hakiki
Pekanbaru, Riau, Indonesia

This work was supported by the Research Institute for Humanity and Nature

ISSN 2192-6336   ISSN 2192-6344   (electronic)
Global Environmental Studies
ISBN 978-981-99-0901-8   ISBN 978-981-99-0902-5   (eBook)
https://doi.org/10.1007/978-981-99-0902-5

This Springer imprint is published by the registered company Springer Nature Singapore Pte Ltd.
The registered company address is: 152 Beach Road, #21-01/04 Gateway East, Singapore 189721, Singapore

Keywords
Peatland restoration · Environmental governance · Global warming · CO$_2$ emissions · Rural community research · Resource governance

Preface

Tropical peatland in Indonesia has drawn global environmental attention since the late 1990s when severe forest fires occurred and smoke from fires caused toxic haze not only in Sumatra and Kalimantan islands but also in peninsular Malaysia and Singapore. Researchers started to identify the carbon emission from fires as one of the most serious causes of global warming. With the severity of the issue, international environmental donors have rushed to support the Indonesian government in creating sustainable peatland governance, and the Indonesian government itself has started to implement various efforts to curb peat fires.

This book is the result of locally based transdisciplinary analysis and action-oriented research on the actual conditions of communities in and around the peatland and the implementation of sustainable peatland governance in Riau Province, Indonesia. Readers will be fascinated with the various approaches we used to understand and depict the local societies and the impacts of peatland governance on them. Readers will also easily understand the challenges that we are facing in addressing the peatland issues. I hope readers will have an interest on the issues and make a commitment to achieve sustainable peatland governance in one way or another in the future.

This research could not have been conducted without the continuous help and support from Riau University and the Peatland Restoration Agency (BRG) of the Republic of Indonesia, especially Prof. Dr. Almasdi Syahza as the head of the Institute for Research and Social Service (LPPM) of Riau University and Dr. Nazir Fuad as the head of BRG. The research is funded by the Tropical Peatland Society Project, "Toward the Regeneration of Tropical Peatland Societies: Building International Research Network on Paludiculture and Sustainable Peatland Management" (RIHN, Project No. 14200117, principal investigator: Osamu Kozan), the Research Support Enhancement Expenses (B) Research Activity Support (RIHN), Grant-in-Aid for Scientific Research (B) (JSPS), "Land Ownership and Peatland Restoration in Indonesia" (Project No. 19H04350, principal investigator:

Kosuke Mizuno), Grant-in-Aid for Early-Career Scientists (JSPS) "Comparative Study of Ethnicity in Eastern Sumatra: Resource Use and the Choice of Livelihoods" (Project No. 20K13293, principal investigator: Takamasa Osawa), and Academic Research Grant from Sompo Environment Foundation "Survival strategies of local people in large-scale peatland fires area in Indonesia" (Principal Investigator: Maho Kasori).

Kyoto, Japan   Masaaki Okamoto
Kanazawa, Ishikawa, Japan   Takamasa Osawa
Banten, Indonesia   Wahyu Prasetyawan
Pekanbaru, Riau, Indonesia   Akhwan Binawan

About This Book

This open access book is one in a series of four volumes introducing peatland conservation and restoration in Indonesia. It focuses on local governance, in particular on regional and local perspectives in Riau, the most peat-destructed province of Indonesia. The book fills a vital gap in the existing literature that overlooks social science and humanities perspectives. Written by authors from different disciplines and backgrounds (including scholars and NGO activists), the approaches to the topic are various and unique, including analysis of GPS logs, social media, geospatial assessments, online interviews (conducted due to the Covid-19 pandemic), and more conventional questionnaires and surveys of community members. The chapters cover an interdisciplinary understanding of peatland destruction and broadly offer insights into environmental governance. While presenting combined studies of established fieldwork methodologies and contemporary technology such as drones and geospatial information, the book underscores the significance of long-term and close collaboration with local residents who confront concrete environmental problems, which leads to better solutions.

Readers will gain a comprehensive understanding of the complexities surrounding peatland conservation and restoration and recognize the significance of locally inclusive approaches that use contemporary but accessible technologies to sustainably govern the globally important resource of peatland. That approach would be useful for other environmentally fragile but important regions and give some ideas to achieve the United Nations' SDGs for (1) No Poverty, (5) Gender Equality, (13) Climate Action, (15) Life of Land.

Contents

Chapter 1
Introduction

Masaaki Okamoto, Takamasa Osawa, Wahyu Prasetyawan, and Akhwan Binawan

Abstract This introductory chapter explains the background of the book. The book was the result of transdisciplinary collaborative research on tropical peatland problems in Indonesia. This chapter describes how the tropical peatlands have emerged as a new development frontier for plantation opening and have experienced serious ecological degradation, causing fire and international smoke damage, and also how the Indonesian government and international organizations have begun to conserve and restore tropical peatlands. The chapter argues that today's peatlands are a political arena involving diverse stakeholders including donors, central government ministries, local governments, environmental NGOs, forestry and plantation companies, and thousands of local communities as peatlands have become a contested space for plantation development and environmental conservation. And it is at the local level, especially the village level in and near the peatlands that the severe conflict of interests has occurred. There has not been much inter-disciplinary and transdisciplinary research on the peatlands at the local level. That is the reason why our research has focused on peatland conservation and restoration efforts at the local level, especially at the village level in this book.

Keywords Peatland restoration · Peatland conservation · Plantation · Riau · Indonesia

M. Okamoto (✉)
Center for Southeast Asian Studies, Kyoto University, Kyoto, Japan
e-mail: okamoto@cseas.kyoto-u.ac.jp

T. Osawa
Institute of Liberal Arts and Science, Kanazawa University, Kanazawa, Ishikawa, Japan

W. Prasetyawan
Syarif Hidayatullah Islamic State University, Tangerang Selatan, Banten, Indonesia

A. Binawan
Perkumpulan Ara Sati Hakiki, Jl. Kayu Putih, Perumahan Athaya IV, Pekanbaru, Riau, Indonesia

M. Okamoto et al. (eds.), *Local Governance of Peatland Restoration in Riau, Indonesia*, Global Environmental Studies,
https://doi.org/10.1007/978-981-99-0902-5_1

1.1   Aim of the Book

This book is part of a planned four-volume series that provides inter- and trans-disciplinary analyses of efforts to conserve and restore tropical peatlands in Indonesia. This book focuses on the governance of such efforts at the local level. In Southeast Asia, particularly in Indonesia, the rapid opening of tropical peatlands for development has resulted in dramatic ecological change and fire and international smoke damage. The carbon and methane emissions associated with peatlands conversion are considered a significant contributor to global warming. The conservation and restoration of peatlands have therefore become a global concern, and the Indonesian government and international aid organizations have begun to act. This book aims to empirically clarify how peatlands policies have been implemented at the local—particularly the village—level in the hope that this analysis can inform improved solutions in the immediate future.

Although a significant amount of research has been conducted on tropical peatlands, most of it takes a natural science approach. Numerous studies have demonstrated, however, that peatland conservation and fire prevention methods based on purely natural science research is often irrelevant when applied in local societies. To achieve sustainable peatland management, humanities and social science approaches are required to duly consider local social, political, economic, and ecological contexts. At the same time, proposals in academic papers and presentations are insufficient to tackle the imminent crisis of peatland destruction and severe fire. Action-oriented research conducted with the continuous involvement of NGOs and local communities, and the implementation of solutions based on such research results, are critical.

In this series on tropical peatlands, we define "trans-disciplinary" as the combination of multi-discipline approaches with the active involvement of local societies. In that vein, this book mobilizes political science, public policy, economics, anthropology, sociology, fish ecology, and fisheries science approaches while also incorporating GPS and drone spatial analyses as well as analyses of the implementation-oriented activities of NGOs working to conserve and restore peatlands.

1.2   Tropical Peatlands as a New Development Frontier

Tropical peatlands in Southeast Asia, especially in Indonesia, have become a frontier for the development of acacia and oil palm plantations. Peat soils are composed of carbonized organic matter resulting from centuries of sedimentation of plant remains in wetlands that have slow bacterial decomposition activity due to low pH waters. Due to their swampy nature, tropical peatlands have been generally considered unproductive and relatively inaccessible. In Indonesia tropical peatland areas were rarely inhabited or cultivated until the 1980s, when they were suddenly subjected to large-scale development plans. As arable land became less and less available in other Indonesian territories, however, the opening of peatlands began to proceed rapidly and massively during the 1990s (Mizuno et al. 2016). A typical example is the one million hectare mega rice project in Central Kalimantan peatlands that was launched at the end of the Suharto authoritarian regime. Although it aimed to transform peatlands into rice fields, the project was a spectacular failure and resulted in large swathes of dried barren fields (Shimamura 2012; Limin et al. 2007; Goldstein 2016). Despite this disastrous result, similar peat swamp projects—especially to develop plantation agriculture—spread throughout Indonesia. In order to prepare peatlands for acacia and oil palm cultivation, extensive channeling systems were built to drain water out of the swamps. This drainage activity dried peat soils below ground, rendering them vulnerable to fire. Fire was often deliberately used to clear peatlands for plantations, but in such desiccated conditions, it often quickly spread on both the dried surface and subsurface, causing increasingly severe smoke damage.

1.3   Loss of Peat and Peat Fires as a Global Environmental Problem

Peat soils store enormous amounts of carbon. Based on a comprehensive analysis of previous studies and data, Page et al. estimate that there are approximately 441,025 km^2 of tropical peatlands globally. Of this, 247,778 km^2, or 56%, is in Southeast Asia, with nearly half (206,950 km^2) spread across Indonesia. In terms of volume, of the world's 1758 Gm^3 of tropical peat, 77%, or 1359 Gm^3, is in Southeast Asia, with 65%, or 1138 Gm^3, in Indonesia (Page et al. 2011, p. 801). Tropical peat stores 89 Gt of carbon, of which 77%, or 69 Gt, is in Southeast Asia. Indonesia boasts the most significant amount, storing 57 Gt, or 65% of the total carbon content of tropical peatlands (Page et al. 2011, p. 809). All available evidence suggests that Indonesia has the most abundant tropical peatlands and carbon fixation globally.

The opening of peatlands for plantations does not automatically reduce their carbon stores, if the peatland is kept wet. Usually, however, plantation companies will dry out peatlands in order to manage plantations more efficiently. As mentioned, dry peatlands are susceptible to fire, which releases vast amounts of carbon stored in the peatlands. In 1997, decreased precipitation caused by the most significant El Niño event on record dried up peat and made tropical forests, including peatlands, more prone to ignition and fire spread. Massive fires in Sumatra, Kalimantan and other parts of Indonesia contributed more than 10% of global carbon emissions that year (Page et al. 2002, p. 61). The fires caused severe smoke damage to neighboring countries and the Singapore and Malaysian governments criticized the Indonesian government's lack of effective fire and smoke control countermeasures.

In 2015, large-scale fires erupted again following an El Niño event. Carbon emissions during the short period from September to October reached 273 Mt. This accounted for 45% of the total annual carbon emissions of Indonesia, Malaysia, and Papua New Guinea, where tropical peatlands are predominantly located (Niwa et al. 2021). These recent experiences, and the reasonable expectation of similar climate scenarios in the future, highlight the need to prevent peatland drying and fires. Maintaining peatland carbon storage capacity is critical to reducing greenhouse gas emissions, the primary cause of global warming.

1.4   The Beginnings and Challenges of Indonesia's Peat Conservation and Restoration Policies

In response to growing international calls for countermeasures against large-scale fires and smoke pollution, in 2009 the Indonesian government, under President Susilo Bambang Yudhoyono, enacted Law No. 32 of 2009.[1] The law required the formulation of environmental conservation and management plans, even down to the district level, and mandated that environmental assessments be conducted when spatial plans were formulated. The law also included a provision prohibiting all burning for land clearing, except for traditional burning of fewer than 2 hectares per household, and imposed penalties on violators. In 2011, President Yudhoyono issued Presidential Directive No. 10, which placed a moratorium on the issuance of new development permits in primary forests and peatlands within conservation, protection, and production forests. In 2013, Yudhoyono extended this moratorium for an additional two years and President Joko Widodo, newly elected in 2015, kept the moratorium in place, making it permanent in 2019 (Jong 2019). The moratorium, however, did not apply to previously permitted projects, to those that had been previously granted extensions, or to projects deemed essential for national development, such as those for geothermal power, oil, natural gas, electricity, rice, and sugarcane. The broad exceptions of the moratorium have meant that peatland clearing has continued within the scope of the law. Illegal peatland development has also continued unabated, as incumbent local heads seeking reelection issue land use permits to companies in exchange for campaign financing (Walhi 2017).

All of this led to the devastating fires and haze of 2015, the worst since 1997. This time, criticism came from beyond neighboring countries and pointed to the Indonesian government's lack of political will to tackle the fires. Indonesia's Ministry of Environment and Forestry blamed hotspots in the northern Malaysian peninsula for causing the haze in Malaysia, but the Malaysian government and the pro-environment international community as a whole dismissed this claim and demanded that the Indonesian government stop the fires and transboundary haze.

The international criticism put Joko Widodo in a tight spot. Due to his popularity and previous success as a small-town mayor and then governor of Jakarta, in 2014 he

---

[1] Yudhoyono dealt with the climate change issue more seriously than any previous president (Sinaga 2020, p. 164). For example, under Yudhoyono, Indonesia hosted COP13 in Bali in 2007, becoming the third country to host the Conference of Parties in Asia.

won the direct presidential election, but only by a slim margin. In 2015, he was not on good terms with the government party and his support base was quite weak (IPAC 2016, p. 1). If he failed to address international criticism stemming from peatland fire, his popularity might have suffered and the movement to remove him would have been emboldened. To demonstrate his determination to address the problem, he established the independent Peatland Restoration Agency (*Badan Restorasi Gambut*, BRG) in 2016. Recognizing the importance of peatlands to global warming, developed countries and international aid agencies stepped in to support the Agency.

With an initial 5-year mandate, the BRG designated 6.1 million hectares of undeveloped wetlands for conservation and approximately 12 million hectares of developed wetlands for restoration.[2] Its target was to restore 2 million ha of peat within five years by "3Rs" approach (rewetting dry peatlands, revegetation to restore tree cover, and revitalization of local economies). In FY 2017, 75 villages were designated as peat conservation villages (*Desa Peduli Gambut*, DPG), or villages where the residents initiate peat restoration activities. As the peat area of these villages amounted to 1.18 million ha, the BRG announced that it had met its goal for FY 2017. However, the designation alone did not indicate actual progress in its peat conservation and restoration efforts on the ground (Majaralhteras.com 2018). In fact, the DPG has not made much progress in restoring peatlands; according to the external evaluator of the program, this is largely due to the low abilities and multiple roles of the people facilitating restoration activities, and to the short period in which they are expected to complete their tasks, only 10 months.

As we have described, peatland fires and smoke damage drew new international attention to Indonesian peatlands. At the same time, global interest in curbing global warming was increasing. Both of these factors led to the creation of Indonesia's Peatland Restoration Agency, which increased awareness of the importance of peat conservation and restoration and gained the support of many international donors. Before the end of its mandate in December 2020, the BRG boasted of several achievements, most notably that it had rewetted 835,288 hectares of peatland surrounding plantation concession areas (achieving 94% of its initial target).[3] The achievements seemed impressive, but the actual restoration process was halfway through. Not BRG but the Ministry of Environment and Forestry had the authority to restore the peatland in the concession areas and claimed that the ministry rewetted around 3.5 million ha of peatland. The ministry did not show any concrete data on the rewetted area, however (Eyes on the Forest 2019; Pramita 2022). And the fire

---

[2] According to the BRG, tropical peatlands in Indonesia cover an area of 18.9 million ha, an area approximately 50 times the total area of Japan. However, the government often changes the official calculation of the total area of peatland; indeed, it is difficult to precisely measure due to inaccessibility and challenges in determining which lands to consider peatland.

[3] The BRG also reported that it had provided technical supervision and assistance to 186 plantation areas amounting to 538,439 hectares (96.9% of the target); it had supported 640 DPG villages with an area of 4.6 million hectares; and that 2295 community groups with approximately 118,576 persons were involved in the DPG program (BRG 2020).

continued even in the peatland outside the concession area which BRG was expected to have the authority to restore. An analysis by an environmental NGO reported that the fire occurred in 69% of the peatland area outside the concession area from January to December 2019. The analysis and other reports suggested that the fires might have occurred because the canal blocking was installed in inappropriate places and some of the canal blocking were broken and unrepaired (Nugraha 2020; Prakoso 2020). Recognizing the work yet to be done, the BRG was transformed into a reformed agency called the Peatland and Mangrove Restoration Agency (*Badan Restorasi Gambut dan Mangrove*, BRGM).

1.5 Tropical Peat Governance Challenges and Recommendations

Today's tropical peatlands have become a contested space: they are a place for plantation development and a global center for environmental conservation. They are a political arena involving extremely diverse stakeholders including donors, central government ministries, local governments, environmental NGOs, forestry and plantation companies, and thousands of local communities. While everyone recognizes the importance of peatlands, not everyone easily agrees with strict rules of peat conservation and restoration. Even though there may be a consensus to conserve and restore peatlands in a certain local area, determining the detailed borders of conservation and restoration zones, and the stakeholders in charge of each zone, is challenged by different actors vying for their concrete interests and survival in that space.

This book aims to elucidate the complicated negotiations, conflicts, and accommodations that occur among various peatland stakeholders. For this, we conducted a detailed study of Rantau Baru Village in Riau Province (see Column 1), offering insights from different districts and villages for comparison with our main study. We chose Rantau Baru village in Pelalawan district, Riau province, Indonesia as the research site for five reasons. Firstly, the peatlands extended in the village hinterland have been rapidly drained for the plantation opening over the past two decades and have been damaged by frequent fires. Villagers have experienced these severe impacts. In that sense, Rantau Baru is at the forefront of the peatland problem. Secondly, the village is an old village, and the villagers have lived in the peat environment for a long time. We can explore customary peatland use and the emerging conflict between villagers and plantation companies on peatland use. Thirdly, we can highlight the importance of the fishery sector for peatland preservation that has not been paid due attention to by previous research and policies. Most villagers in RB have been dependent on the sector, and peatland preservation is the key to the sustainable management of the community-based fishery. Fourthly, we have really felt and recognized the desperate need of villagers to tackle various peatland issues and to come up with sustainable management of peatland and the coexistence between environmental protection and economic affluence. And finally, we have received strong support from the Pelalawan district government and village government for our trans-disciplinary study in this village.

Although the term "transdisciplinary" is used in a variety of ways depending on the era and field of study (Bernstein 2015), this book orients toward concrete engagement with the peat problem, commitment to the local communities, and collaboration among diverse stakeholders. The researchers contributing to this book come from diverse academic and practical backgrounds. They include political scientists, sociologists, economists, ecologists, anthropologists, and local environmental NGO officials, all of whom focus on how peat problems occur and should be solved in the local space. Inquiry and resolution both require a multi-layered understanding of local social conditions and livelihoods as well as trust and collaborative relationships with residents and local governments. The researchers were therefore not detached from the research subject, the local space, or the inhabitants. Rather, the studies in this book emerged through mutual exchange, discussion, and practice about peatlands and their protection. This transdisciplinary approach, emphasizing the exchange of different kinds of knowledge, has an important role to play in analyzing the current state of governance of peatlands and addressing the peat problems in each local space.

Both top-down and bottom-up approaches to peat conservation and restoration are necessary, but the top-down approach continues to dominate in Indonesia. At the same time, the voices and demands from the communities within and around tropical peat areas remain muted. This book therefore emphasizes the significance of actual peatland use and spatial perceptions of existing customary communities, and strongly advocates for bottom-up practices.

The transdisciplinary analyses of the chapters reveal that peatlands should not be the only focal point when considering peatland conservation and restoration. Conservation of river ecosystems and sustaining fishery livelihoods are also critical to the future of peat. Although many previous studies have highlighted the importance of bottom-up approaches and local perspectives in sustainable environmental governance, we do not consider such terms as fixed, but rather as dynamic. Utilizing the strengths of the humanities and social sciences, this book presents how peat conservation can be realized amid the transformations occurring in local societies. At the same time, we utilize the natural sciences in order to describe the rapid changes occurring in the natural environment, which cannot be understood through local experience alone. Thus, we aim to support the many Indonesian people who seek to establish sustainable peat conservation societies that are grounded in sound science and appropriate for local communities.

1.6   Structure of the Book

The book is comprised of eleven chapters and a conclusion. Three columns explain various technical aspects of our transdisciplinary research, including our data gathering in the main research site, the process of drone mapping, and our online community research methodology.

Chapter 2 reveals the complexities of the politics of mapping forests and peatland at the national, provincial, and local levels. Several cases demonstrate how multiple stakeholders jockey for power and protect their interests in the mapping process and how such processes have failed to create a consensus map at each level. The authors conclude that while pro-environment mapping is not easily accepted, the rubber-stamp mapping of illegal forest and peatland opening is no longer accepted uncritically, as it once was.

With Chap. 3, we move to the main part of the book, the transdisciplinary study of Rantau Baru Village in Riau Province. Chapters 3–5 provide sociological and ecological analyses of the basic conditions of the human inhabitants and the environment in and around the village. Chapter 3 is a study of land use governance in Rantau Baru, where lands and rivers in the traditional territory have been historically treated as ancestral common space, but where, in recent years, peatlands have been sold to outsiders. Describing the transformation of land use from the past to the present, this chapter explores the causes of the commodification of traditional territory. Chapter 4 documents the relationships between scientific and local names of fishes found in the area, and infers recent degradation in freshwater ecosystems in the mid-Kampar River Basin based on a survey conducted in Rantau Baru. It also uses comparative review of studies and methods of effective freshwater protection measures in other developing countries in order to assess the potential of establishing effective freshwater protection measures in our study area. Chapter 5 uses a GPS logger to analyze fishing patterns and practices in Rantau Baru, demonstrating the essential function of the submerged forests of the peat environment in fishing livelihoods. Presenting the case of a nearby village, the author proposes fishing tourism as a potential sustainable livelihood.

Chapters 6–10 provide critical analyses of Rantau Baru and other villages from the perspectives of anthropology, gender, public policy, economy, and geography. Chapter 6 questions the emphasis placed on indigenous local knowledge (IK) and local wisdom (*kearifan lokal*) in recent peatland restoration efforts, highlighting the potential risks of researcher over-emphasis of isolated local wisdom in traditional knowledge and practices. The author suggests the need to investigate the dynamism, interaction, and transformation of knowledge beyond the framework of local areas, which can result in a better understanding of local realities and build a broader network of cooperation. Chapter 7 investigates the differentiated knowledge and roles of both men and women in peatland management in Rantau Baru Village. It finds that although men are significantly more knowledgeable about peatland ecology and management than women and peatland agricultural activities are dominated by men, gender roles are more evenly distributed in fishery activities. Women and men play complementary roles in household maintenance and reproduction, but women do not participate nearly as much as men in the public sphere of sociopolitical activities. The study provides new insight into the community's knowledge of peatland dynamics according to gender, and the potential roles of both male and female community members in peatland restoration.

Using the case of Pelalawan District, Chap. 8 investigates how village governments have utilized their budgets for peatland restoration and fire prevention since the enactment of the 2014 Village Law and to what extent local communities have been involved in the project planning process. The author provides a rather grim picture of the law's impacts on peatland restoration, but also describes how some villages have launched interesting programs even with small budgets. Chapter 9 assesses the Willingness to Pay (WTP) for environmental conservation of aquatic and peatland ecosystems in Rantau Baru. It uses a contingent valuation method to measure how villagers value each ecosystem and an ordinary least square method for estimation. It finds that WTP for conserving fishing areas is closely associated with household expenditures, while WTP for conservation of peatlands is associated with education and weakly associated with household expenditures. Using maps created in collaboration with the villagers, Chap. 10 discusses the position of Rantau Baru as an *adat* community and the village's current predicament. While various laws assert the importance of the rights of *adat* communities, the maps clearly show legal and illegal encroachment on village territory by oil palm companies and the resulting degradation of peatlands. The authors point out that respecting the rights will lead to improved livelihoods and better management of peat environments.

Chapters 11 and 12 study peatland communities in different parts of Riau Province. Analyzing a multi-ethnic local community in Pelalawan District, Chap. 11 finds that peatland conservation policies and related livelihood improvement programs involve only a small segment of richer non-peatland landholders, and suggests that existing conservation programs may thus accelerate existing economic disparities. Chapter 12 examines the peatland management of local communities in Kepulauan Meranti District in Riau that participate in a social forestry scheme. The chapter argues that the introduction of social forestry and ecosystem services valuation is not sufficient to achieve meaningful results; the active involvement of community members in the ecosystem services valuation process (including the mapping) is required to achieve sustainable peatland management.

In the conclusion, we propose that the convergence of a long-term future vision of sustainable peatland governance among the government, academia, and communities is paramount, and we dedicate our efforts to this goal.

References

Bernstein JH (2015) Transdisciplinarity: a review of its origins, development, and current issues. J Res Pract 11(1):art R1

BRG (Badan Restorasi Gambut) (2020) Siaran pers: tugas dan nakhoda baru BRG. https://brgm.go. id/siaranpers/tugas-dan-nakhoda-baru-brg/. Accessed 10 Sep 2022

Eyes on the Forest (2019) Kebakaran gambut masih menghantui, restorasi gambut sudah sejauh mana? http://eyesontheforest.or.id/reports/kebakaran-gambut-masih-menghantui-restorasi-gambut-sudah-sejauh-mana. Accessed 1 Dec 2022

Goldstein J (2016) Carbon bomb: Indonesia's failed Mega Rice Project. Arcadia no 6. https://doi. org/10.5282/rcc/7474

IPAC (Institute for Policy Analysis of Conflict) (2016) Update on the Indonesian military's influence. IPAC Report no 26, IPAC, Jakarta

Jong HN (2019) Indonesian ban on clearing new swaths of forest to be made permanent. Mongabay. com, 10 June. https://news.mongabay.com/2019/06/indonesian-ban-on-clearing-new-swaths-of-forest-to-be-made-permanent/. Accessed 11 Oct 2022

Limin SH, Jentha, Ermiasi Y (2007) History of the development of tropical peatland in Central Kalimantan, Indonesia. Tropics 16(3):291–301. https://doi.org/10.3759/tropics.16.291

Majaralhteras.com (2018) Laporan tahunan BRG 2017: kesadaran masyarakat melindungi lahan gambut meningkat. Majaralhteras.com, 6 Feb. https://majalahteras.com/laporan-tahunan-brg-2017-kesadaran-masyarakat-melindungi-lahan-gambut-meningkat. Accessed 15 Sep 2022

Mizuno K, Fujita MS, Kawai S (eds) (2016) Catastrophe and regeneration in Indonesia's peatlands: ecology, economy and society, Kyoto CSEAS series on Asian studies, vol 15. NUS Press, Singapore; Kyoto University Press, Kyoto

Niwa Y, Sawa Y, Nara H et al (2021) Estimation of fire-induced carbon emissions from Equatorial Asia in 2015 using in situ aircraft and ship observations. Atmos Chem Phys 21(12):9455–9473. https://doi.org/10.5194/acp-21-9455-2021

Nugraha I (2020) Empat tahun BRG: daya dan upaya pulihkan gambut negeri. Mongabay, 28 Jan. https://www.mongabay.co.id/2020/01/28/empat-tahun-brg-daya-dan-upaya-pulihkan-gambut-negeri/. Accessed 1 Dec 2022

Page SE, Siegert F, Rieley JO et al (2002) The amount of carbon released from peat and forest fires in Indonesia during 1997. Nature 420:61–65. https://doi.org/10.1038/nature01131

Page SE, Rieley JO, Banks C (2011) Global and regional importance of the tropical peatland carbon pool. Glob Change Biol 17(2):798–818. https://doi.org/10.1111/j.1365-2486.2010.02279.x

Pramita D (2022) Adu klaim restorasi gambut. Tempo, 30 July. https://majalah.tempo.co/read/ lingkungan/166553/menguji-klaim-keberhasilan-retorasi-gambut. Accessed 1 Dec 2022

Prakoso A (2020) Nasib restorasi gambut Indonesia. Pantau Gambut, 28 Oct. https://en. pantaugambut.id/publications/the-fate-of-indonesias-peat-restoration. Accessed 1 Dec 2022

Shimamura T (2012) Nettai deitan shicchi no gainen (Overview of tropical peatland). In: Kawai S, Mizuno K, Fujita M (eds) Nettai baiomasu shakai no saisei (Recovery of tropical biomass society). Kouza, seizon kiban ron, vol 4. Kyoto University Press, Kyoto, pp 104–121

Sinaga LC (2020) Assessing the commitment of Indonesian Government towards climate change policy: the Yudhoyono Presidency 2004–2014. Politica 11(2):163–182. https://doi.org/10. 22212/jp.v11i2.1752

Walhi (2017) Moratorium 25 tahun menghentikan deforestasi dan menyelesaikan konflik. Walhi, 11 May. https://www.walhi.or.id/index.php/moratorium-25-tahun-menghentikan-deforestasi-dan-menyelesaikan-konflik. Accessed 18 Aug 2022

Chapter 2
Contentious Politics of Mapping for (De)forestation in Indonesia: From the National to Provincial and Community Levels

Masaaki Okamoto, Made Ali, and Kazuo Watanabe

Abstract Democratization and decentralization have accelerated the dispersion of power, even in the cartography of Indonesia. Different actors from the national to the provincial and community levels have started to create maps that reflect their own purposes and interests. As far as forest and peatland areas are concerned, some maps aim to reconfirm the reality of deforestation and the opening of peatland for plantations. Others aim to reclaim the forest and peatland. Without clear and binding regulations on the mapping, a plurality of maps has prevailed at all levels in Indonesia. This is increasingly reinforced by the fact that map-based land use has become the norm of spatial planning, making the creation of a legally binding map quintessentially important for any stakeholder. The cases from the national, Riau provincial, and community levels in this chapter demonstrate that all the mapping processes have failed to create a consensus map: while pro-environment mapping is not easily accepted, the rubber-stamp mapping of illegal forest and peatland opening is no longer simply accepted as before.

Keywords Mapping · Riau · One Map Policy · Peatland

2.1 Introduction

Mapping, or cartography, is political. This view has become quite common since the introduction of critical cartography by Harley (1989). Drawing on an eclectic combination of Foucault's discourse and Derrida's rhetoricity of texts, Harley persuasively elucidates the existence of power even in a purely scientific map.

M. Okamoto (✉) · K. Watanabe
Center for Southeast Asian Studies, Kyoto University, Kyoto, Japan
e-mail: okamoto@cseas.kyoto-u.ac.jp

M. Ali
Jikalahari, Jl. Kamboja, Delima, Tampan, Pekanbaru, Riau, Indonesia

© The Author(s) 2023
M. Okamoto et al. (eds.), *Local Governance of Peatland Restoration in Riau, Indonesia*, Global Environmental Studies,
https://doi.org/10.1007/978-981-99-0902-5_2

13

Drawing a border line on a modern map might lead to a unilateral confirmation of ownership. If that border is claimed by a state, it can become an officially sanctioned border without any involvement of the real owners of the territory or the communities living in the border area. In that sense, the transformative power of a map and the lines on it reside not in the map, but rather, as Scott (1998, pp. 87–88) says, "in the power possessed by those who deploy the perspective of that particular map," the most powerful of which is the state. A powerholder can transform land ownership or utilization by claiming land based on the map that he created. Or, to borrow from Winichakul (1997, p. 110), a map is a model for what it purports to represent. Therefore, every state tends to monopolize map-making power and cartography is nationalized. The state guards its knowledge carefully: maps have been universally censored, kept secret, and falsified (Harley 1989, p. 12).

Recently, however, with the new governance ideas of decentralization and participation, and the development of technology such as Global Positioning System (GPS), satellite images, and drones, a movement by both local governments and communities to create locally initiated maps has emerged. Maps created by communities with the help or facilitation of academics and/or NGOs often echo the counter and participatory map movement to revise, discredit, or negate official maps by the state or to visualize the discrepancies between official maps and actual land utilization in order to achieve a bottom-up spatial justice (Peluso 1995; Radjawali et al. 2017).

This might be the common knowledge about the political history of mapping, especially in Western societies. While the Indonesian case also follows this trend in general, its mapping history naturally has its own contextual characteristics. The Indonesian state, especially under the authoritarian Suharto regime (1966–1998), monopolized the mapping power, but different ministries in Jakarta made their own authoritative maps without coordination, reflecting their own interests and the interests of Suharto cronies' and other powerholders. The plurality of official maps weakened the authority of each map, accelerated massive land grabs by powerholders, marginalized traditional customary communities by stripping them of access to forest resources, and resulted in environmental destruction. Commencement of the democratization and decentralization era transformed this. As mapping and planning power became democratized and decentralized, bottom-up and/or participatory mapping and planning have been introduced and have played a role in countering centralized mapping and actual power. Local governments have created their own maps and planning to serve locally entrenched vested interests, while local communities and NGOs have engaged in participatory mapping to achieve environmentally conscious or traditional community-conscious mapping and planning. Today, the relationship between the reality and the map, and among different maps, has become more complex than during the Suharto period. While it was difficult to apply Scott's (1998) understanding of mapping power even during the Suharto era, today it is far more difficult.

The politics of mapping and land grabbing are such important issues in Indonesia that numerous works have been published, including on the following: the birth and impact of "political forest" as state-designated forest (Peluso and Vandergeest 2001,

2020; Kelly and Peluso 2015), the large-scale land grabbing during the authoritarian regime (McCarthy 2000; Brockhaus et al. 2012), the One Map Policy attempt to rectify the plurality of maps (Nuhidayah et al. 2020), the consistent power of the Ministry of Forestry over land planning and management (Barr et al. 2006a; Gellert and Andiko 2015; Maryudi 2015), the changing forest planning and mapping policies (Wardojo and Masripatin 2002; Santoso 2003), how decentralized planning and local elections accelerated deforestation (Suwarno et al. 2015), the political economy of the conflicts between the central government and local governments concerning spatial planning (Setiawan et al. 2016; Setiawan et al. 2017 on Central Kalimantan; Suprapto et al. 2018 on Riau), and the positive and negative impacts of counter-mapping and participatory mapping in different areas (Pramono et al. 2006 on West Kalimantan; Widianingsih and Morrell 2007 on Solo, Central Java; Wollenberg et al. 2008 on Malinau, East Kalimantan, Radjawali et al. 2017; Dewi 2016; de Vos 2018 on Sambas, West Kalimantan; Tilley 2020).

The focus of these previous works tends to be on the dynamics of mapping at one level, whether it be national, local, or community. They therefore do not adequately elucidate the different mapping power dynamics and the various dynamic interactions between the mapping and the reality in the field. Building on these previous works, this chapter describes various mapping initiatives that are born from and carry different motivations, from the changing national mapping policies and the provincial spatial planning process to specific cases of participatory planning implementation in peatland-dominant Riau, Indonesia. In all the mapping processes at these three levels, conflict among powers and interests, and between the pros and cons of creating environmentally or peat-conscious maps, results in the absence of a consensus map at any level. This chapter argues that the lack of such a consensus map is a sign of the severity of existing conflicts and stakeholders' rising awareness of the power of mapping. This study uses text data analysis of legal and NGO documents and online news as well as spatial data analysis of satellite and drone images.

2.2   Centralized Mapping for Deforestation

The 32-year Suharto authoritarian regime (1966 to 1998) achieved a rather high economic growth partly by exploiting the country's natural resources, such as oil and gas, coal, and forests. This accelerated deforestation, especially on Sumatra and Kalimantan islands, and revealed a weak commitment to environmental conservation and forest-area community sustainability. Land-use planning under Suharto produced centrally driven, poorly implemented plans, and the forest estates were severely degraded during the Suharto period (Jepson et al. 2002; Wollenberg et al. 2008). According to Wollenberg et al. (2008), there were two phases of national land-use planning during the Suharto era: centralized planning through the early 1990s and limited decentralized planning by the late 1990s.

In 1967, the Suharto regime issued Basic Forest Law No.5 and embarked on a territorial strategy that designated 143 million ha, or three-quarters of the nation's total land area, as a state "forest zone" (*kawasan hutan*) (Barr et al. 2006b, p. 1). As far as the "Outer Islands," such as Sumatra and Kalimantan, were concerned, most land areas were categorized as *kawasan hutan*. For example, the total land area of Riau Province (Sumatra) was designated *kawasan hutan*. This large-scale designation aimed to bring foreign investment into the forest sector (McCarthy 2002, pp. 869–870). That is why forest land use planning came to have a significant influence on land use planning as a whole, with the Department of Forestry (or the Ministry of Forestry since 1981) becoming the sole government agency in charge of land-use planning by 1992 (Chakib 2014, pp. 14–15). The ministry was so powerful that it was referred to as "the Golden Ministry" in reference to the formal and informal incomes available to its employees (Kelly and Peluso 2015, p. 487). The main mission of the ministry was not to protect forests, but rather to utilize or exploit them. Weak concern for forest protection was clearly illustrated by the fact that of the 30,000 civil servants in the ministry, only 477 were employed in the forest protection and conservation directorate (McCarthy 2000, p. 95).[1]

In 1984, at the request of the Ministry of Home Affairs, the Ministry of Forestry produced a map of forest use called Consensus-Based Forest Land Use Planning (*Tata Guna Hutan Kesepakatan*, TGHK) (at a scale of 1:500,000). The map classified forest land according to function, as either conservation forest (*hutan konservasi*), protection forest (*hutan lindung*), production forest (*hutan produksi*), or convertible production forest (*hutan produksi yang dapat dikonversi*). Much of this designation was done with little regard for conditions on the ground, such as the existence of customary communities. For example, 96.4% of the land area of Riau Province was categorized as forest area. A few years after a new TGHK forest use map was updated and revised to a scale of 1:250,000, it became the country's base map, despite ongoing inaccuracies and problems (Chakib 2014, p. 14; Suprapto et al. 2018, pp. 197–198).

In 1992, the Suharto regime enacted a new law on the spatial planning process to "increase efficiency of spatial utilization, spatial quality, harmony of spatial utilization with environment, harmony in regional growth, development equalization, national unity and integrity as well as social welfare" under the Coordinating Board for National Spatial Use Management (*Badan Koordinasi Penataan Ruang Nasional,* BKPRN) and the National Development Planning Board (*Badan Perencanaan Pembangunan Nasional,* BAPPENAS). The law initiated a partial decentralization of mapping power by obliging the provinces to design new spatial plans, obliging districts and cities to have their own plans in line with the provincial plans, and by giving citizens the right to be informed about the spatial plan and to participate in the planning process.

---

[1] One of the most striking facts is that even as of 2019, the Ministry of Forestry (now the Ministry of Environment and Forestry) still had 125.93 million ha under its authority as forest area.

There was a synchronization process (*paduserasi*) between the provincial spatial policies and TGHK in order to create a spatial plan for each province (*Rencana Tata Ruang Wilayah Provinsi*, RTRWP).[2] The extent of the decentralization of spatial planning was quite limited, however. Mapping was concentrated at the provincial level, and the finalization of the integrated maps was contentious because of disharmony and incompatibility with the TGHK maps. The Ministry of Forestry largely stalled and resisted the efforts of the planning boards. Top-down mapping prevailed (Chakib 2014, p. 15; Wollenberg et al. 2008). This did not mean that the spatial plan proposed by the Ministry of Forestry was smoothly implemented, however. A number of factors hampered the implementation: (1) inaccurate and inconsistent maps, produced under various spatial planning policies; (2) poor coordination between the Ministry of Forestry and other ministers; (3) poor coordination among district, provincial, and central bodies; (4) presidential decrees that took precedence over existing plans; (5) a lack of local government capacity or will; (6) vested political and business interests; (7) a lack of financial resources; or (8) simply the lack of a plan (Wollenberg et al. 2008).

The National Coordinating Agency for Survey and Mapping (*Badan Koordinasi Survei dan Pemetaan Nasional*, Bakosurtanal) directly under the president was assigned the duty to coordinate the planning and implementation of a national survey and mapping, but it was too weak compared to other ministries and agencies—and Suharto's cronies—to create an integrated spatial plan and implement it. Different official maps produced by different ministries and agencies coexisted with no coordination: *Bakosurtanal* had a topographical map, the Ministry of Forestry had a forestry map and a forest product utilization map, the Geographical Survey Institute had a plantation map, the Ministry of Energy and Natural Resources had a map of mineral and coal concessions, and the Ministry of Labor and Immigration had a map of immigration areas. All these official maps were created with such a rough scale (of 1:250,000) that they were not practical for detailed land use planning.

The result of this top-down and uncoordinated rough mapping and spatial planning process and feeble implementation of plans was large-scale land grabbing by timber conglomerates and later by the pulp and oil palm plantation companies of Suharto's cronies and those with strong connections to powerholders in Jakarta (Barr et al. 2006b, p. 1). Local networks of power and interests joined in the grabbing at a smaller scale (McCarthy 2002). With cross-ownership, hidden deals, and silent partnerships, ownership of forest concessions was anything but transparent. In late 1998, the Ministry of Forestry and Plantation, the successor of the Ministry of Forestry, revealed that just twelve companies controlled virtually all of Indonesia's forest concessions, with three groups controlling more than 2 million ha each (McCarthy 2000, pp. 105–106). These actors not only controlled the concession areas on the maps, but also carried out illegal logging in areas that had not been approved by the ministry. Organized syndicates of illegal loggers had also long been

---

[2]Riau Province (later divided into the two provinces Riau and Riau Islands) and Central Kalimantan Province failed to achieve synchronization between the provincial spatial plan and TGHK.

active in most timber-producing provinces. By the end of the Suharto era, Indonesia had lost roughly one third of its forest cover since 1967 (Barr 2006, p. 28). The map prepared by the Ministry of Forestry had been used to suppress the customary communities' historical entitlement to the forest to benefit the cronies.

2.2.1 One Map Policy and the Dispersion of Mapping Power

The fall of Suharto and the commencement of the democratization and decentralization era introduced new institutional rules of the politico-economic game, dispersing power from the executive to the legislative and from the national to the local levels. Mapping power has also been dispersed, with different actors initiating map making in an uncoordinated way, causing the plurality of maps to become even more severe. The most challenging issue for a democratic Indonesian government in the field of mapping and spatial planning, therefore, is to address the plurality of official maps. In 2007, a new law on spatial planning introduced more transparent and participatory bottom-up planning procedures, but that did not solve the plurality problems. Different ministries in Jakarta insist on using the maps that reflect their own vested interests. It was only in 2009 that the central government began to rectify the uncoordinated mapping and planning conditions. That year, President Susilo Bambang Yudhoyono promised to reduce CO_2 emissions by 26% by 2020 in the G20 meeting in Pittsburgh, USA. Acknowledging the discrepancies in forest areas according to the Ministry of Forestry's map and the Ministry of Environment's map in a cabinet meeting in December 2010, President Yudhoyono ordered related ministries to create a base map. This marked the start of the One Map Policy in Indonesia (Karsidi 2016).

The Geospatial Information Agency (*Badan Informasi Geospasial*, BIG), the successor of *Bakosurtanal* was put in charge of the One Map Policy. After the fall of the 32-year authoritarian regime, Indonesia not only began to democratize, but also to decentralize. Local governments, especially district and city governments, were given wider authority, and began overissuing permits for plantations and concessions for mineral resources to private companies based on inaccurate maps of the permit and concession areas (Setiawan et al. 2016). Rampant issuance of permits and concessions accelerated deforestation; decentralization also accelerated illegal logging and deforestation (Smith et al. 2003; Abood et al. 2014). The hugely challenging task before BIG was to create a base map that integrated all the maps, including those used to issue local permits and concessions, into a single authoritative map.

First, BIG calculated the total area of Indonesia as 192.257 million hectares. It then overlaid the map produced by the Ministry of Energy and Mineral Resources with maps produced by the Geographical Survey Institute, the Ministry of Forestry, and the Ministry of Labor and Immigration. In doing this, it found that the maps— and the associated authority of the entity that created each map—overlapped in a total area of 20.862 million hectares, or 11% of the total land of Indonesia. Table 2.1

Table 2.1 Mining concession and the plantation permit areas where authorities overlap (Karsidi 2016, p. 27)

	Total area (Ha)	Overlapping area (Ha) hectare	Overlapping area as percentage of total percentage
Sumatera	480,793.3	37,298.7	7.8%
Jawa-Bali	135,218.3	328.2	0.2%
Kalimantan	544,150.1	123,932.8	22.8%
Maluku	78,896.5	543.9	0.7%
Nusa Tenggara	67,290.4	1,271.6	1.9%
Sulawesi	188,522.4	12,646.1	6.7%
Papua	110,928.7	27,653.0	24.9%
Indonesia	1,913,578.7	208,625.0	10.9%

details the total hectarage of mining concession and plantation permit areas on seven islands, as well as the percentage of land where authorities overlap in these areas. The table clearly demonstrates how different ministries have created maps according to their own interests without any concrete and serious coordination between the ministries.

The overall purpose of the One Map Policy is to collect and overlay 85 thematic maps created by 19 ministries and 34 provincial governments to create a base map to a scale of 1:50,000 (Sarbini et al. 2017). Recognizing the importance of one map to Indonesia's economic development, Joko Widodo prioritized the One Map Policy by issuing Presidential Decree No. 9/2016 on the acceleration of the One Map Policy (Anwar 2018). The decree regulates that Indonesia will create a base map by 2019 with input from environmental NGOs such as WWF (World Wildlife Fund) and Wetlands International. But, the One Map Policy faces technical obstacles to the collection and management of geospatial data, such as less precise and less detailed old maps of the agencies, a lack of geospatial information and guidelines, varying technologies and capabilities at the local levels, and administrative obstacles stemming from the conflicting vertical and horizontal interests of the different layers of government and their respective agencies. Line ministries believe that land cover maps should be produced by their own ministries rather than BIG. For example, the Ministry of Forestry asserts that its knowledge of the condition and classification of the forests is based on their own groundwork, rather than that of BIG, which depends only on interpretation of data via digital information (Nuhidayah et al. 2020, pp. 381–382).

As far as peatland is concerned, the Indonesian government has created a new map for the conservation of peat. Just before the end of his presidential term, President Yudhoyono issued government regulation No. 71/2014 on the conservation and management of peat ecosystems. In 2015, his successor Jokowi was forced to demonstrate—at least superficially—his strong will to protect the forests, including peatlands, after the devastating fires that occurred in peat areas in that year. In January 2016, Jokowi merged the Ministry of Forestry and the Ministry of Environment into the new Ministry of Environment and Forestry and established the independent Peatland Restoration Agency (*Badan Restorasi Gambut*, BRG)[3]. In December 2016, he revised the above government regulation to further advance peat conservation. The new government regulation (No. 57/2016) designated the Peat Hydrological Unit (*Kesatuan Hidrologi Gambut*, KHG) as the unit of peat area protection and divided peat areas into two categories based on their function: conservation or cultivation. Based on this regulation, the Ministry of Environment and Forestry issued a peat map that demarcates 865 KHGs in a total area of approximately 25 million ha (with roughly half designated as conservation areas and half as cultivation areas) across Indonesia.

Unexpectedly, by mid 2019, BIG did create a base map, (which included the KHG areas) from the 85 existing maps. However, access to this map is limited to central and local government stakeholders and it was not consensually created with the acceptance of all the ministries, agencies, and local governments. This base map is just a starting point for discussion among the stakeholders and it is constantly changeable and renewable. The One Map is not a fixed map for land use planning in Indonesia and it will not invalidate existing mining concessions or plantation permits that are not in accordance with the map. BIG expects that the ministries and agencies will follow the One Map when they renew concessions and permits. But each concession or permit is valid for a long period of time; for example, coal mining concessions are valid for eight years and oil palm plantation permits for 25 years. If the Jokowi government waits for the renewal of concessions and permits, Indonesia will not have one officially sanctioned map in the coming years and the plurality of maps will not change soon. While the Ministry of Environment and Forestry strongly intends to keep the designated forest area as large as possible, the Ministry of Energy and Mineral Resources aims to legalize mining concession areas in forest areas, and local governments are willing to create their own maps to expand non-forest areas for economic gain. In this sense, the One Map Policy that began with Yudhoyono's CO_2 reduction pledge and accelerated under the Jokowi government might not have an immediate impact on stalling deforestation or the opening of peatland for plantations, but still, it may have a chance of becoming a scientific and political tool to stimulate an environmentally conscious discourse in the long run.

2.3   Spatial Planning to Justify Deforestation in Riau

In connection with the central government's initiative to create one unified map, it has become essential for local governments, such as those of provinces, districts, and cities, to create spatial plans that correspond to the era of decentralization. Specifically, after the law on spatial planning was enacted in 2007, each province, district, and city was required to create its own spatial plans. Some local governments

---

[3] The presidential regulation on BRG stipulated that BRG would end its duties in 2020. Therefore the president created a new institution called the Peatland and Mangrove Restoration Agency (Badan Restorasi Gambut dan Mangrove, BRGM) in 2021.

struggled to reconcile the wide gap between actual land utilization and the official maps, and with the overlap among maps produced by central government ministries and local governments (Potter and Badcock 2001; Barr et al. 2006a; Setiawan et al. 2016). The following section will examine the case of Riau.[4] Although various actors are involved in the mapping politics at the Riau provincial level (Suprapto et al. 2018, p. 203), this chapter focuses on the Ministry of Forestry, the provincial government, and environmental NGOs because these are the main and visible actors that openly contest the legality and illegality of provincial mapping and spatial plans.

Riau Province is one of the most natural resource-rich provinces in Indonesia, with its oil resources, oil palm plantations, and acacia plantations. Therefore, it has achieved a high economic growth while accelerating environmental destruction and deforestation. Corporate interests have so dominated land utilization that the spatial mapping process has continued to reflect their interests. The two largest timber company groups, APP (Asia Pulp and Paper) and APRIL (Asia Pacific Resources International Holdings Limited), have been dominant players since the Suharto period and now control more than 2 million of the 9 million ha of forest area in Riau, while 45% of APP's and 70% of APRIL's concessions are in peatland areas (Nugraha 2018). In addition, the number of oil palm plantation companies with tens, hundreds, and thousands of hectares has increased incessantly, especially since decentralization, making Riau the province with the largest oil palm plantation area in Indonesia (of 3.4 million ha) (Hariandja 2020).

As far as forest area is concerned, the provincial government's standpoint is distinctly different from that of the Ministry of Environment and Forestry (Sinabutar et al. 2015; Suwarno and Situmorang 2017). The discrepancy between actual forest area in Riau and the forest area on the map is stark. At the same time, the history of spatial planning revision has been an unceasing effort to legalize illegal forest use by tailoring the mapped forest to the actual forest by expanding the areas designated as non-forest and convertible production forest.[5]

Following the central government's enactment of the spatial planning law in 1992, the Riau provincial government issued a provincial bylaw on spatial planning in 1994. The Ministry of Forestry did not approve of the bylaw, because it allocated a large portion of land as non-forest area, thus diminishing the Ministry's power (Potter and Badcock 2001, p. 10). As explained in footnote 1, two versions of provincial spatial planning have existed ever since the failure of the synchronization effort between RTRWP and TGHK in Riau Province. That is why the term "illegal forest" in the above paragraph and in the following sentences might be a bit tricky. Following TGHK, the Ministry of Forestry and environmental NGOs viewed the

---

[4]For a detailed review of the prolonged process of the Riau's provincial spatial planning until 2018, see Made et al. (2018). This chapter is partially reliant on it.

[5]According to the Ministry of Forestry, "Convertible Production Forest is a production forest that can be changed or designated into non-forest status by the release of forestry land or by exchange through a ministerial decree (Pusat Humas Kementerian Kehutanan 2011)." In other words, the Ministry has the authority to change the status of a convertible production forest to a non-forest area.

opening of forest areas as illegal, while the provincial government viewed it as legal according to its own RTRWP. The National Land Agency (*Badan Pertanahan Nasional*) did not oppose the view of the provincial government.

Table 2.2 shows the changes in the area designated as forest land in Riau by the Ministry of Forestry from 1986 to 2016. In 1986, almost all of Riau was designated as forest land, but the post-*Reformasi* government kept decreasing the forest area. According to the Ministry of Forestry's map, in 1986 non-forest area accounted for only 1.3% of the total area of the province,[6] which was far from the actual situation in Riau. Although oil palm plantations had already illegally cleared expansive areas of forest, those areas were still categorized as forest. The provincial government requested the Ministry of Forestry to reflect the actual situation or legalize the illegal operations in the spatial plan by increasing the area designated as non-forest. The request, or demand, was partially met by the ministry after decentralization, and the non-forest area thus increased from 1.3% in 1986 to 39.8% by 2016.

2.3.1 Legalizing Deforestation in the Decentralization Era

The first post-*reformasi* initiative from Riau Province to legalize illegal plantations in the spatial plan came in 2009, when the Riau provincial governor, Rusli Zainal, decided to revise the unapproved 1994 spatial plan and requested the Ministry of Forestry to change the status of previously designated forest areas to non-forest, or to "de-forest." In 2011, the Ministry of Forestry decided to deforest around 1.75 million ha in Riau Province. In 2012, Rusli requested the Ministry of Forestry to de-forest a further 3.5 million ha of forest area. Faced with the governor's request, the Ministry of Forestry created an integrated team comprised of bureaucrats from related ministries, professors, Riau provincial bureaucrats, and logging company association representative to evaluate the governor's request and make a recommendation in 2009. After receiving revised provincial requests three times, the team recommended to de-forest around 2.7 million of the requested 3.5 million ha (Tim Terpadu RTRWP Riau 2012; Jikalahari 2018, pp. 16–17). However, the Ministry of Forestry did not follow this recommendation from the team, and instead issued a decision to de-forest 1.62 million ha in 2014. The environmental NGO network, Jikalahari found that this decision legalized the status of approximately 77,000 ha of plantations operated by 104 of the 378 illegally operating oil palm enterprises in the province (Fitria 2017a).[7]

---

[6] After the decentralization, Riau Province was divided into two provinces, Riau and Riau Islands Province. The percent of non-forest area in 1986 was the ratio of non-forest area in Riau Province, not including the land area of the future Riau Islands Province.

[7] Fitria's analysis was based on the results of a survey oil palm plantation in Riau that was conducted by a special committee of the provincial parliament in 2015. The survey found that out of the 513 oil palm plantation companies in Riau, only 135 (26.3%) were operating legally, even after the Ministry of Forestry had expanded the non-forest area in 2011. The remaining 378 companies (73.7%) were illegally cultivating oil palm on forest land.

Table 2.2 The changing forest and non-forest areas in Riau Province (Made et al. 2018, p. 37)

	Forest Utilization Planning Map (1986) (excluding the area in today's Riau Archipelago Province)		Ministry of Forestry Decision VII/7651 (2011)		Ministry of Forestry Decision II/878 (2014)		Ministry of Forestry and Environment Decision 903 (2016)[a]	
Forest area	**8,865,823.00**	**98.7%**	**7,121,344.00**	**79.3%**	**5,499,693.00**	**61.2%**	**5,406,992.00**	**60.2%**
Protected forest (Hutan lindung)	271,841.00	3.0%	213,113.00	2.4%	234,015.00	2.6%	233,910.00	2.6%
Conservation forest (Kawasan suaka alam, Kawasan pelestarian alam)	438,835.00	4.9%	617,209.00	6.9%	633,420.00	7.0%	630,753.00	7.0%
Limited production forest (Hutan produksi terbatas)	2,663,960.00	29.6%	1,541,288.00	17.2%	1,031,600.00	11.5%	1,017,318.00	11.3%
Production forest (Hutan produksi)	1,336,907.00	14.9%	1,893,714.00	21.1%	2,331,891.00	26.0%	2,339,578.00	26.0%
Conversion production forest (Hutan produksi konversi)	4,154,280.00	46.2%	2,856,020.00	31.8%	1,268,767.00	14.1%	1,185,433.00	13.2%
Non-forest area	**119,000.67**	**1.3%**	**1,863,479.67**	**20.7%**	**3,485,130.67**	**38.8%**	**3,577,831.67**	**39.8%**
Total area	8,984,823.67	100.0%	8,984,823.67	100.0%	8,984,823.67	100.0%	8,984,823.67	100.0%

[a]The Riau Provincial Spatial Planning Law Draft in 2018 followed this decision

Not satisfied with the Ministry of Forestry's 2014 decision, the provincial government demanded more illegally operating plantations in forest areas be legalized. In 2015, the provincial governor appealed to the National Ombudsman of Indonesia for possible abuse of authority by the Ministry of Forestry, which had ignored the team's recommendation and had "caused legal uncertainty to deliver the public service." The Ombudsman issued a recommendation to the Ministry of Forestry to revise its decision in order to expand the non-forest area for residential areas, business and government areas, defense infrastructure and facilities, and the development needs of the national and local interests. It also suggested creating a "holding zone" (or an area the spatial utilization of which has not been decided yet: *kawasan yang belum ditetapkan perubahan peruntukan ruangnya*) for the area which was still designated as forest area according to the ministry's decision but was no longer in the forest area according to the team's recommendation.

Following the Ombudsman's recommendation, the Minister of Forestry issued a new decision in 2016 that de-forested around 90,000 ha of land. Again, this amount was far less than the provincial government's request, and so in 2016 it submitted a draft bylaw on provincial spatial planning to the provincial parliament, aiming to deforest a further million ha of forest area. If accepted, the proposed bylaw would change the status of the oil palm plantation areas of 32 companies from illegally grabbed forest areas to non-forest areas.

The provincial government and parliament agreed that the lack of spatial planning had caused an investment slump in Riau. After receiving the draft bylaw on spatial planning, the parliament immediately set up a special committee to analyze its relevance in September 2016, aiming to pass the bill in one month (Rozi 2016). The committee realized that converting a forest area into a non-forest area was legally impossible without permission to release a forest area (*izin pelepasan kawasan hutan*) from the Minister of Forestry. It determined that this legal problem would be solved by following the Ombudsman's recommendation to create a "holding zone" for approximately 406,000 of the one million ha of the provincial government-proposed non-forest area. The committee decided to exclude roughly 600,000 ha from the holding zone because these areas were under the control of large companies without permission. On the other hand, the ex-head of the committee justified the creation of the holding zone for 406,000 ha to protect the customary communities (*masyarakat adat*) who had lived in the forest areas for years (see Binawan and Osawa, Chap. 10,) and to develop national strategic projects, such as toll roads and dams (Bertuahpos 2018). The bylaw regulated that approximately 322,000 ha of the holding zone was designated as "peoples' farms" (*perkebunan rakyat*). The parliament approved the committee's draft bylaw and enacted it in September 2017.

2.3.2 Counter-Mapping in Riau and the Persistent Deforestation Process

The bylaw was filled with irregularities and ambiguities, however. First, the justification to create the 406,000 ha-holding zone turned out to be misleading at best, or outright false. Environmental NGOs, including Eyes on the Forest and Jikalahari, used portable GPS units to conduct a field investigation, or counter-mapping, of 12% of the 32,000 ha designated as peoples' farms in the holding zone. They found, and mapped, that these areas were not for small farmers, but rather for four oil palm plantation companies, 10 businessmen, and three cooperatives in partnership with the plantation companies, each holding 50 to 12,000 ha of land (Fitria 2017b; Eyes on the Forest 2018).

The province's spatial planning did not consider the existence of customary community land at all and, for the most part, neglected the central government policy to protect the peat forest. As mentioned earlier, the Jokowi government established BRG and issued the regulation to protect 865 KHGs comprised of 25 million ha (with 50% designated as conservation area and 50% as cultivation area). The regulation stipulates that Riau Province has approximately 5 million ha of peat ecosystem, with 2.47 million ha designated as a conservation area and 2.57 million ha designated as a cultivation area. But the newly passed provincial spatial planning bylaw only designated approximately 22,000 ha as a peat conservation area. This was less than 1% of the peat conservation area designated in the central government's regulation No. 57/2014.

Another complicated issue is the term "holding zone" itself. The provincial government used the term to refer to forest areas with no clear land-use objective based on Presidential Instruction No. 8/2003. The instruction regulates that the provincial governor can propose to the Minister of Forestry to designate a forest area that the province is requesting to change the status of, but to which the Ministry has not yet agreed, as a "holding zone." This means that the status of a holding zone is, by definition, legally ambiguous. This ambiguity may be the reason why the provincial government later stopped using the term "holding zone" and started to use the term "outline" instead to refer to the same 406,000 ha of forest area. The provincial bylaw unilaterally defines the term "outline" as the border delineating currently designated forest areas that are planned to be developed for uses other than forest. If we follow this definition, the "outline" area has a clear land-use objective other than forest use. The problem is that the provincial government's justification for changing the term from holding zone to outline is based on a 2017 decision of the Ministry of Home Affairs, which was made to follow Government Regulation No. 8/2013. Strangely, neither the ministry's decision nor the government regulation mentions the term "outline" at all. The term outline, therefore, has no legal foundation, making it legally unjustifiable for the provincial government to utilize the "outline" forest area for non-forest needs without the approval of the Ministry of Environment and Forestry.

After the Riau provincial government submitted its spatial planning bylaw to the Ministry of Home Affairs in October 2017, the Ministry, together with the Ministry of Environment and Forestry, BIG, BAPPENAS, and the Ministry of Agrarian and Spatial Planning, evaluated the bylaw. In its evaluation report in November 2017, the Ministry of Home Affairs requested the provincial government to revise 26 points in the bylaw and conduct a kind of environmental assessment called a strategic study on the environment (*kajian lingkungan hidup strategis*, KLHS) and to submit the KLHS to the Ministry of Environment and Forest.[8] The Ministry of Home Affairs, BIG, BAPPENAS and the Ministry of Agrarian and Spatial Planning soon agreed with the revised bylaw, but the Ministry of Environment and Forest was not satisfied with the KLHS submitted by the provincial government. The main reason for the Ministry's rejection of the KLHS was the discrepancy between the ministry's and the provincial government's designated forest and peat conservation areas (Yugo 2018a; Suparto 2019, pp. 91–92).

The Ministry of Environment and Forest requested a further revision of the KLHS to be submitted within a year, by April 2018. Strangely, when the provincial government secretary requested the Ministry of Home Affairs to issue a registration number for the spatial planning bylaw without submitting the KLHS to the Ministry of Environment and Forest, the Ministry of Home Affairs issued the registration number for the bylaw—thus approving it—in late April (Yugo 2018b). The acting provincial governor and parliament greeted this issuance with excited expectation of an influx of investment to the province. The provincial government's division head of investment and integrated service predicted that 53 trillion Rupiah from 171 domestic companies and 160 foreign companies would be invested, mainly in the plantation and hotel sectors, following the issuance of the spatial planning bylaw (Advertorial DPRD Provinsi Riau 2018; Analisadaily 2018).

After the provincial government and parliament enacted Bylaw No. 10/2018 in May 2018, environmental NGOs in Riau harshly criticized the spatial planning process for being full of irregularities, including the unfinished revision of the KLHS (Jikalahari 2018). Jikalahari and Walhi submitted a constitutional complaint on the bylaw to the Supreme Court (*Mahkamah Agung*) in August 2019 (Fitria 2019). Two months later, the court decided in favor of the plaintiffs, ruling that some articles of the bylaw were unconstitutional and issuing a correction order to the provincial government, including cancelling the use of the term "outline," nullifying the province's unilateral decision to open forest areas for non-forestry use, and ordering the province not to ignore the central government's peat restoration policy (Mahkamah Agung 2019).

This lawsuit—and victory—of the environmental NGOs indicates that civil society counter mapping may be able to successfully contain the spatial justification of the deforestation process by the local government and business actors. But it is doubtful if the provincial government will simply follow the court's decision and

---

[8]The central and local governments are required to conduct a KLHS for any development plan in order to follow the sustainable development principle, in accordance with Law No.32/2009.

invalidate all the permits of the illegally operating plantation companies. The illegal plantations are too dominant to neglect. A survey conducted by Eyes on the Forest in 2019 reveals that 47% of oil palm plantations are in the forest area and 39% of plantations outside the forest area do not have business permits, meaning that only 14% of oil palm plantations Riau are operating legally outside the forest (Eyes on the Forest 2021, p. 1).[9]

After becoming provincial governor in 2019, Syamsuar established a special taskforce to tackle the problem of illegal oil palm plantations in August of that year, boasting that the government would take measures against 1.2 million ha of illegal plantations. By January 2020, the taskforce had identified around 80,000 ha of illegal plantations and found 32 plantation companies with around 58,000 ha of illegally owned plantations, but no measures have been implemented (Tim Publikasi Katadata 2020; Fitria 2021). At the same time, the provincial government is distorting the court's correction order by proposing a status quo plan in which the deforested area in the outline/holding zone is kept as it is until the long-term land contracts for logging and plantations in the zone expire (Dewi 2021; Gunawan 2021).

The Riau case illustrates that the central government is far from monolithic, and different stakeholders at the national and local levels are aggressively engaged in the official mapping and counter-mapping process. The provincial governor and parliament have prioritized economic development through the opening of land for plantations and pressured the Ministry of Environment and Forestry to accept the province's own version of a spatial map that legalizes the illegally opened forest area, and the Ombudsman has supported the provincial government's stance. But, environmental NGOs have successfully used counter-mapping to problematize the pro-plantation provincial mapping process and appealed to the court, winning a lawsuit. This may mean that democratization and decentralization have produced a dynamic mapping process and the environmental NGOs' counter mapping and legal strategy have borne fruit, at least for the moment. This dynamism will not end soon, however, and pro-plantation actors both at the national and provincial levels will have other strategies to assert their deforestation will.[10]

---

[9]The widespread illegal operation of oil palm plantations is not limited to Riau Province. A government joint team established to enforce the law against hundreds of plantation and mining firms operating illegally in Central Kalimantan Province found in 2011 that out of a total of 967 plantation and mining firms in the province, only 76 companies had secured permits to convert the forest into business endeavors. Of 325 plantation companies with a total area of 4.6 million hectares, only 67 had obtained permits from the Forestry Ministry. In the mining sector, of the 615 registered companies in the province, only nine held permits to convert forests in an area of 30,000 ha. Mining companies, mostly small-scale coal mining companies, were operating in a 3.7 million ha area (Simamora 2011).

[10]For example, a controversial 812-page Omnibus law to facilitate job creation by synchronizing as many as 79 laws, including the laws on spatial planning, forest and environment, passed the national parliament in October 2020. It aims to weaken the decision-making power of ministries and strengthen that of the president. It contains articles on illegal business operations in forest areas. The articles have the potential to legalize the illegal opening of forest areas for oil palm plantations

2.4 Participatory Mapping at the Village Level

The provincial spatial planning process was tarred by the strong economic motivation of the local government and business sector to legalize deforestation, but the deforestation drive has not simply prevailed. This is because a counter drive has been taking place at the community level to engage in a bottom-up mapping process.

The bottom-up mapping movement to break up the state's monopolistic mapping power began in Canada in the 1970s. The importance of bottom-up mapping was realized for the first time in Indonesia in 1990, during the final days of the Suharto authoritarian regime. Decentralization and democratization, and access to cheaper surveying and mapping technology, such as portable GPS trackers and drones, accelerated the bottom-up and participatory, or counter, mapping movement across Indonesia to reclaim customary land that the Suharto regime forcefully plundered, promote environmental justice (Radjawali et al. 2017), and to define village borders.

In 2017, of the 83,172 *desa* (villages) and *kelurahan* (the smallest administrative unit in an urban area) in Indonesia, only 14.6% had definitive borders. Realizing the need for definitive borders and to end border conflicts in order to accelerate village development, the Ministry of Home Affairs issued a ministerial decree on village borders in 2006. Furthermore, the central government, after strong pressure from villages, enacted a law on villages in 2014 that clarified village authority and drastically increased village budgets. The law obligates each village to create an annual and a mid-term development plan, including a spatial one. Such a plan requires a clearly defined village area and, therefore, in 2016 the Ministry of Home Affairs revised the above ministerial decree in order to accelerate the determination of borders with the support of NGOs. Even though some local governments have started to adopt the participatory mapping method at the village level, the central government has not officially acknowledged a participatory map, maintaining that the accuracy of the maps created by GPS trackers is not high enough and the participatory mapping actors have no qualification as surveyors. But these maps become a reference for future official map making.

In view of these various developments, the authors of this chapter conducted a participatory mapping case study to determine the borders and clarify land utilization in a village in Riau Province. The process utilized satellite maps, GIS trackers, and drones, and was conducted in collaboration with a local NGO called Hakiki and the Pelalawan District government under the Future Earth Trans-disciplinary Project called "Building a Sustainable Governance of Smallholders' Oil Palm Plantations in Indonesia." The purpose of the mapping was to incubate a spatial discussion platform for the villagers to think about the future possibility of village development that is not totally dependent on oil palm plantations. As Chap. 10 by Binawan and Osawa elucidate, the lack or ambiguity of village borders in Riau Province have

---

and other purposes (Mawan 2021). Three individuals and three civil organizations submitted a petition on the unconstitutionality of the law to the Constitutional Court. The Court accepted the petition and decided that the law was conditionally unconstitutional. This process clearly suggests that the current central government has a strong intention to legalize illegally opened plantations.

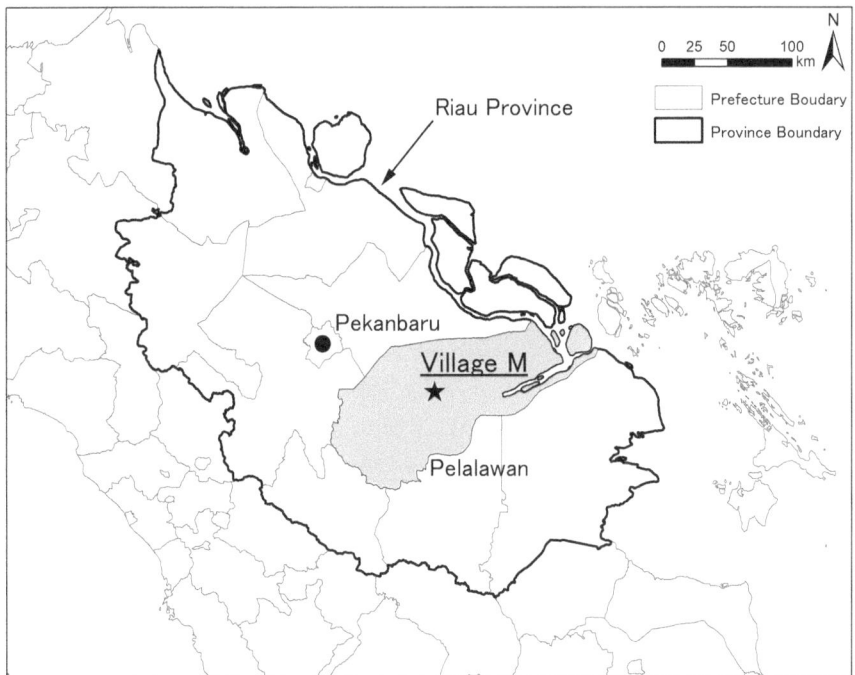

Fig. 2.1   Village M in Riau Province

created the disincentive to have an idea or vision to empower the village as a unit and also opened more space for companies and businessmen to establish oil palm plantations illegally in forest and peat areas.

After discussion with the Institute for Research and Community Service (*Lembaga Penelitian dan Pengabdian kepada Masyarakat*, LPPM) of Riau University and the Pelalawan District government, we chose Village M in Pelalawan District as the research site. The village is approximately 100 km southwest of Pekanbaru, the capital of Riau Province, as shown in Fig. 2.1. The village head welcomed our idea to create a detailed village map for the future village plan partly because the village had no clear borders with neighboring villages and it needed a village map for the mid-term development plan, as mentioned above. With the consent of the district government, we agreed to conduct a participatory mapping process by involving the villagers.

The village is divided into six hamlets (RW). Four of these, RW1, 2, 3, and 6, are located in a peat swamp area (see Fig. 2.2), and two, RW4 and 5, are located in a hilly area with extensive mineral soil. Previously, the peatland was intractable and foreign, even to those living near peatland areas. Villagers sometimes extracted timber and non-timber products from the peatlands, but they did not use the peatland as they did the mineral soils (see also Osawa, Chap. 6). Peatlands emerged as a last frontier of Riau Province for oil palm plantations especially in the late 1990s. Since

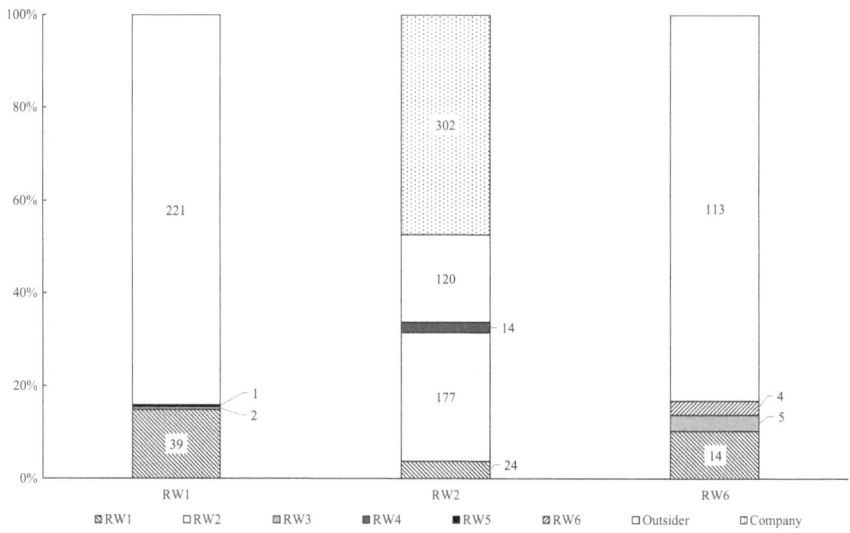

Fig. 2.2 Ownership ratio in RW 1, 2, and 6

then, those with capital have been investing massive amounts of money to convert peatlands into plantations. As mentioned in the previous sections, plantations were established not only in non-forest peatlands, but also illegally in forest peatlands. In 2015, due to the El Niño phenomenon, large scale fires in peat swamps in Sumatra and Kalimantan caused serious smoke pollution in neighboring countries.

We can clearly see the land grabbing of peatlands occurring at the micro-level of M Village. This began around 2000, and the conversion to oil palm plantations proceeded rapidly. These lands were developed by people with capital from outside the village. But this does not simply mean that there is a sharp discrepancy or dichotomy between poor villagers and land-holding outsiders, because village-scale land grabbing was carried out in collaboration with the village officers. In the case of M Village, some villagers themselves helped outsiders grab peatland for their economic benefit. Although land grabbing in the peatland frontier brought some formal and informal money to the village officers and common villagers, that land grab also resulted in serious peat fires and damage to the villagers themselves. Ironically, outsiders and "insiders" are in a complicit relationship when it comes to the peat fire calamities.

2.4.1 Village Boundary Determination

The first step in the mapping process was to determine the village and RW borders. Based on the WorldView satellite image taken on April 2, 2015 with a ground resolution of 50 cm, we interviewed village leaders, neighborhood leaders, and other influential people, as well as ordinary residents, from February 2016 to get a general idea of the village boundaries. Our experience has shown that it is not easy for residents who are not used to looking at maps to identify the actual location from the map, but when looking at high-resolution satellite imagery, most residents are able to recognize their houses, cultivated land, and landmark facilities, structures, and trees in the village. In this process, we found that most of the residents could recognize their houses, farmland, landmark facilities, and specific structures and trees. Satellite images taken within a year of the interviews were enlarged and printed on A0-sized sheet of paper for the residents to see, which enabled them to get a rough idea of the village boundaries. In addition, we went around the village with a GPS device with the residents to confirm the coordinates of the boundaries.

No major problems were encountered in determining either the RW borders or the village's boundaries with neighboring villages.[11] Following confirmation with officials from the neighboring villages, we were able to draw up a village boundary map as shown in Fig. 2.3, which delineates the boundaries of each village and each RW. Based on this village boundary map, the area of each hamlet from RW1 to RW6 was calculated to be 1436 ha, 1106 ha, 3388 ha, 676 ha, 2401 ha, and 581 ha, respectively, for a total village area of 9588 ha.

2.4.2 Visualizing Land Ownership in Peatlands

Once the boundaries of the village and its hamlets had been demarcated, we attempted to determine the status of the ownership and use of the peatlands that were spread across the hamlets. The exact distribution of peatlands in the village is difficult to ascertain. However, we know from interviews with local residents that the peatland was not yet cleared in the early 2000s, so we extracted the forest area between the river and the village from the LANDSAT satellite image taken on August 15, 2002 and assumed that this area was peatland. Since this study was a pilot study and the survey period and budget were limited, the survey area was limited to the entire area of RW2 and RW6 and the northeastern part of RW1. We followed the same survey method that we used to identify the boundaries of the hamlets and the village: interviews with residents based on satellite imagery and a field survey.

Figures 2.2 and 2.4 show that most of the peatland in each hamlet is not owned by the residents, but by outsiders. In RW1, for example, 221 ha of land is owned by outsiders, while only 39 ha is owned by RW1 residents. In RW2, the amount of land owned by outsiders and residents was 120 ha and 24 ha, respectively, and in RW6, 113 ha and 14 ha, respectively. We also found that almost all of the land owned by outsiders was used for oil palm plantations. Of these, the largest area was owned by a

---

[11] No regulation obligates a village to demarcate RW borders. It seems rare that a village has clear RW borders.

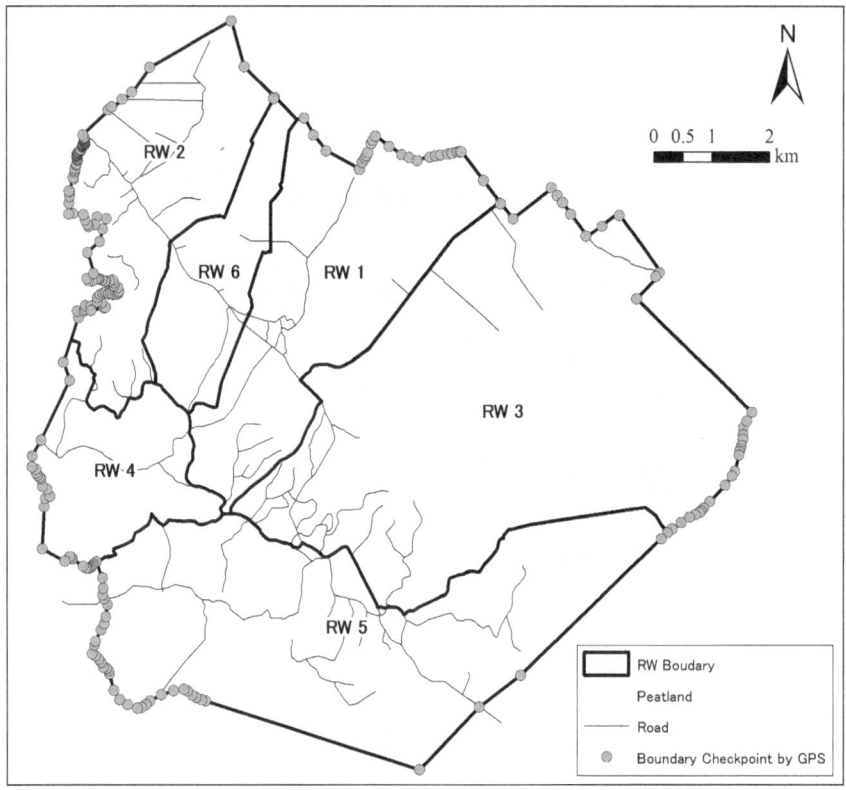

Fig. 2.3 Village and RW borders in M village

company with an address in Pekanbaru, which occupied 175 hectares, or 70% of the peatland area in RW2.

Although the above situation was well known among the local residents, there was no spatial and quantitative information on the situation. The visualization of the realities of land use and ownership in the village resulted in the stagnation of the mapping process.

2.4.3   Revealing the Political Nature of Maps

Because Hakiki had a trusting relationship with the local government secretary of Pelalawan District and we had formed a friendly relationship with the M Village head, we were able to establish the boundaries with neighboring villages and between the RWs relatively smoothly using satellite imagery and drones. However, as the land ownership of the peat swamp became clearer, it became more difficult to continue the survey, as the map showed that the peat swamp was owned by people

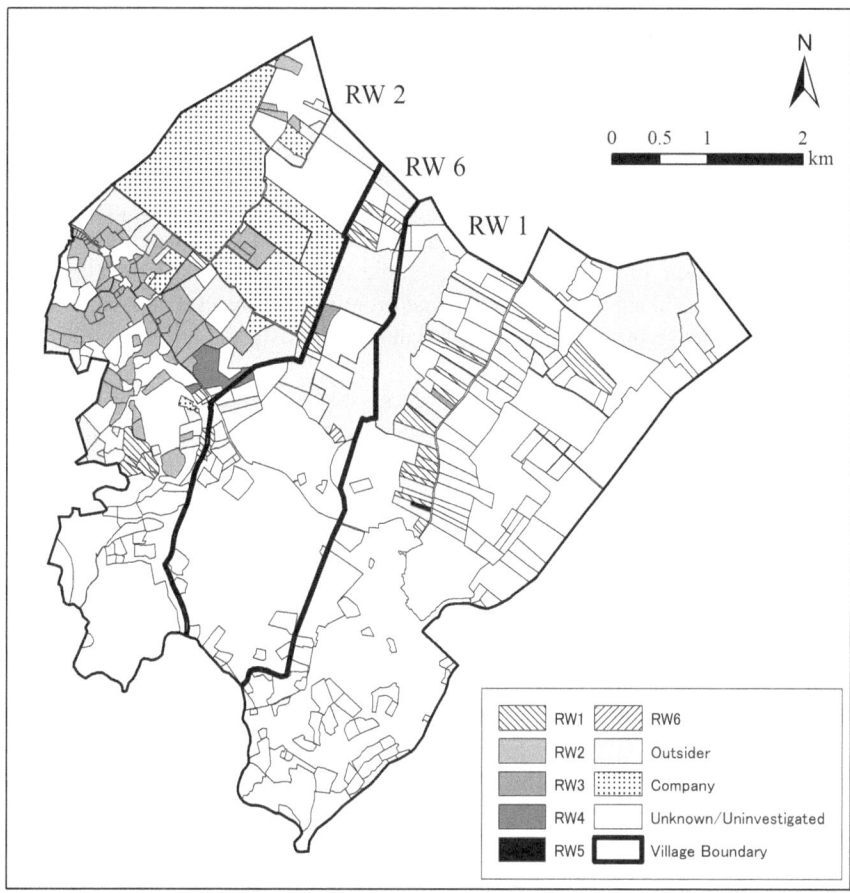

Fig. 2.4 Land ownership in RW 1, 2, and 6 in M village

outside the village even though it was in the forest conservation area. Not only did it show that government officials had illegally given permission for development, but it also revealed that some of the outsiders were locally influential actors.

As this case illustrates, bottom-up or participatory mapping is not a panacea. Even if it is a bottom-up approach, no map can be made without taking into consideration the hierarchical relationships that exist in villages and/or customary communities. The voices of formal and informal leaders may prevail. If a map is made solely by marginalized people and the NGOs supporting them, it will take a long time for that map to be recognized and legitimized, and if the map is simply exposed to the villagers, it may cause more visible, serious, and violent conflicts within the community. In the case of M Village, the village boundary could be smoothly fixed on the map after consultation with the neighboring villages. The problem arose when clarifying the land ownership within the village boundaries. The village head, who had agreed to create a village map, had not really thought through the implications of mapping power. He did not anticipate that the bottom-up mapping process would reveal information on land (legal and illegal) borrowing and selling, which had been decided only by a small elite of the village. Some villagers were involved in opening the peatland opening for outsiders and knew the actual status of peatland utilization, but the socio-political impact would be significant once it was visible as a map, which is what the village head was understandably afraid of. The participatory mapping process at the village level therefore resulted in a halfway map with no village recognition. In the process, however, the village head recognized 1000 ha of natural forest as village communal land and promised to protect it. The map, which includes the pledge-protected communal land, now exists and it can be utilized at any time in the future.

A similar stagnation tends to happen during the map-making process in larger villages outside of Java and Bali in Indonesia, where outsiders complicit with village officials illegally open peatland areas for oil palm plantations, sometimes with the involvement of villagers. The problem here is the definition, or meaning, of illegality. Establishing a plantation in a peat area within the state forest is illegal because it is against the forestry regulations. But illegality in this sense is too common in Indonesia, as 3.4 million ha of oil palm plantations, or around 20% of some 16.8 million ha of total oil palm cover in Indonesia, is located inside the forest zone. Approximately 1.2 million ha of smallholder plantations are also inside the forest zone (Fakultas Kehutanan UGM 2018; Bakhtiar et al. 2019). Simply reversing all the "illegal" plantations to the forest is unworkable, but simply legalizing all the "illegal" plantations will only further accelerate the deforestation. Here emerges ideas to incorporate the social forestry scheme under the Jokowi government and to introduce an agroforestry model of oil palm plantations mixed with other tree crops (Sumardjono et al. 2018; Bakhtiar et al. 2019). If the central government is forced to address this issue with a clear-cut policy, then currently unrecognized maps, such as the one we created in M Village, might be meaningful for future village plans.

2.5 Conclusion

Mapping is a political act. During the Suharto authoritarian regime, the mapping power held by the central government was significant, but not fully centralized in one institution, such as the Ministry of Forestry. Different ministries produced and used different maps for their own interests with weak coordination among them. They imposed these maps without considering the real life of the communities or the negative environmental impacts of doing so. The maps preceded, transformed, and corrupted the reality. Different actors claimed specific territories as concession, plantation, conservation, or other areas, using different "official" maps to advance their interests. Maps thus served the interests of central government and related corporate actors. In turn, then, these vested interests preceded, transformed, or corrupted the "official" maps.

The fall of the Suharto regime and the emergence of new technology have expanded the political opportunity structure for local actors to counter centrally decided mapping. The deconstruction, democratization, and decentralization of mapping power is underway. The One Map Policy launched by Yudhoyono and continued by Joko Widodo is an ambitious endeavor to consolidate all centrally defined and locally created maps into a single authentic and official base map to a scale of 1:50,000. Although BIG created a base map in 2019 to rectify the plurality of "official" maps, it has not been accepted by all the stakeholders as a "One Map." As far as forest and peatland areas are concerned, the Ministry of Environment and Forestry and BRG, or its successor agency, BRGM (see footnote 3), aim to keep the forest area as large as possible by promoting forest conservation as a means to stop global warming. At the same time, the Ministry of Agriculture, the Ministry of Energy and Mineral Resources, and local governments prioritize plantations and natural resource exploitation, demanding the expansion of non-forest areas. It is hard to imagine that these agencies will reach any consensus on one base map any time soon. The creation of one map itself can be, however, at least one small but important step toward an environment-conscious discourse prevailing.

The politicization of mapping is also quite clear in Riau Province's spatial planning and M Village's map-making processes. Still today, Riau Province has had no spatial planning because of the lingering conflicts among various stakeholders. On the one hand, the provincial governor and parliament have consistently promoted expansion of the deforested area in the spatial plan and have received a certain level of support from the central independent organization of the Ombudsman. On the other hand, local environmental NGOs have used counter-mapping to problematize the process of legalizing illegal plantations and to illustrate the destructive nature of deforestation and peat fires. These NGOs have sometimes succeeded in obtaining the support of Jakarta stakeholders, such as the Supreme Court. The M Village participatory mapping process—in which and environmental NGO used GPS devices and drones—was quite successful in determining the village borders and in obtaining a pledge to conserve communal land. The process stagnated, however, when it came to visualizing peatland utilization. The mapping process clearly revealed the disproportionate ownership of peatlands by outsiders and thus had the potential to invoke villager anger. The local NGO found it difficult to continue creating the map against the will of village elite, who supported the status quo. This process demonstrates the emerging awareness of the political power of mapping at the village level. Although the M Village map (and others like it) currently remains unfinished, it can be used in the future.

As these mapping and spatial planning processes, and the actual land grabbing history, show, democratization, decentralization, and the introduction of IT have precipitated an important dispersion of mapping power, empowering the spatial justice movement. While the actors supporting pro-plantation mapping are powerful at every level, there is still a chance to achieve spatial justice and a more sustainable environment by combining mapping and judiciary powers in Indonesia.

References

Abood SA, Lee JSH, Burivalova Z (2014) Relative contributions of the logging, fiber, oil palm, and mining industries to forest loss in Indonesia. Conserv Lett 8(1):58–67. https://doi.org/10.1111/conl.12103

Advertorial DPRD Provinsi Riau (2018) Akhirnya, Mendagri tandatangani RTRW Riau: dampak positif bakal dirasakan Masyarakat. Situs Riau, 19 Apr. http://situsriau.com/. Accessed 18 Sep 2021

Analisadaily (2018) Kemendagri setujui raperda RTRW. Analisadaily.com, 26 Apr. https://analisadaily.com/. Accessed 18 Sep 2021

Anwar MC (2018) Kebijakan satu peta a la Jokowi, begini penjelasan lengkapnya. CNBC Indonesia, 11 Dec. https://www.cnbcindonesia.com/. Accessed 14 Oct 2021

Bakhtiar I, Suradiredja D, Santoso H et al (eds) (2019) Palm inside: resolving the oil palm invasion inside forest zone. Kehati, Jakarta

Barr C (2006) Forest administration and forestry sector development prior to 1998. In: Barr C, Resosudarmo IAP, Dermawan A et al (eds) Decentralization of forest administration in Indonesia: implications for forest sustainability, economic development and community livelihoods. Bogor, CIFOR, pp 18–30

Barr C, Resosudarmo IAP, Dermawan A et al (2006a) Decentralization's effects on forest concessions and timber production. In: Barr C, Resosudarmo IAP, Dermawan A et al (eds) Decentralization of forest administration in Indonesia: implications for forest sustainability, economic development and community livelihoods. CIFOR, Bogor, pp 87–107

Barr C, Resosudarmo IAP, McCarthy et al (2006b) Forests and Decentralization in Indonesia: an overview. In: Barr C, Resosudarmo IAP, Dermawan A et al (eds) Decentralization of forest administration in Indonesia: implications for forest sustainability, economic development and community livelihoods. CIFOR, Bogor, pp 1–17

Bertuahpos (2018) Holding zone di RTRW, ini penjelasan mantan ketua pansus DPRD Riau. Bertuahpos, 26 April. https://bertuahpos.com/. Accessed 8 Sep 2021

Brockhaus M, Obidzinski K, Dermawan A et al (2012) An overview of forest and land allocation policies in Indonesia: is the current framework sufficient to meet the needs of REDD+? For Policy Econ 18:30–37. https://doi.org/10.1016/j.forpol.2011.09.004

Chakib A (2014) Civil society organizations' roles in land-use planning and community land-rights issues in Kapuas Hulu Regency, West Kalimantan, Indonesia. Working P1. https://doi.org/10.17528/cifor/005426

Dewi R (2016) Gaining recognition through participatory mapping? The role of Adat land in the implementation of the Merauke Integrated Food and Energy Estate in Papua. Indonesia. Austrian J South-East Asian Stud 9(1):87–106. https://doi.org/10.14764/10.ASEAS-2016.1-6

Dewi RS (2021) 1,2 juta hektare perkebunan ilegal di kawasan hutan Riau dieksekusi. GoRiau, 25 Aug. https://www.goriau.com/. Accessed 20 Sep 2021

Eyes on the Forest (2018) 'Legalisasi' perusahaan sawit melalui *holding zone* dalam rancangan peraturan daerah rencana tata ruang wilayah Provinsi Riau (RTRWP) 2017–2037: area tak dibebankan izin di 17 kebun sawit bukanlah lahan peruntukan rakyat. Eyes on the Forest, Pekanbaru

Eyes on the Forest (2021) Omnibus law bukan legalisasi otomatis untuk perkebunan sawit ilegal: investigasi EoF pada kebun sawit ilegal dan penjualan tandan buah ilegal di Riau. Eyes on the Forest, Pekanbaru

Fakultas Kehutanan UGM (2018) Press release: menyikapi polemik tanaman sawit di dalam kawasan hutan Indonesia. Fakultas Kehutanan UGM, Yogyakarta

Fitria N (2017a) Kertas posisi: RTRWP Riau untuk rakyat, bukan untuk segelintir pemodal dan monopoli korporasi. Jikalahari, 4 Aug. https://jikalahari.or.id/. Accessed 7 Sep 2021

Fitria N (2017b) Holding zone dalam ranperda RTRW 2017–2037 untuk korporasi dan cukong sawit. Jikalahari, 18 Oct. https://jikalahari.or.id/. Accessed 21 May 2021

Fitria N (2019) Judicial review RTRWP Riau untuk keselamatan warga, ruang kelola rakyat Riau dan perlindungan lingkungan hidup dan kehutanan. Jikalahari, 8 Aug. https://jikalahari.or.id/. Accessd 27 May 2021

Fitria N (2021) Dua tahun Gubernur Syamsuar: lamban realisasi komitmen LHK. Jikalahari, 19 Feb. https://jikalahari.or.id/. Accessed 21 Sep 2021

Gellert PK, Andiko (2015) The quest for legal certainty and the reorganization of power: struggles over forest law, permits, and rights in Indonesia. J Asian Stud 74(3):639–666. https://doi.org/10.1017/S0021911815000613

Gunawan A (2021) Dinas LHK Riau bakal eksekusi 1,2 Juta hektare perkebunan ilegal. Bisnis Sumatra, 25 Aug. https://sumatra.bisnis.com/. Accessed 20 Sep 2021

Hariandja R (2020) Kebun sawit Riau terluas, tak jamin rakyat sejahtera dan berdaulat pangan. Mongabay, 20 May. https://www.mongabay.co.id. Accessed 15 Oct 2021

Harley JB (1989) Deconstructing the map. Cartographica 26(2):1–20. https://doi.org/10.3138/E635-7827-1757-9T53

Jepson P, Momberg F, van Noord H (2002) A review of the efficacy of the protected area system of East Kalimantan Province. Indonesia. Nat Areas J 22(1):28–42

Jikalahari (2018) Anotasi hasil evaluasi ranperda RTRWP Riau 2018–2038. Jikalahari, Pekanbaru

Karsidi A (2016) Kebijakan Satu Peta (One Map Policy): roh pembangunan dan pemanfaatan informasi geospasial di Indonesia, 2nd edn. Sains Press, Cibinong

Tim Publikasi Katadata (2020) Sengkarut lahan perkebunan sawit Riau. Katadata, 29 Jan. https://katadata.co.id/. Accessed 21 Sep 2021

Kelly AB, Peluso NL (2015) Frontiers of commodification: state lands and their formalization. Soc Nat Resour 28(5):473–495. https://doi.org/10.1080/08941920.2015.1014602

Made A, Okto YS, Fitria N (2018) Public review rancangan peraturan daerah rencana tata ruang dan wilayah Provinsi Riau 2017–2037. Witra Percetakan, Pekanbaru

Mahkamah Agung (2019) Putusan Nomor 63 P/HUM/2019

Maryudi A (2015) The political economy of forest land-use, the timber sector, and forest certification. In: Romero C, Putz FE, Guariguata MR et al (eds) The context of natural forest management and FSC certification in Indonesia. Occasional Paper 126, CIFOR, Bogor, pp 9–34

Mawan A (2021) Was-was masa suram era UU cipta kerja. Mongabay, 16 Sep. https://www.mongabay.co.id/. Accessed 21 Sep 2021

McCarthy JF (2000) The changing regime: forest property and *reformasi* in Indonesia. Dev Change 31(1):91–129. https://doi.org/10.1111/1467-7660.00148

McCarthy JF (2002) Turning in circles: district governance, illegal logging, and environmental decline in Sumatra, Indonesia. Soc Nat Resour 15(10):867–886. https://doi.org/10.1080/08941920290107620

Nugraha I (2018) Potret 5 tahun kebijakan konservasi hutan APP dari pantauan organisasi lingkungan. Mongabay, 20 Feb. https://www.mongabay.co.id. Accessed 10 Sep 2021

Nuhidayah L, Davies PJ, Alam S (2020) Resolving land-use conflicts over Indonesia's customary forests: one map, power contestation and social justice. Contemp Southeast Asia 42(3):372–397

Peluso NL (1995) Whose woods are these? Counter-mapping forest territories in Kalimantan, Indonesia. Antipode 27(4):383–406. https://doi.org/10.1111/j.1467-8330.1995.tb00286.x

Peluso NL, Vandergeest P (2001) Genealogies of the political forest and customary rights in Indonesia, Malaysia, and Thailand. J Asian Stud 60(3):761–812. https://doi.org/10.2307/2700109

Peluso NL, Vandergeest P (2020) Writing political forests. Antipode 52(4):1083–1103. https://doi.org/10.1111/anti.12636

Potter L, Badcock S (2001) The effects of Indonesia's decentralisation on forests and estate crops in Riau Province: case studies of the original districts of Kampar and Indragiri Hulu. CIFOR, Bogor

Pramono AH, Natalia I, Janting Y (2006) Ten years after: counter-mapping and the Dayak lands in West Kalimantan, Indonesia. Unpublished

Pusat Humas Kementerian Kehutanan (2011) Indonesian forestry. Pusat Humas Kementerian Kehutanan, Jakarta

Radjawali I, Pye O, Flitner M (2017) Recognition through reconnaissance? Using drones for counter-mapping in Indonesia. J Peasant Stud 44(4):817–833. https://doi.org/10.1080/03066150.2016.1264937

Rozi F (2016) Rapat paripurna jabawan pemprov: maksimalkan pembentukan raperda RTRW, DPRD Riau bentuk pansus. GoRiau, 5 Sep. https://www.goriau.com/. Accessed 20 May 2021

Santoso H (2003) Forest area rationalization in Indonesia: a study on the forest resource condition and policy reform. World Agroforestry Centre (ICRAF), Bogor

Sarbini N, Wibisono Y, Sampurno DTW (2017) One Map Policy as national priority fundamental development. And its implementation. Powerpoint presented at an international workshop on dilemmas of One Map Policy and participatory mapping in Indonesia, Kyoto University, Kyoto, 7 August 2017

Scott JC (1998) Seeing like a state: how certain schemes to improve the human condition have failed. Yale University Press, London

Setiawan EN, Maryudi A, Purwanto RH et al (2016) Opposing interests in the legalization of non-procedural forest conversion to oil palm in Central Kalimantan, Indonesia. Land Use Policy 58:472–481. https://doi.org/10.1016/j.landusepol.2016.08.003

Setiawan EN, Maryudi A, Purwanto RH et al (2017) Konflik tata ruang kehutanan dengan tata ruang wilayah (Studi kasus penggunaan kawasan hutan tidak prosedural untuk perkebunan sawit Provinsi Kalimantan Tengah). Bhumi 3(1):51–66. https://doi.org/10.31292/jb.v3i1.226

Simamora AP (2011) 967 forestry firms under govt scrutiny. The Jakarta Post, Feb 2. https://www.thejakartapost.com/. Accessed 27 Sep 2021

Sinabutar P, Nugroho B, Kartodihardjo H et al (2015) Kepastian hukum dan pengakuan para pihak hasil pengukuhan kawasan hutan negara di Provinsi Riau. Jurnal Analisis Kebijakan Kehutanan 12(1):27–40. https://doi.org/10.20886/jakk.2015.12.1.27-40

Smith J, Obidzinski K, Subarudi et al (2003) Illegal logging, collusive corruption and fragmented governments in Kalimantan, Indonesia, Int For Rev 5(3):292–302. https://doi.org/10.1505/IFOR.5.3.293.19138

Sumardjono MS, Simarmata R, Wibowo RA (2018) Penyelesaian masalah penguasaan dan pemanfaatan kawasan hutan untuk perkebunan sawit rakyat. Kehati, Jakarta

Suparto (2019) Problematika pembentukan peraturan daerah (Perda) tentang rencana tata ruang wilayah Provinsi Riau. Bina Hukum Lingkungan 4(1):79–96. https://doi.org/10.24970/bhl.v4i1.87

Suprapto S, Awang SA, Maryudi A et al (2018) Kontestasi Aktor dalam Proses Revisi Rencana Tata Ruang Provinsi (RTRWP) di Indonesia (Studi Kasus: Revisi RTRW Provinsi Riau). Jurnal Wilayah dan Lingkungan 6(3):193–214. https://doi.org/10.14710/jwl.6.3.193-214

Suwarno A, Hein L, Sumarga E (2015) Governance, decentralisation and deforestation: the case of Central Kalimantan Province, Indonesia. Q J Int Agric 54(1):77–100. https://doi.org/10.22004/ag.econ.206297

Suwarno E, Situmorang AW (2017) Identifikasi hambatan pengukuhan kawasan hutan di Provinsi Riau. Jurnal Analisis Kebijakan Kehutanan 14(1):17–30. https://doi.org/10.20886/jakk.2017.14.1.17-30

Tilley L (2020) "The impulse is cartographic": counter-mapping Indonesia's resource frontiers in the context of coloniality. Antipode 52(5):1434–1454. https://doi.org/10.1111/anti.12634

Tim Terpadu RTRWP Riau (2012) Haisl penelitian terpadu perbahan Kawasan hutan dalam revisi rencana tata ruang wilayah Provinsi Riau. Slide presented at the Ministry of Forestry, Jakarta, 5 December 2012

de Vos R (2018) Counter-mapping against oil palm plantations: reclaiming village territory in Indonesia with the 2014 village law. Crit Asian Stud 50(4):615–633. https://doi.org/10.1080/14672715.2018.1522595

Wardojo W, Masripatin Nur (2002) Trends in Indonesian forest policy. Policy trend rep 2002. IGES, Kanagawa, pp 77–87

Widianingsih I, Morrell E (2007) Participatory planning in Indonesia: seeking a new path to democracy. Policy Stud 28(1):1–15. https://doi.org/10.1080/01442870601121320

Winichakul T (1997) Siam mapped: a history of the geo-body of a nation. University of Hawaii Press, Honolulu

Wollenberg E, Campbell B, Dounias E et al (2008) Interactive land-use planning in Indonesian rainforest landscapes: reconnecting plans to practice. Ecol Soc 14(1):Article 35

Yugo O (2018a) Mendagri membina dan mengawasi pemda Riau terkait penyelenggaraan RTRWP Riau. Jikalahari, 23 Apr. https://jikalahari.or.id/. Accessed 20 June 2021

Yugo O (2018b) Direktur PHD Dirjen Otda Kemendagri melakukan maladministrasi perihal nomor register ranperda RTRWP Riau 2017-2037. Jikalahari, 30 Apr. https://jikalahari.or.id/. Accessed 18 Sep 2021

Chapter 3
Selling Peatland for the Future: History, Land Management, and the Transformation of Common Land in Rantau Baru

Takamasa Osawa and Akhwan Binawan

Abstract Rantau Baru, an old fishing village on the bank of the Kampar River in Indonesia, is surrounded by peat hinterlands. The village territory has been recognized by previous local states for hundreds of years, and the villagers have managed it as ancestral common space based on a matrilineal system and headmanship. However, since the 1990s, acacia and oil palm companies have encroached on the peatlands of the traditional territory. In this situation, many villagers have either sold or plan to sell peat hinterlands in the village territory. How has their ancestral territory transformed into tradable land, and why have they chosen to sell it? What is the relationship between the traditional values of customary space and the adoption of the perspective of land as a commodity? Based on historical research on local land governance and a present-day household survey of land use and attitudes toward peat space, this chapter argues that the privatization of peatlands has transformed a once-common space into a commodity. Villagers sell peatland to actualize its potential amid anxiety and economic difficulty to contribute to the stable future of their descendants.

Keywords Rantau Baru · Regional history · The Petalangan · Traditional village governance · Land management · Transformation of common land

3.1 Introduction

Chaps. 3–10 focus on the administrative village (*desa*) of Rantau Baru (Pangkalan Kerinci Sub-district, Pelalawan District) and examine its peatland governance using an inter/transdisciplinary approach. This chapter provides an overview of the history

T. Osawa (✉)
Institute of Liberal Arts and Science, Kanazawa University, Kanazawa, Ishikawa, Japan
e-mail: tosawa@staff.kanazawa-u.ac.jp

A. Binawan
Perkumpulan Ara Sati Hakiki, Jl. Kayu Putih, Perumahan Athaya IV, Pekanbaru, Riau, Indonesia

© The Author(s) 2023
M. Okamoto et al. (eds.), *Local Governance of Peatland Restoration in Riau, Indonesia*, Global Environmental Studies,
https://doi.org/10.1007/978-981-99-0902-5_3

and present situation of Rantau Baru and analyzes the relationship between people and land from the perspective of social and cultural anthropology.

Since the 1990s, the Indonesian government has implemented policies to regulate or mitigate the expansion of oil palm and acacia industries in peat environments to prevent peatland degradation and fire. In 2011, it imposed a moratorium on the issuance of new licenses for acacia and oil palm plantations in dryland forests and peatlands under the international framework of "reducing emissions from deforestation and forest degradation and the role of conservation sustainable management of forests, and enhancement of forest carbon stocks in developing countries" (REDD+) (Murdiyarso et al. 2011; Sloan et al. 2012). The types of large-scale concessions granted during the previous few decades were halted for an initial period of two years with the hope of slowing deforestation and the subsequent degradation of peatland. This policy was amended and extended repeatedly and became permanent in 2019. In 2016, the government established the Indonesian Peatland Restoration Agency (*Badan Restorasi Gambut*, BRG). The BRG adopted the strategies of rewetting, revegetation, and revitalization (3R), paying special attention to locals' lives.

However, in Riau, the oil palm industry continued to encroach on peatland because moratorium policies contain many loopholes (Jong 2019). First, such policies do not protect all forests and peatlands, and second, a protected area may be amended through lobbying by local authorities (Jong 2019). Furthermore, at the village level, peat hinterlands continue to be bought and sold, and oil palms continue to be planted on peatlands. Changes in the regulation of the oil palm industry in the 2000s made it possible for small landholders to either work on oil palm plantations as worker-cum-owners or receive revenues from plantations as unearned income (Kawai 2021, pp. 40–41), which facilitated the speculative hoarding and selling of land. For example, in this manner, the Ngaju Dayaks in Kapuas District of Central Kalimantan have sold peat hinterlands from which they used to collect forest products, such as rattan, to buyers related to the oil palm industry (Lubis 2013, pp. 48–50; see also Jong 2021). Capital continues to define the borders and conditions of peatlands that the local community once used. Similarly, Rantau Baru villagers have sold peat hinterland in the village territory to buyers since the mid-2000s.

However, in Rantau Baru, where village lands have been inherited from the villagers' ancestors over at least several centuries, and villagers have managed the land, river, and resources in the village territory as a common space based on a matrilineal system and headmanship. The boundaries of the village were determined in the precolonial era and recognized by the Pelalawan kingdom, which ruled the region between the 18th and mid-20th centuries. The village territory can thus be considered an inalienable ancestral territory (Chou 2010). Despite this, even Rantau Baru villagers have sold peat hinterland. How has their ancestral territory transformed into a tradable commodity, and why have they chosen to sell the hinterland? What is the relationship between the traditional values of customary space and the adoption of the perspective of land as a commodity?

In describing the formation of the Rantau Baru village and its history of land use, this chapter examines how and why village peatlands have been divided, privatized, and sold. First, we summarize the history of the Kampar River Basin based on past records and previous historical studies and analyze the position of people in history and explain that their territories were recognized by previous states. Second, based on the results of unstructured interviews and questionnaire surveys, we provide an overview of the ethnicity of villagers, the village matrilineal system and headmanship, and the environment in the territory. Finally, we examine the processes through which the village's customary space (particularly peatland) was divided, documented, and eventually sold to outsiders.

Although the villagers aspire to own oil palm gardens, this is difficult for them to accomplish due to environmental and economic reasons. Selling peatland is an alternative means that enables them to actualize the potentiality of peatland space, which contributes to the stable future of their descendants. A deeper understanding of such behaviors and ways of thinking among local residents is necessary to prevent peatland degradation, drying, and fires in a bottom-up fashion.

3.2 Historical Background: The 29 *Pebatinan* Under the Malay Kingdoms

3.2.1 Brief History of the Pelalawan Area

Although the history of Rantau Baru stretches back centuries, the story of how the village began has not been passed down through generations in a clear manner.[1] However, it is well known that the village was one of the 29 *pebatinan*, or administrative districts, designated by the Pelalawan kingdom in the precolonial era. The old name of the village, Bokol Bokol, is also mentioned in a colonial record written in the 1860s (Faes 1882, p. 518). In this section, we provide an overview of the history of the Kampar River Basin, examine the process through which the territory was formulated, and elucidate the status of local people vis-à-vis the state before and after the proclamation of Indonesian independence in 1945. To this end, we largely depend on the works written and edited by Tenas Effendy et al. (Effendy 1997a, 1997b, 2002; Effendy et al. 2005). Tenas Effendy was a leader of traditional culture (*tokoh adat*) in Riau and had a profound influence on cultural policies from the 1990s to the early 2010s. After the 1960s, he traveled to villages in the area,

---

[1] According to a report on Rantau Baru written by the BRG, people moved from West Sumatra to the present-day area of Rantau Baru in the 17th century (BRG 2019, p. 35). However, we could not find any evidence or records to confirm this.

collecting and recording cultural practices and oral histories.[2] He also wrote and edited many historical articles on Pelalawan District. However, his descriptions often do not mention any sources, and in some cases, he would present his own views without credible sources (see also Masuda 2012, p. 241). In presenting the following historical overview, we recognize the limits of these materials.

Settlements and states in eastern Sumatra were typically established and developed along several large rivers and their tributaries. Historically, eastern Sumatra, including the Kampar Basin, was influenced by the Malay kingdoms of Srivijaya, Malacca, and Johor, which depended on maritime trade overseas. Several local kingdoms established centers on the shores of the estuaries of large rivers that contain headwaters from the Barisan Mountains and pour into the Malacca Strait. These kingdoms locally controlled their basins by following or allying with the Malay maritime kingdoms. Settlements relied on large rivers and their tributaries as trading routes for crops, spices, minerals, and forest products produced and harvested in the upstream and midstream regions (Bronson 1978; Kathirithamby-Wells 1993). These regions were also a destination of out-migration (*merantau*) of the Minangkabau, who lived in the highland areas of the Barisan Mountains. Since ancient times, people of eastern Sumatra have continuously migrated to lowland areas along the rivers. The resulting settlements and states that developed on the riverbanks enabled communication among the Malays from coastal areas, the Minangkabau from the highlands, and the indigenous people who lived in the forest areas (Andaya 2008, pp. 81–100; Barnard 2003; see also Osawa 2022). The population of this region can thus be characterized by its mixed nature (or *kacuness*) (Barnard 2003, pp. 2–3). Hamidy (1987, p. 22), a local anthropologist who conducted fieldwork in several settlements of the 29 *pebatinan,* notes that while the people seem to have various origins, their status and identity have been formulated through the everyday interactions among the communities.

At the end of the 14th century, one of the local kingdoms, the Pekantua kingdom, was established and began to exert power over the Kampar Basin (Effendy et al. 2005, p. 67). Malay merchants from Malacca and the eastern coast of Sumatra, who traveled to the kingdom's center, and the people of the kingdom came into contact with Islam (Masuda 2012, p. 242). Around the turn of the 16th century, the Malacca kingdom, aiming to extend its influence into the Kampar basin, attacked and occupied Pekantua. By accepting the royal family of Malacca as its ruler, the Pekantua kingdom was permitted to continue. As the Malacca kingdom had adopted Islam in the 15th century, the Pekantua king was named a Sultan (Effendy et al. 2005) and Islam gradually spread throughout the basin. During the next two hundred years, the center was moved several times around the Kampar estuary before finally settling in the mid-18th century in Sungai Rasau, or present-day Pelalawan. It was then that the kingdom was named Pelalawan (Effendy et al. 2005, pp. 73–75).

---

[2] In this region, the history of each village is memorialized in the form of poetry singing or *tombo* (Effendy 1997b, pp. 21–22). In Rantau Baru, the detailed origin of the villagers has not been handed down.

Around the turn of the 19th century, Pelalawan was invaded by the Siak kingdom. However, the polity of the Pelalawan kingdom was maintained by accepting the brother of the Siak king as a ruler (van Anrooij 1885, pp. 268–269; Barnard 2003, p. 139, 160; Effendy et al. 2005, pp. 77–79).

In the mid-19th century, when the Dutch government extended its direct control over the Indonesian archipelago, both the Indoragiri and Siak kingdoms were subsumed under Dutch control. In 1879, the Pelalawan kingdom was incorporated into the colonial government structure (Effendy et al. 2005, p. 88). Although Pelalawan was placed under its suzerainty, the Dutch colonial government recognized that in the "Outer Islands," such as Sumatra and Kalimantan, local authorities had continuously ruled their territories based on their customary laws, or *hukum adat*. Therefore, Pelalawan remained a regional self-governing kingdom (*Zelfbesturende Landschappen*) (Masuda 2012, p. 247). During the Japanese occupation (1942–1945), although the Pelalawan kingdom was organized as a district (*gun*) of Eastern Sumatra Province (*shu*), the Japanese military government maintained the Dutch administrative organization, allowing the kingdom's structure and governance to continue. In 1945, the Pelalawan king officially handed sovereignty of the kingdom to the Republic of Indonesia (Effendy 1997a, p. 633, Effendy et al. 2005, pp. 91–92; Masuda 2012, pp. 246–249).

Throughout the Pekantua and Pelalawan eras, kings controlled the Kampar Basin by appointing the heads of each area in the region. According to Masuda (2012, p. 399, quoting an unpublished manuscript written by Effendy and Jaafar[3]), Sultan Alauddin Riayat Syah II, who ruled the Pekantua kingdom between 1528 and 1530, formulated the rights and obligations of each customary law (*adat*) of local communities. Additionally, he decided that each *adat* would be managed by *batin,* or the heads of *pebatinan*, and designated areas to each *batin* as the customary land (*tanah wilayat*) (see Masuda 2012, pp. 242–243). At the turn of the 19th century, these customary areas were organized into 29 *pebatinan* that still exist today (Effendy et al. 2005, p. 42, 82). Syarif Jaafar, who ruled Pelalawan from 1866 to 1872, divided the kingdom into four regions and dispatched a head (*orang besar*) to each region, which comprised several *pebatinan*. He confirmed the right of the customary land of each *pebatinan* by issuing land certificates (*Gran Sultan/ Selat Keterangan Hutan Tanah*). Following this example, his successors also issued land certificates to each local area in the kingdom's territory in compliance with the wishes of local heads until the kingdom became part of the Republic of Indonesia (Effendy 1997a, pp. 632–633; Effendy et al. 2005, pp. 86–87, 120–121). Effendy (1997a, p. 632) describes village governance under the Pelalawan kingdom in the following manner: "given the freedom to organize their law and custom as well as having rights to ownership, management, use and care of the land, the Petalangan people [the people of the 29 *pebatinan*] were able to lead their way of life unhindered, adhering to the model and the values laid down by the *adat*."

---

[3]Effendy T, Jaafar HTS (1982) Selayang pandang latar belakang sejarah peskuan batin-batin di kecamatan Pangkalan Kuras. Unpublished.

Fig. 3.1 A map of the 29 *pebatinan* created during the Dutch colonial era and preserved in Tenas Effendy's house (Reprinted from Masuda 2012 with the permission of the author)

In a map depicting the names and boundaries of the 29 *pebatinan* made by the Dutch colonial government (Fig. 3.1), Rantau Baru is named *Kebatinan Rantaubaroe* and is in almost the same location as the present-day village. Masuda (2012, p. 248, 286–297) deduces that this map was created between 1916 and 1938 based on the division of colonial administrative areas.

Following Indonesian independence, the Pelalawan kingdom was reorganized as an administrative district, or *wilayah kewedanaan*, and the king became its head (*wedana*). The four regional heads of the kingdom, or *orang besar*, became sub-district heads (*camat*), and the *batin* were appointed as village heads (*penghulu*) (Effendy 1997a, p. 633, 2002, p. 368).[4] In 1947, land certificates from all 29 *pebatinan* were gathered to determine the boundaries of the administrative villages. However, when the Dutch military reinvaded the Indonesian archipelago, soldiers stationed in Pelalawan and the surrounding area experienced fighting and damage. In the ensuing disorder, the certificates were lost and remain missing today (Effendy 1997b, p. 58; Masuda 2012, pp. 254–255).

From the late 1940s to the mid-1960s, a village was called a *kepenghuluan*. If the territory of a *kepenghuluan* became too unwieldy, it was divided into several new

---

[4]In 1950, the Pelalawan area was made part of the new administrative district of Kampar. In 1999, Kampar District was divided and the new Pelalawan District was established.

kepenghuluan, and new *penghulu* were appointed as heads of the new villages (Masuda 2012, p. 263). In 1965, when the central government integrated local administrative systems,[5] these *kepenghuluan* were reorganized as *desa*. In these processes, the village heads were gradually replaced by people from outside the village because most existing heads did not fulfill the education qualifications that were newly required by the central government. As a result, the authority of the *batin* was separated from the official administration (Effendy 1997a, p. 633, 2002, pp. 366–369).

3.2.2 The Position of the People under the State

While Effendy's works describe the Pelalawan kingdom's rule as being void of any major problems for the people, colonial records describe a slightly different relationship between the kingdom and communities. According to a record written by a Dutch person who visited Pelalawan (Faes 1882) in the late 19th century, the people that he met (probably rulers and officials at the center of Pelalawan) identified themselves as the descendants of the Johor kingdom, while he found that they were influenced by Minangkabau culture (1882, p. 496).[6] In Pelalawan, the cultivated land was owned privately, and the king only had the right to tax crops and forest products. However, the king levied fines upon the people as he wished, and if the fines were not paid, the offenders were tortured (Faes 1882, p. 500).[7] In the face of such harsh conditions, the people of several *pebatinan* fled and the overall number of *pebatinan* in the kingdom gradually decreased. Although Dutch colonial officials acknowledged 29 *pebatinan* in 1811, the number decreased to 22 by 1865, and only 11 remained at the end of the 19th century (Faes 1882, pp. 518–519). Faes (1882, p. 496, 519) does not explain the details of this decline, noting that it was caused by the population reduction as a result of heavy taxation and corvée imposed by the Pelalawan kingdom after usurpation by the Siak kingdom. Masuda (2012, p. 243) supposes that the reduction of *pebatinan* in the mid-19th century was caused by their temporary move upstream to hinterland forests to avoid taxation. As the 29 *pebatinan* have been maintained to date, Masuda's supposition seems plausible.

Other colonial records indicate that the people of the 29 *pebatinan* occupied a marginal position in the state system. The main people of the *pebatinan* were once

---

[5] *UU No. 19 Tahun 1965 tentang Desapraja sebagai Bentuk Peralihan untuk Mempercepat Terwujudnya Daerah Tingkat III di Seluruh Wilayah Republik Indonesia.*

[6] The population of the kingdom in 1880 was approximately 6000 people (Faes 1882, p. 496).

[7] Faes (1882, pp. 504–505) points out that the revenue of the kingdom largely depended on import and export duties, harbor dues, and transit duties in addition to licenses for timber logging in the territory and revenues from the sale of opium. Given the general characteristic that the states in this region largely depended on mediating trade between the Minangkabau highlands and the maritime trading kingdoms of the coastal areas, governing the territory and people might have been rather less important for the Pelalawan rulers.

called the Talang (*Orang/Suku Talang*) or the Petalanagan, and these names remain in their *adat* (mentioned later). They were regarded as different from the Malays from the coasts of eastern Sumatra and the Malay Peninsula, that is, coastal Malays (*Melayu Pasisir*, see also Kasori, Chapter 11), who comprised the majority of eastern Sumatra.[8] Nieuwenhuijzen (1858), who visited the capital of the Siak kingdom, distinguished the Suku Talang (Talang tribes), non-Muslims living in forest areas, from the king's subjects (*hamba raja*), who were Muslims living in settlements controlled by the kingdom. Indeed, the king's subjects, or ordinary citizens of the kingdom, were typically coastal Malays.[9] However, the categorization of the people during this era was ambiguous. For example, Nieuwenhuijzen (1858) classified as Suku Talang not only tribal groups living in the forest areas of the Kampar Basin, but also those living along the tributaries of the Siak River (present-day Sakai).

Van Anrooij (1885), a Dutch official who visited the Siak kingdom in the latter half of the 19th century, distinguished the people living in the Siak-Pelalawan border area from other tribal groups. According to his record, the boundaries were called "*pertalangan,*" and the people who lived in this area were the Talang (*Orang Talang*) (van Anrooij 1885, pp. 294–295).[10] These people were nominal Muslims who lived in the forest, depended on swidden cultivation (van Anrooij 1885, p. 295), and were categorized as "king's folk," or *rakyat raja*, a class beneath the king's subjects (*hamba raja*).[11] While food sharing and marriage between the Talang and the king's subjects had previously been avoided, inter-class marriages increased at the end of the 19th century (Van Anrooij 1885, p. 324). Although van Anrooij acknowledges that he is not aware of the system through which people were classified into these two different classes, he infers that the degree of their belief in Islam was the essential criterion (1885, p. 324). According to Wilkinson's dictionary (1959), "Orang Talang" refers to the people living on the eastern coast of Sumatra, who had "have become Moslems and [followed] the Adat Minangkabau but [retained] many of their primitive customs and beliefs" (1959, p. 1154).

The "backward" or "tribal" image of the Talang visible in the Dutch records seems to reflect the perspective of the coastal Malays, who originated in the coastal areas and were the majorities and/or rulers in the region. As the people of the 29 *pebatinan* lived in relatively upstream forest areas and depended on swidden

---

[8] In 1995, the population of the Petalangan was approximately 58,000 people, while the population of the Malay, including those in Riau, in 2000 was 1,790,000 people (Masuda 2012, p. 62).

[9] In the Siak hierarchy, Minangkabau, who supported the establishment of the Siak kingdom and their descendants, had an exceptional independent high position in the kingdom as "four lineages" (*empat suku*) (van Anrooij 1885; Barnard 2003; Nieuwenhuijzen 1858). While the Talang have a matrilineal system similar to that of the Minangkabau, their position was much lower than the Minangkabau.

[10] As these border areas were covered by thick forests, there were no clear boundaries between the kingdoms (van Anrooij 1885, pp. 294–295).

[11] In contrast to Nieuwenhuijzen, van Anrooij (1885) distinguishes the Talang from other tribal groups and asserts that they were in the position of *rakyat tantera*, which is one rank below *hamba raja* but above *rakyat banang*, which included tribal groups, such as the Sakai.

cultivation, the coastal Malays would have considered them "backward" or "tribal" minorities in the peripheral areas of the state. However, the actual relationship between these people and coastal Malays must have been heterogeneous in each local community. Indeed, some communities would have accepted the obligations of taxation and the services of the state.[12] In this process, as they converted to Islam and married coastal Malays, they came to be regarded as "Malays of the Pelalawan kingdom." However, while communicating with the kingdom and other communities, some would have lived in the upstream forests of the tributaries. When their relationship with the kingdom deteriorated, they moved to more upstream forests around the boundaries of the kingdom and lived there. As Hamidy (1987, p. 22) notes, the culture and people of this region can be characterized by its mixed nature, and their identity and positions were formulated through everyday interactions in history. In this situation, their territories were recognized by the Pelalawan kingdom based on the *pebatinan* system, while the communities maintained their autonomy to a large extent.

3.3 Rantau Baru Village and Land Management Based on *Adat*

3.3.1 Rantau Baru Village and its People

Rantau Baru was one of the 11 *pebatinan* that continued throughout the 19th century (Faes 1882). The present-day main settlement of Rantau Baru village is located on the northern banks of the Kampar River, a few kilometers downstream of the confluence with the Bokol Bokol River, a narrow tributary that flows from the northern hinterland to the Kampar. "Rantau Baru" can be translated as "new frontier," and the present-day main settlement is rather new. The old settlement, called Malako Kocik (small Malacca), was situated on the south bank of the Kampar and faced confluence. Although the villagers do not know the exact year, several generations ago, a plague spread through the old settlement and its people moved to the new one. Currently, the site of Malako Kocik is used as a public cemetery for the village, and the name is used for a sub-village of the present-day main settlement. Several other old names, such as "Kampung Tolok" and "Kampung Tuo," have also been handed down; however, the places and periods from which they originate are unknown, as the site of the settlement would have repeatedly moved around the confluence. While many *pebatinan* were divided into several administrative villages after the independence of Indonesia, Rantau Baru was not divided. It is unclear when the villagers adopted Islam.

---

[12] The Petalangan community, for example, played an important role in the rituals of the Pelalawan kingdom as members of the royal orchestra, or *nobat* (Turner 1997).

There are two settlements in the administrative village of Rantau Baru. The main settlement is composed of the administrative sub-villages (or *dusun*) of Sepenjung (Dusun 1) and Malako Kocik (Dusun 2). These sub-villages have no clear boundaries and are typically referred to collectively as Rantau Baru. This chapter refers to this collective settlement as "the main/traditional settlement." According to the village census in 2020, the population of this settlement was 581. The other settlement is the Sei. Pebadaran (or Dusun 3), which is eight kilometers north of the main settlement. This settlement was formed in 2004 with district government support (mentioned later) and had a population of 206 in 2020.

The main language in Rantau Baru is a Malay dialect that is understood by other Riau Malay and Indonesian speakers. Villagers call their language "rural Malay" (*bahasa Melayu kampong*), as opposed to Riau Malay or Indonesian. According to historical records, people in this area were called Talang (as mentioned above), while academic papers written after the 1990s often refer to them as Petalangan. There are several hypotheses regarding the meaning of *talang*. In Dutch records describing the history of the Pelalawan kingdom, *talang* means "district" (*distriten*) (Faes 1882, p. 500). Effendy (2002, p. 364; see also Wilkinson 1959, p. 1154) implies that the term *talang* is related to "middleman" or "trader". In another article, however, he suggests that the name is derived from the local use of bamboo (*talang*) to mark settlements and obtain water (Effendy et al. 2005, p. 115). Hamidy (1987, pp. 22–23) and Masuda (2012, p. 60, 285) assert that *talang* refers to a settlement or space in the forest.

Although the villagers know that they have been referred to as Talang in the past, they never identified themselves as Talang or Petalangan during the duration of our study.[13] One middle-aged villager explained that Talang/Petalangan implies "primitive" or "living in the forest with no clothes." His parents' generation is particularly reluctant to be called these names and always identify themselves as Malays (*Orang Melayu*) or inland Malays (*Melayu Daratan*) when they want to differentiate themselves from coastal Malays.

Such discrepancies between self-identification and identification by others are common in eastern Sumatra. This is a result of the process of state expansion between the 18th and 19th centuries, during which the state designated ethnic names for the people living in peripheral areas (Osawa 2022). For example, indigenous people living in the upstream areas of the Mandau River are referred to as the Sakai in colonial records and Indonesian government documents (e.g., van Anrooij 1885), but they called themselves the Batin (*Orang Batin*) in the past and only adopted the government's categorization and began calling themselves Sakai in the 1960s (Porath 2003, p. 4). Indigenous people living in the Bengkalis Island are referred to as the Utan (*Orang Utan*) in colonial records and Indonesian government documents. Reluctant to be called this name, they negotiated with the district

---

[13] Masuda, an anthropologist who has worked in the nearby village of Betung, Pangkalan Kuras Sub-district has also never heard villagers refer to themselves as Talang or Petalangan (personal communication).

government to recognize them with their preferred name, the Suku Asli (Osawa 2022). "Talang" seems to have been a name used by the Pelalawan administration to collectively refer to various peoples in the kingdom. While it is uncertain whether local people accepted this name during the Pelalawan era, given that they never call themselves the Talang today, we can assume that the name was a given one, not one of self-identification.

In contrast to Talang, the name "Petalangan" has been found in academic works since the 1990s. The use of this term would be associated with a rise in support for indigenous communities facing large-scale deforestation in Sumatra caused by the development of oil palm and acacia industries during this period (see Okamoto et al., Chap. 2). Land grabbing threatened the lives of indigenous communities living around the forest, such as the Orang Rimba in Jambi province. In the 1990s, NGOs like WARSI tried to support the Orang Rimba, but it was extremely risky to claim their land rights under Suharto's authoritarian regime (Li 2001, pp. 666–672). Therefore, they linked their support to nature conservationism, which was a popular cause among national and international donors, critics, and academics (Li 2001, pp. 656–658). Instead of demonstrating land rights, they emphasized the importance of natural conservation. In this process, they differentiated the Orang Rimba from "ordinary" rural communities, emphasized their way of life as harmonious and symbiotic with the nature, and urged the government to conserve the rainforest and their lives (Li 2001). In Riau, Tenas Effendy and his networks were the main agents of this approach. They established an NGO, received support from donors, and claimed the need to protect the rainforests and the lives of people living in the forest environment in Pelalawan (Masuda 2009). It is highly probable that during this process, Effendy and his network began calling local people ethnic "Petalangan." While emphasizing their position as a version of the Malays, they underlined their position of having lived harmoniously with the forest environment. In this context, many studies have explored local traditions and the use of resources in a manner that is harmonious with natural surroundings (e.g., Effendy 1997a, 2002; Hamidy 1987; Turner 1997).[14]

While neither the term Talang nor Petalangan is used by Rantau Baru villagers to describe their ethnicity, these terms are used to refer to their *adat* (traditions, customs, and customary law). Their *adat* handed down by the people living in the 29 *pebatinan* is called "*Adat Melayu Petalangan*" (Petalangan Malay *Adat*). This is regarded as distinct from "*Adat Melayu Pasisir*" (Coastal Malay *Adat*), which was passed down by Malays living in downstream areas. In this context, Rantau Baru villagers may identify themselves as "inland Malays" (*Orang Melayu Darat*) to distinguish themselves from coastal Malays. While *adat* generally involves practices and rules, such as ceremonies, rules of inheritance, management of forest and river resources, and magic, the *Adat Melayu Petalangan* is distinctly characterized by its

---

[14]Neither the term Talang nor Petalangan is registered on the national list of politically and geographically isolated traditional communities (*Komunitas Adat Terpencil*) in Riau, which identifies people in marginalized and tribal positions.

matrilineal social system, in contrast with the non-lineal, or bilateral, system of *Adat Melayu Pasisir*. In *Adat Melayu Petalangan*, houses, gardens, and other properties are inherited via maternal lines, and the lands in the village territories are managed by exogamous matrilineages or *suku*.[15] There are generally several *suku* in each settlement. The total number of *suku* in the 29 *pebatinan* is unknown (Masuda 2012, p. 92).

3.3.2 Traditional Local Governance: Roles of the Suku and Adat Leaders

Three exogamous *suku* exist in Rantau Baru: Suku Melayu Datok Tuo (or Melayu Tuo), Suku Meliling, and Suku Melayu Datok Mudo (or Melayu Mudo).[16] According to the villagers, the ancestors of Melayu Tuo and Meliling were the first to settle in the area, and the ancestors of Melayu Mudo later joined the settlement. Melayu Tuo and Meliling possessed the most land, while Melayu Mudo possessed a section. The traditional title of the head, Datok Sati, called Batin Bokol Bokol in the past, is bequeathed to a male of the Melayu Tuo, while the title of Datok Sari Koto and Datok Mangku, which are regarded as the deputy headman, are bequeathed to the males of the Meliling. Nowadays, Datok Sati lives in the town of Sorek, which is dozens of kilometers away from Rantau Baru, and Datok Sari Koto and Datok Mangku manage the ceremonies and procedures related to *adat* on his behalf. In the main settlement, the three *suku* each possess land for homesteads. Although the boundaries are quite vague, many Meliling women have homesteads in the southeast part of the village, while the Melayu Mudo have homesteads in the northwest and Melayu Tuo in the area between the two. According to our questionnaire survey, although several Javanese and Malays from outside the village live in the main settlement, in every case, either the respondent or their spouse belongs to one of the three *suku*.

Effendy (1997b) notes that following independence, many *batin* from the *pebatinan* were replaced by village heads from outside of the villages. However, Batin Bokol Bokol maintained his role as the administrative village head. After Independence, the Batin Bokol Bokol became the *penghulu,* and his successor also took the role of *penghulu.* When the national law of village governance was implemented in 1965, the title of the administrative village head changed from

---

[15]While their matrilineal system would have been influenced by the Minangkabau, which is the largest matrilineal society in the world, the *Adat Melayu Petalangan* is distinct from that of the Minangkabau to a certain extent (Masuda 2012, p. 92, 116).

[16]When people who belong to other *suku* from different villages are admitted to Rantau Baru through marriage, their *suku* is replaced by the three *suku* based on the historical relationship between them. For example, when a man belonging to Suku Dayung marries a Meliling woman, he is regarded as a member of Melayu Datuk Mudo because Dayung is identified with Melayu Datok Mudo in the village.

penghulu to *kepala desa*. After the introduction of village head elections in the 1980s, one of the two *adat* leaders described above (Datok Sati, Datok Sari Koto, and Datok Mangku) was chosen as the *kepala desa* by the villagers. In 1998, a villager who did not play the role of the *adat* head was selected as the *kepala desa*.

In addition to the three aforementioned *adat* leaders, the three *suku* have their own heads. These six heads are called *ninik mamak*.[17] The three *ninik mamak* of the *suku* are positioned under Datok Sati, Datok Sari Koto and Datok Mangku; they settle disputes within their own *suku* and act as representatives for all members of their *suku*. Each *suku* head is assisted by (1) two men (called *ketuo anak jantan*) who represent the male children of the *suku*, (2) two women (called *ketuo anak betino*) who represent the female children of the *suku*, and (3) two males belonging to other *suku* (called *ketuo semondo*, or "head of one hut"). As marriage between members of different *suku* is regarded as a marriage between the *suku*, *suku* heads and their assistants manage the engagement and wedding ceremony procedures in the village. These people also play important roles in the management of natural resources such as rivers, forests, and land. Each year, *ninik mamak* hold a bid meeting to establish fishing rights for that year in dozens of Kampar tributaries and oxbow lakes, with *ninik mamak* and *ketuo semondo* constituting a committee to manage the meeting. The funds from bid sales are used for various welfare initiatives, such as managing and repairing mosques and providing grants to children who have lost either of their parents.

The bid meeting represents villagers' attempts to manage the space and resources in the river. This is a typical fishing village in this region; its villagers largely depend on fishing in the Kampar River and its tributaries for their livelihoods (see also Nakagawa and Nofrizal et al., Chaps 4 and 5). According to Masuda (2012, pp. 213–223), in the village of Betung at the midstream of the Nilo River, commercial fishing using raft huts was developed after the 1970s. In Rantau Baru, raft huts have not been used at all, and it is unclear when the bid meeting began or when commercial fishing became the main livelihood. According to the villagers, while fishing has been an important source of their livelihood, they depended more on swidden cultivation twenty years ago (mentioned later) and fished only during agricultural off-time and off-season. The hauls were both consumed in houses, processed, and sold to the cities of Kerinci and Pekanbaru via buyers who visited the village (see Nofrizal et al., Chap. 5). Commercial fishing in Rantau Baru would have been developed earlier than in Betung, as access from the cities to Rantau Baru was much easier than Betung through the mainstream of the Kampar. After swidden cultivation declined around twenty years ago, commercial fishing became the main livelihood in the village.

---

[17] "*Ninik*" or "*nenek*" means "grandparent," and "*mamak*" means "mother's brother."

3.3.3 Management of Sialang Areas and Swidden Cultivation

The *suku* manage the *sialang* trees and the surrounding forests. *Sialang* trees are often especially big and can be regarded as landmarks. Bees build their hives in these trees, which are protected alongside the surrounding forests as *sialang* areas (*kepung sialang*). It is prohibited to cut down the trees in these areas. In Rantau Baru, *sialang* areas are concentrated on the bank of the Kampar with mineral soils, and each *suku* possesses one *sialang* area and manages many *sialang* trees. During our visit to the village, a man living in a neighboring village cut down a *sialang* tree possessed by Melayu Mudo. The *ninik mamak, ketuo anak jantan,* and *ketuo anak betino* of Melayu Mudo repeatedly held meetings as representatives of the *suku*. After negotiations, it was decided that the man had to pay compensation. It is said that *sialang* trees and the surrounding forests are possessed and inherited by *suku*, and the yields of honey and beeswax should be shared among the *suku* members. While such *sialang* trees still exist, trees that are protected and grown by individuals may be possessed by individuals and inherited by their close kin. In these cases, the yields are shared between the owner and the honey collectors rather than by the *suku*. Such *sialang* trees have become more common in recent years. Nevertheless, the trees and the surrounding forests are associated with identity and cosmological order for the people around the midstream Kampar, in which *sialang* trees are identified with human beings (Effendy 2002). In 2021, the umbrella organization of Malay communities across Riau that conducts the activities for the protection and prosperity of Malay tradition and culture, the Association of Malay *Adat* (*Lembaga Adat Melayu Riau*), required the district head (*bupati*) of Pelalawan to properly protect *sialang* forests from deforestation by industries around the regions of the midstream Kampar based on the local *adat*.

In terms of land management, the *suku* managed the plots for the slash-and-burn cultivation of rice on mineral soil, which was conducted on the south bank of the Kampar until approximately twenty years ago. During the dry season, which begins in March and April, villagers would open the forests and burn them, build a hut to live in for four or five months, and cultivate rice and several types of vegetables (such as cucumbers) for their own consumption in the dry field. They continuously cultivated a plot for two or three years before leaving it and opening a new plot. Some families planted rubber and oil palm trees at the corner of the plot and carried the yields to the town of Kerinci. They were allowed to use the land until the trees were blasted. The villagers could open any forests with the agreement of *ninik mamak* and *suku* members, with the exception of *sialang* areas and the plots in which other villagers had already planted trees. However, as an important *adat*, the lands around a plot that had been opened by the members of a *suku* were meant to be cultivated by those of the same *suku*. It was prohibited to cultivate plots next to the plots of different *suku* in a mosaic-like manner. While slash-and-burn cultivation sustained their livelihood together with fishing until approximately twenty years ago, it suddenly declined. This is because while villagers had been able to predict the seasonal floods of the Kampar in the past, irregular floods increased and the crops were repeatedly spoiled.[18] At present, most of the land for the field has become a secondary forest, while a part is used for rubber and oil palm gardens.

The river, land, and forests in the realm of the *pebatinan* in the past and the administrative village of Rantau Baru in post-independence Indonesia were controlled by the three matri-lineages of the *suku*. This means that the space in the village was possessed not by individuals but by *suku* as a common space. According to the villagers, although *ninik mamak* have the authority to control the space as representatives of their *suku* to a certain extent, it is necessary for them to agree with *suku* members if they want to yield or sell the space in the village territory to outsiders.

3.3.4 Floods and Resettlement Programs

A large part of the village area is situated in the floodplain of the Kampar, and the land suffers from seasonal floods during the flood season between November and March. The scale varies from year to year, but the relatively low land, including most of the main settlement, is covered in water every year. Therefore, the houses in the settlement are built on piles that are approximately two meters high. During the floods, the villagers use boats for transportation, the primary school in the village is closed, and youths who travel to junior high or high schools in Kerinci by motorcycle or omnibus during the dry season must board outside the village. The people here live in this environment because, according to the villagers, the haul from fishing in the Kampar tributaries increases just after the rainy season. However, because of the seasonal isolation and difficulty regarding sedentary cultivation, the village's population has been in flux and many people have migrated to other places. Many people who were born in the village moved to the town of Sorek to ensure a stable livelihood (*untuk cari makan*) and engaged in trade and other labor. Most of them have maintained a kin relationship with the villagers and often return to the village. We were told that when a mosque in the village needs to be repaired, the people who live in different places significantly contributed to its repair through donations.

Both the government and the villagers regard floods as the main reason for the village's delayed development. During the last three decades, two large-scale resettlement programs have been implemented in Rantau Baru. The first program involved resettlement to present-day Kiyap Jaya Village, which is upstream of the Bokol Bokol River. This program was led not by the government, but by a network of *batin* headmen. The upstream hinterland of the Bokol Bokol River was the realm of Pebatinan Sekijang Mati (one of the 29 *pebatinan*) and became an administrative

---

[18] The reason for these irregular floods could be related to the charging and discharging of water by a hydroelectric dam (PLTAK Koto Panjang) at the upstream of the Kampar Kanan River, which began its operation in 1997.

village of Kijang in post-independence Indonesia. At the beginning of the 1990s, a man with the title of Datuk Sekijang who had spent his childhood in Rantau Baru sympathized with the village's situation of flooding every year. Therefore, he decided to cede 500 hectares of unused forest land in his *kebatinan* to Rantau Baru after making agreements with the *ninik mamak* of Rantau Baru and his communities. As a result, 100 households, which were selected by drawing lots, moved to a new place, received five hectares of land each, and began cultivating oil palms. At first, the new settlement was called "Rantau Baru Atas" (or Upstream Rantau Baru; the old settlement was called "Rantau Baru Bawah" or Downstream Rantau Baru). In 2005, the settlement became a new administrative village of Kiyap Jaya alongside several settlements that belonged to the village of Kijang. While the settlement is considered under the control of Datok Sati in terms of *adat*, the *suku* heads are different from those of Rantau Baru. Almost all villagers in present-day Rantau Baru have relatives in the village of Kiyap Jaya.

The district government led the second resettlement program in the early half of the 2000s. In this project, the district government built 100 houses made of brick and concrete and prepared gardens of 500 square meters for each house in the hinterland, where the influence of the floods was minimal. These homesteads were provided to villagers in the main settlement. The village office was also built in this area, which was called Dusun 3. However, the program was unsuccessful. First, as Dusun 3 was far away from the Kampar, which many people depended on for their livelihood, many families who received a new homestead returned to the main settlement just after they moved. Second, the soft ground of peat soil was unsuitable for the houses, and in many houses, the floor was tilted and the whole structure was distorted. These houses were dismantled. Third, although some houses could be used, the settlement was still influenced by floods. Consequently, many people returned to their old houses in the main settlement, which resulted in many empty houses. Therefore, the village office is implementing a policy to lend empty houses to migrants who have moved to the village to work in oil palm plantations. In our questionnaire survey, we obtained 45 responses from 49 houses in the Dusun 3. We found that 29 respondents and their spouses did not have any kinship with the *suku*. Their ethnic backgrounds were Javanese, Bataks, Nias, and Malays born outside of Rantau Baru.

One villager who was born and brought up in Rantau Baru and worked in the village office stated that "Rantau Baru is the eldest, but the most undeveloped village in Pelalawan District." While it is uncertain whether this village is the oldest in the region, it is certain that people have lived around the confluence of the Kampar and Bokol Bokol for several centuries and controlled the land, river, and resources based on their matrilineal social system. However, the underdevelopment of this village in relation to other villages in Pelalawan should be given attention. Based on the questionnaire survey, Prasetyawan (Chap. 9) notes that the expenditure per capita of this village undergoes the poverty line in Riau. The marshy peatland and flood-plains that cover most of the village area have prevented them from changing their livelihood from fishing to being owners of oil palm gardens that usually provide better incomes and have been the main livelihood in neighboring villages. In such a situation, the relationship between the villagers and their ancestral land has transformed.

3.4 Development of Peatlands and Change in the Relationship

3.4.1 Construction of a Road and Plantations and Change in the Peat Environment

While the people in Rantau Baru have maintained the village territory based on their matrilineal system, the space that they manage only comprises the Kampar River, its tributaries, and the mineral banks along the river. The main stream of the Kampar, an oxbow lake, and the Bokol Bokol River are common fishing grounds among the villagers. The tributaries and creeks, which are only two-three meters in width, are managed at a bid meeting once a year. Mineral banks are used for the *sialang* areas, which are owned by the *suku*, and fields, which are opened on the bank with an agreement among the *ninik mamak* and *suku* members. However, peat hinterlands, which are regarded as the historical and administrative territories of the village, have not been managed by *suku* in a concrete manner. Until the mid-1980s, the peatland was covered by a thick rainforest and a swampy and marshy area, which prevented people from entering the area and using its resources. The peat swamp forests were only used for the occasional logging of timber and harvesting of honey and beeswax, as the village only possessed a few *sialang* trees (not *suku,* according to the villagers) in the swampy forest (see also Osawa, Chap. 6). The peat swamp forests were only used during the flood season because the flood water provided access to the inside of the forest via boats. The main settlement was surrounded by swampy forests, and there were no paths that connected the settlement to other places. Therefore, transportation was completely dependent on the waterlines. However, this situation changed at the end of the 1980s when the acacia and oil palm industries were introduced to the village.

The first turning point occurred in the mid-1980s when Asia Pulp and Paper, a large acacia company, obtained a concession of the vast land in Kampar District (present-day Pelalawan District was a part of Kampar District) and constructed a large industrial road, Jl. Korridor Riau Andalan Pulp and Paper (Jl. RAPP), which crosses the northern hinterland of Rantau Baru. Following this, a path connecting the main settlement with the road was constructed, which connected Rantau Baru with the town of Kerinci and other villages by land routes. In the 1990s, two vast oil palm plantations, operated by the companies of Pusaka Megah Bumi Nusantara and Langgam Inti Hibrindo, were constructed in the peat areas northwest and southeast of the village (see Binawan and Osawa, Chap. 10). They constructed canals and water gates in the swamp forests, controlled water, and created grounds that were suitable for growing oil palms. Although the areas of the oil palm plantations partly overlapped with the village territory, the villagers did not take action against it or negotiate.

The construction of industrial roads and oil palm plantations changed the landscape of the peat hinterland. Since approximately 1995, peat swamp forests have suffered from fires during the dry season and transformed into grasslands. Due to these fires, a few *sialang* trees and their forests were burned. According to the villagers, while there were occasional small-scale fires in the peatlands before the 1990s, both the scale and frequency dramatically increased after the mid-1990s. The villagers believe that the main reason for this is the increase in the number of outsiders entering the peatlands. After the construction of Jl. RAPP, many anglers living in Pekanbaru and Kerinci began to fish the tributaries and canals. They fished throughout the night and built fires to avoid mosquitoes. Carelessness with these fires or cigarettes is believed to have been the cause of peatland fires. However, the villagers rarely criticized road construction or oil palm plantations as the causes of fires.

We asked some villagers why they had not taken actions against the construction of the road and the oil palm plantations that encroached on the village territory. According to a male village official, these projects were impelled by the district governments, so they had no choice in the matter. Moreover, the village has opened to the outside world because of the road and the villagers who work in the company. Therefore, they do not have the courage (*tidak berani*) to complain or criticize the work of the companies or the government, regardless of whether they cause fires.

The construction of Jl. RAPP stimulated the dramatic transformation of the natural and social landscape of the peat hinterland. It introduced oil palm industries to the peat hinterland of the village territory, which the villagers rarely used. Then, fires frequently occurred, and the landscape of swampy forests was lost. It is noteworthy here that Rantau Baru is one of the villages targeted by BRG's Peatland Care Village (*Desa Peduli Gambut*, DPG) program. Although a program facilitator visited the village in 2019, they did not actively educate the villagers or conduct any collaborative projects. Additionally, the district government has not conducted any particular programs in the village for restoring degraded and dried peatland. The impact of the government's peatland restoration policies has been minimal in Rantau Baru (see also Osawa, Chap. 6).

3.4.2 Expansion of Oil Palm Plantations and the Spread of Land and Compensation Letters

At the end of the 1980s, a path (Jl. Lama, mentioned later) connected the main settlement and Jl. RAPP. While it connected the village with the outside world via a land route, it also enabled the villagers to access the peat hinterland more easily. As a result, since the mid-2000s, peat hinterland in the traditional village territory of the *tanah wilayat* began to be documented and commodified.

In Indonesia, oil palm industries were typically developed after the mid-1970s under collaboration schemes between large-scale companies and their contracted farmers (*Perusahaan Inti Rakyat*, PIR scheme). The PIR scheme was modified several times, and in the latter half of the 1990s (around the collapse of Suharto's regime), it became possible for small- to medium-sized enterprises to bring in oil palm industries (Kawai 2021; Nagata 2021, p. 80). In the 1990s, for various reasons (such as the rise of cultivation technology and the increase in support from local governments other than the PIR scheme) the area of oil palm plantations that was owned by local farmers began to rapidly increase across Indonesia (Kawai 2021, pp. 37–38; Nagata and Arai 2013). The PIR scheme was revised again in the mid-2000s, which enabled the small landholders to choose to either work on their land as worker-cum-owners or receive the revenues of their plantations as unearned incomes (Kawai 2021, pp. 40–41). In accordance with the relaxation of regulations and the expansion of the oil palm industry, the demand for land rapidly increased, and land in rural areas has become the target of dealings in Riau.

In Riau, deals with land after the 1980s were often made by producing a set of two letters—land letters (*Surat Keterangan Tanah*, SKTs) and compensation letters (*Surat Keterangan Ganti Rugi*, SKGRs)—which are issued by administrative village offices and signed by the sub-district head (Dethia et al. 2020; Mujiburohman et al. 2015). As a large part of Riau was categorized as state forest (*hutan negara*) based on the Basic Forestry Law in 1967[19] and its successor, the Forest Law in 1999[20] (see Okamoto et al., Chap. 2), the basic title of the land is attributed to the Indonesian state (Dethia et al. 2020, pp. 425–426; Gusliana et al. 2019; Fitzpatrick 2007, p. 138). Therefore, when people want to claim their right to land use in such areas, they need to obtain permission from the government. In the 1970s, the right was recognized by obtaining slashing permission (*izin tebas tebang*).[21] If the objective area was less than two hectares, the head of the sub-district had the authority to issue permission by considering the opinion of the village head (Mujiburohman et al. 2015, pp. 170–171). However, the permit did not issue a right over the land that people once had slashed and cultivated but did not use at the time of applying for permission (Mujiburohman et al. 2015, pp. 172–173, 180–181). Therefore, village offices began to issue SKTs, explaining that the letter holder could open and use the land. In the 1980s, oil palm companies began encroaching on land close to settlements, and the issuing of SKTs rapidly prevailed across Riau as a countermeasure (Mujiburohman et al. 2015, pp. 179–180). An SKT explains that the land has been cultivated by the letter holder; it is acknowledged by land users around the land concerned, the heads of neighborhood associations (*Rukun Tetangga/Rukun Warga*,

---

[19] *UU Nomor 5 Tahun 1967 tentang Ketentuan–ketentuan Pokok Kehutanan.*

[20] *UU Nomor 41 Tahun 1999 tentang Kehutanan.*

[21] The cutting permission corresponded to the opening permission (*izin membuka tanah*) in the Ministerial Regulation of Ministry of Home Affairs No. 6 in 1972 regarding the Delegation of Authority for Granting Land Rights (*Peraturan Menteri Dalam Negeri Nomor 6 tahun 1972 tentang Pelimpahan Wewenang Pemberian Hak Atas Tanah*) (Mujiburohman et al. 2015, p. 170).

RT/RW), and the village head; and reported to the sub-district head (Mujiburohman et al. 2015, pp. 183–184). Thus, SKTs became a general document showing the right to use land in Riau.

In contrast, the SKGR is a proof of the transfer of land use in state forest areas (Dethia et al. 2020; Mujiburohman et al. 2015). This transfer means that the transferee must produce the new SKT and pay "compensation" (*ganti rugi*) to the transferor to use the land for their own purposes (Dethia et al. 2020, p. 426). Additionally, the letter is used to mortgage land (Dethia et al. 2020). The letter form is issued by village offices, filled in by both the transferor and transferee, acknowledged by the heads of neighborhood associations, and signed by the heads of the village and sub-district, which is similar to the SKT (Dethia et al. 2020, p. 427). The detailed legal standings of SKTs and SKGRs are incomplete and controversial; Dethia et al. (2020, p. 427) states that SKGRs in Riau are not related to the legal control of the land (cf. Mujiburohman et al. 2015). Nevertheless, land is transferred through the SKT and SKGR, and people understand this procedure as selling and buying land (Dethia et al. 2020, p. 429). The latter has enabled Rantau Baru villagers to sell the right of land use in the village territory, which is situated in the state forest area.

3.4.3 Compartmentation of the Land

It is unclear where Rantau Baru villagers obtained the slashing permissions before the introduction of SKTs or when the village office first issued SKTs. However, some households asked the village office to issue one and held an SKT before 2000. At first, these letters were issued for the land that the households continuously used for several years, such as their homesteads and the gardens on the Kampar bank, where they planted rubber trees and oil palms. While these lands were owned by the *suku*, they could obtain the letters for the lands that they controlled when the *ninik mamak* agreed.

The village office of Rantau Baru began to sell peat hinterlands in the village territory after the mid-2000s. During this period, vast peatlands south of Dusun 3 and the southern hinterland of the Kampar were sold to oil palm companies via brokers. As these deals were conducted at the village head's discretion without the agreement of the village community. Thus, the details of the processes are unknown. However, these were unused peatlands far from the main settlement, and the villagers did not complain to the head or resist their decisions. Similarly, a neighboring village, Pangkalan Kerinci Barat, sold land southwestward of Dusun 3 to brokers by issuing a SKT and SKGR, even though some of the land overlapped with the traditional territory of Rantau Baru. The overlapped area remains disputed between Rantau Baru and Pangkalan Kerinci Barat (see also Banawan and Osawa, Chap. 10). The new landholders are companies or urban residents who have planted or will plant oil palm trees on the land. For their plantations, they employ Javanese and Batak migrants who have experience in cultivating oil palm and offer relatively low salaries. According to a villager, Malays in Rantau Baru do not want to be employed in these plantations, as the salary is much lower than the income obtained through fishing.

In addition, the village office comparted other areas of peatland and distributed them to the villagers. The path connecting the main settlement and Jl. RAPP is called the Jl. Lama, and the office provided land along Jl. Lama to villagers in 2004. Under this policy, each household received one hectare (40 × 250 m) of land. In addition, in the mid-2000s, the village office launched a project to construct a new 8 km long path, Jl. Baru, between the main settlement and Dusun 3. Although the construction is still ongoing, the constructed part has already made it possible for people to access peat hinterlands. In 2012, the village office comparted land along Jl. Baru and provided two hectares (60 × 333 m) of land for each household. In these processes, the households that did not have land in the village and wanted to possess it could receive the lands with priority. The villagers who had land around the Kampar bank or who did not want to have it did not receive the land. In both cases, the village office issued SKTs when the villagers asked for them. If the villagers obtained the SKTs, they could sell or mortgage the land. While almost all land is covered by thick peat, some parts contain mineral soils. As many villagers wanted to receive mineral land, the locations of the land were determined via a lottery. Much of the land distributed by the village office has been sold to urban residents, including Chinese, Javanese, and Batak people in Pekanbaru. According to the villagers, if they obtain the land letters, they do not require the agreement of the *suku* to sell the land.

In this way, peatland in the hinterland that had not been used productively and had no clear boundaries was comparted and sold to outsiders or distributed to the villagers. Some people planted oil palm seedlings on the land. However, little was produced, mainly because of floods and fires. In the years in which large floods occurred, most of the land was submerged, destroying the plantations. In the late 2010s, for example, a Batak in Pekanbaru bought ten hectares of land along Jl. Lama and planted several thousand oil palm seedlings. Due to a large flood in 2018, however, all except for three seedlings of the thousands withered and died. In addition, frequent fires have burned oil palm seedlings and peat soils. The fires in 2016 and 2019 were on such a large scale that they burned a wide area of hinterland, and many oil palm trees (including adult trees) were damaged or died. If excavators are used to dig deep ditches or canals to manage the water and soil properly, it would be possible to produce oil nuts, even on peatlands in the floodplains. However, a sizable investment is needed, and the villagers do not have enough capital to modify the land.

3.4.4 Land Registration and the Commodification of Land

Peatlands along Jl. Lama and Jl. Baru possessed by the villagers or sold to outsiders have not been used productively.

Table 3.1 Land type and land use in Rantau Baru (Source of data: Based on responses from questionnaire survey)

	Years they had the land	Valid answers	State of soil			Use of the land				Gender registered on SKT		
			Basically mineral	Basically peat soil	Not known	Oil palm	Not used in productive ways	Already sold	Not known	Male	Female	Did not have the letter/ Not known
Homestead	1950~2020	89	68 (76%)	19 (21%)	2 (3%)					42(47%)	39 (44%)	8 (9%)
	(Only Dusun 1&2)	70	67 (96%)	3 (4%)	0					30 (43%)	38 (54%)	2 (3%)
Along Jl. Lama	around 2004	31	5 (16%)	25 (81%)	1 (3%)	8 (26%)	9 (29%)	14 (45%)	0	20(65%)	9 (29%)	2 (6%)
Along Jl. Baru	around 2012	61	4 (6%)	56(92%)	1(2%)	10 (16%)	30 (49%)	21 (34%)	0	46(75%)	11 (18%)	4 (7%)
Kampar Riverbank	2000-2010	20	16 (80%)	4(20%)	0	15(75%)ᵃ	5(25%)	0	0	10(50%)	6 (30%)	4 (20%)
Other place	2005-2018	23	10 (43%)	12(52%)	1(5%)	17(74%)	4(17%)	0	2(9%)	17(68%)ᵇ	6 (24%)ᵇ	2 (8%)

ᵃ The figure includes one case of a rubber garden.
ᵇ The figures include two households that had two SKTs with the names of a man and woman registered.

ᵃ The figure includes one case of a rubber garden
ᵇ The figures include two households that had two SKTs with the names of a man and woman registered

Table 3.1 compares the land types and uses in homesteads along Jl. Lama and Baru, on the Kampar Riverbank, and in other places (including outside Rantau Baru), according to responses to our questionnaire survey.[22] The amount of peatland along Jl. Lama (81%) and Jl. Baru was very high (92%). Moreover, 45% of the land along Jl. Lama and 34% along Jl. Baru had already been sold at the time of the survey. Furthermore, 29% (Jl. Lama) and 49% (Jl. Baru) of people had not used land productively.[23] Although we confirmed with several respondents that oil palm was cultivated, they noted that, "The trees are still growing before producing the nuts" and "the trees were heated by the fires and are almost dying." According to one villager, few villagers (and outsiders) can continuously harvest fresh fruit bunches using the land around Jl. Lama and Jl. Baru. However, the land on the bank that was previously used for slash-and-burn cultivation and the land in other places, including outside the village, is typically productive, with part of this land comprising peatlands.

[22] The land situation was only investigated with the people who answered "Yes" to the following question: "Have you or your family received land (including land that has been sold) or possessed land until now around (respective place)?"

[23] The figures of "not in productive way" are the total of the answers that chose the options of "Grassland" (*Sumak*), "Forest" (*Hutan*), and "Not used for anything" (*Tidak gunakan untuk apa saja*).

Here it is worth noting the gender listed on the SKTs. According to the *adat*, the land and houses are owned by women, as previously mentioned. However, it is more common to register land under a man's name on the SKT. Table 3.1 shows that even for the homesteads of the traditional settlement (Dusun 1 and 2), only around half of the respondents (54%) registered these under a woman's name. Of the land distributed along Jl. Lama in 2004 and Jl. Baru in 2012, little is registered under a woman's name (29% and 18%, respectively). The percentages of land on the Kampar riverbank and other places registered to women were also relatively low (30% and 24%, respectively).

Even though the villagers emphasized that the houses and land should be possessed by women in this village, in recent years the tendency is to register land under male names. When asked about the reason why male names are registered on SKTs, respondents answered that "the registration is only nominal" ("*Itu nama saja*") and emphasized that the land and houses are still possessed by women. They then pointed to the Indonesian national custom of registering the husband of a nuclear family as the "head of the household" (*kepala keluarga*) on the family register. This is the same in Rantau Baru, where the husband's name is registered as *kepala keluarga,* unless the husband is deceased or the couple has divorced. While anyone can be registered on an SKT, if the husband is registered, trust in the SKT rises and it can be used to borrow money and mortgage the land more easily than an SKT with the names of other family members. In addition, SKTs under the husband's name are said to be easier in terms of procedures when selling land.

This demonstrates that new attitudes toward the relationship between individuals and land more or less differ from that of their *adat*. According to the *adat*, land in the village territory should be managed and inherited by the *suku* and women, and even now, some land (such as *sialang* areas) is managed as common spaces that are inalienable and inherited from ancestors. However, some of the space is comparted and distributed to households by land registration and sold to outsiders. This tendency emerged not only in Rantau Baru, but also in other places in the "decentralized" era of Indonesia. When describing land reform in Bali in the same era, MacRae (2003) noted that ancestral land can become a commodity by obtaining the legal title and documents to the land. The government title casts off the land from customary restrains, which opens "the door to its inevitable commodification" (2003, p. 159). In Rantau Baru, the issuance of SKTs and SKGRs by the village office has reduced the significance of *suku*'s common land, and land is dealt with as though it is the property of an individual or household, rather than an inalienable property inherited from ancestors.

This tendency is especially prominent in peatlands. While peatlands are a part of the village territory inherited from the ancestors, it was typically an unmanaged area. Osawa (Chap. 6) points out that the villager's recognition of peatland is characterized by its externality from the settlement. Peatlands have neither been used for their livelihood nor for rituals. For Rantau Baru villagers, they depended on water transportation in the past, and the peatland did not constitute an essential landscape. This contrasts with the Orang Suku Laut in Riau Islands province, who regarded the sea route as inherited from their ancestors (Chou 2010). Although there were some *sialang* trees in this area, they were burned by fire. As a result of the fires and documentation, peatland came to be something completely alienable in their landscape. This tendency is clearly seen in the contrast between the peatland and riverbanks. While the peatland has been sold by the village office and villagers, the land on the riverbanks has been regarded as an inherited area and has not been sold to outsiders. Peatland thus came to be less important for the villagers than the riverbanks, which remain an essential space.

3.4.5   Selling Peatland as an Alternative to Being Farmers

Today, the area of the land owned by village individuals, *suku*, and the village office amounts to approximately three thousand of the eight thousand hectares of the total village area (see Banawan and Osawa, Chap. 10). Most of the remaining five thousand hectares was ceded to oil palm companies through industrial concessions or sold to outsiders by the village office. However, part of the land was distributed to the villagers along Jl. Lama and Jl. Baru and sold to urban residents by individuals. Why do individuals sell peatland to outsiders through this process? According to unstructured interviews with villagers regarding their attitudes toward selling land,[24] reasons for selling land are related to their economic situation and uneasiness about and aspirations for their livelihoods.

The first case is that of a woman (in her forties) who works everyday as a fisher in the Kampar River. Her family received peatlands along Jl. Baru in 2012 and sold it to a policeman living in Pekanbaru in 2013. When asked what the money was used for, she replied, "For my daughter." This daughter had graduated from a high school in Kerinci and the money was used to pay for her wedding ceremony. The cost of the ceremony was approximately Rp40,000,000. The bridegroom had paid the Rp20,000,000 toward the ceremony as the bride wealth (*antaran*), the bride, and her family needed to pay the remainder. Therefore, they sold the land. She stated that the cost of weddings and bride wealth has increased significantly, and it is now more expensive than before. We asked her whether she knew how the neighbors had used the money they obtained from selling their land. She answered, "It varies from house to house. For example, it may be used to send children to schools in Kerinci." In the face of rising prices and the necessity to educate children, villagers often need a significant amount of money. Selling peatland is a feasible way to obtain money in a relatively short period.

However, peatlands are sold not only because of villagers' short-term need for cash, but also because of their long-term anxiety about and aspirations for their future livelihood. Rantau Baru villagers feel uneasy about the uncertainty of fishing.

---

[24]While we conducted many unstructured interviews both in direct and online manners (see Dewi et al., Column 3), here we select and describe two cases that seem to represent the villager's ways of thinking and their economic situation.

According to our questionnaire survey, which included questions about recent fish catches, 120 out of 148 respondents answered that the total catch has been decreasing (81%). Through unstructured interviews with villagers, Dewi (Chap. 7) describes a woman who hopes that her daughters will earn a living through agriculture, using the knowledge they acquire in university. Such aspirations are common among the villagers. By engaging in agriculture, they can diversify their livelihoods from an exclusive dependence on fishing to a combination of fishing and agriculture, making them more adaptable in the future.

"Agriculture" here means possessing an oil palm garden. It is a general view among villagers that their village is undeveloped, and this underdevelopment is often emphasized vis-à-vis the situation of Kiyap Jaya. A man in his forties states that the people who migrated to Kiyap Jaya about thirty years ago are now more well-off. They own several hectares of oil palm gardens, live in large houses made of bricks, and have cars that only a few Rantau Baru villagers possess. Well-off villages that succeeded in the oil palm business surround Rantau Baru. Thus, Rantau Baru villagers also aspire to own oil palm gardens.

This is so with the second case, a family that has come to own oil palm gardens around the main settlement in recent years. We interviewed the head of the family, who was a man in his sixties. His primary occupation is boatbuilding; he also lends his boats to anglers who come to the village from cities every weekend. His family was relatively well off in the village. In 2000, he and his wife bought three hectares of land on the southern bank from other villagers. In 2010, they bought 0.4 hectares of land around his house, which had been another villager's homestead. They planted oil palm seedlings on these lands and harvested the yields. He stated that he bought the land "because we want to keep land for our children and grandchildren." He stated that as the broad area of village land was occupied by oil palm companies and sold to outsiders, the land available for villagers' livelihoods became narrower. His daughter lives in the village and has a child. The land on which they planted oil palms has the potential to provide a stable livelihood for his children and grandchildren.

However, this case was exceptional. The land on the riverbank is rather narrow, and most parts have been controlled, as they are owned by *suku*. Even if empty lands exist, it is economically difficult for ordinary villagers to buy them. Therefore, they tried to create a garden in the peat area. Although they planted oil palm seedlings on the peatland distributed by the village, almost all of them were damaged by floods and fires. While it might be possible to cultivate by introducing proper land improvement and agricultural techniques, this seems to be extremely difficult for ordinary villagers, because their knowledge of cultivation and the capital and labor that can be invested in peatland are limited.

Although they aspire to own oil palm gardens, it is difficult to accomplish this because of various factors. In this situation, selling peatland can be an alternative means of responding to their uneasiness and aspirations for their uncertain future. Although selling is rather more unstable than possessing oil palm gardens, it provides a temporary infusion of cash to better provide for their families. When money is used to invest in children's future, such as marriage and education, the children might be able to choose their future by themselves. Selling land allows them to draw out the productive potentiality from the unused lands and respond to the uncertain future. Long (2009) describes a similar case of a Malay family in the Riau Islands. He describes a woman's decision to sell land significantly below the market rate to Batak tenants, even though it had been inherited from her husband's ancestors and had legal certificates. Although she wanted to use the land productively, the land was far from her home and had been occupied by many Batak families. As this case also illustrates, although selling land is not ideal, it can be seen as an alternative way to liquidate land productivity. This contributes to her aspiration to overcome economic difficulties and become a mother and grandmother in the future (2009, pp. 73–75).

As an alternative means to possessing oil palm gardens, peatlands can be commodified. However, not all ancestral territories have been commodified equally. Land is essential for villagers' livelihoods. The land on the Kampar banks has a clear connection with their ancestors, history, and identity, and is difficult to commodify. In contrast, peatland has a less clear connection with ancestors and identity, it is not essential for their lives and livelihoods, and it cannot contribute to their livelihoods within a short period. Peat hinterland has been a part of their territory since the era of the Pelalawan kingdom, and villagers have recognized it as such. However, its importance is relatively low and it is modified based on individual choices. The ancestral territory has been compared, classified, and ranked. This process has been triggered by a combination of the explosive expansion of oil palm industries and increased aspirations, hope, desires, and anxiety for the future among the villagers.

3.5   Analysis: Landscape and Resistance

In Rantau Baru, while peatland was regarded as a part of the *pebatinan* and as a common space inherited by the members of the three *suku*, the space did not have boundaries within it, and its use was limited. However, with the introduction of oil palm cultivation, the elimination of *sialang* trees, land deals by the village office, and the proliferation of individual land registrations, a large portion of the peatland space was divided and distributed. In this process, individuals have classified and ranked the value and meaning of ancestral spaces. Peatland has come to be largely regarded as alienable, and some villagers have sold it for their future prospects. Thus, Rantau Baru demonstrates the transformation of peatland from a "space" to a "place." According to Filippucci (2016), a landscape is "something constructed by humans in the course of their daily lives and interactions, both physically and also symbolically, by being invested with meaning, memory, and value." "Place" emerges from everyday life, is subjective, and has a "foreground actuality," while "space" is separate from everyday life and has a "background potentiality," although the two cannot be completely detached (Hirsch 1995, p. 4). In other words, although a landscape is not something that carries monolithic meaning and value, it can be transformed into a "space" and "place." The meaning and value of places and spaces are interchangeable in and through the communication between insiders and outsiders (Hirsch 1995, p. 13). This transformation can be accomplished by new interpretations that are added to the existing imagination (Ingold 2016, p. 3). In Rantau Baru, while the *suku* space contained a "background potentiality," the actualization of the potential of peatland to benefit life, livelihoods, and identity was limited. Selling peatland to outsiders activated the "background potentiality" of the peatland into an "actuality" that could contribute to the futures of their children and grandchildren, thus maintaining and reproducing the Rantau Baru community.

Tsing (2005) connected the increased encroachment on the forest and locals' reactions or resistance to it in Kalimantan to the collapse of Suharto's regime. Tsing suggested the concept of "friction," in which the universal and the local create frictions that drive globalization forward. While Rantau Baru's recent history exhibits this to some extent, it also differs from Tsing's case. The people of Rantau Baru have not demonstrated against or resisted encroachment on their ancestral peatland, because they already more or less depend on the capital and infrastructure of the oil palm and acacia industries. Moreover, the encroachment is of peripheral peatland, which is less useful for their lives and livelihoods. Rather, the villagers see the "universal" as a means to realize their aspirations, desires, and hopes and to mitigate their anxieties about the future. Here, the values of the universal and the local are not in opposition; instead, they complement each other. The friction here is rather weak and can be found in individual choices.

For more than thirty years, peatland in Riau has been a frontier for acacia and oil palm industries, and many of the forests and peatlands have been transformed into acacia and oil palm plantations. In addition, the moratorium policy that has been implemented prohibits the exploitation of peatland to a certain extent. Nevertheless, the peripheries of the forests and peatlands must be minutely territorialized by industries without salient resistance from the locals. Although Rantau Baru villagers aspire to own oil palm gardens, environmental and economic constraints make this difficult. Selling peatland is a means to actualize the potentiality of peatland space, which contributes to the stable future of their descendants.

3.6 Conclusion

The recent commodification of ancestral common lands in Rantau Baru provides important lessons for peat restoration policies. Despite BRG promotion of community-led restoration and use of peatlands (as described in Chaps. 2 and 6), bottom-up peat restoration activities have limitations. In Rantau Baru, land has been commodified through the land letters issued by the village office. This implies that giving greater discretion to local communities may encourage legal and illegal land sales and the expansion of oil palm plantations, which result in peatland fires. While a bottom-up approach is essential for sustainable peat restoration, appropriate top-down approaches that restricts the expansion of oil palm plantations and provides villagers with benefits through peatland conservation are also needed.

It is significant that these land sales are conducted by villagers who experience changes in local economic and environmental conditions and seriously look to the community's future. In this chapter, we describe villagers desire to become oil palm farmers and the sale of land as an alternative to this path, both of which seem to be a reflection of their anxiety about the instability of fishing income and decreasing fish resources. In other words, the sale of land and the active introduction of oil palm plantations may be deterred if villagers are able to earn stable incomes and understand the significant risks of converting peatlands into oil palm plantations.

Continuous communication with villagers is necessary to understand the multifaceted risks of oil palm plantations. Villagers have a deep understanding of their community environment and are concerned for its future. For this reason, the most desirable form of communication is not a one-sided "promotion of understanding," but rather open discussions among officials, researchers, and villagers about the current situation and the future of the village. While seeking acknowledgment from all parties that the expansion of oil palm plantations will lead to further fires (Binawan and Osawa, Chap. 10) and a decrease in fish stocks (Nakagawa, Chap. 4), it is necessary to support livelihoods that are compatible with the local environment and villager aspirations.

Income stability cannot be achieved solely through the restoration and improved use of peatlands. Livelihoods in Rantau Baru are heavily dependent on fisheries resources; therefore, increasing fish stocks by improving the coastal environment (Nakagawa, Chap. 4) and the introduction of high value-added fishing methods (Nofrizal et al., Chap. 5) should be considered. It is also possible to achieve a future less dependent on oil palm plantations by establishing a peatland management system that contributes to the village economy. A combination of these proposals should be used to effectively improve the livelihoods of Rantau Baru and conserve the peatlands.

In any case, achieving sustainable peatland restoration requires a deep understanding of the history of the community, its economic and cultural situation, and the members' aspirations for the future. While recognizing the risks of developing peatland, peatland restoration activities should be carried out through continuous communications.

References

Andaya LY (2008) Leaves of the same tree: trade and ethnicity in the Straits of Melaka. University of Hawai'I Press, Honolulu

van Anrooij HAH (1885) Nota omtrent het rijk Siak. Tijdschr Indische Taal Land Volkenkd 30: 259–390

Barnard TP (2003) Multiple centres of authority: society and environment in Siak and Eastern Sumatra, 1674–1827. KITLV Press, Leiden

BRG (Badan Restorasi Gambut) (2019) Profil desa peduli gambut: Desa Rantau Baru, Kecamatan Pangkalan Kerinci, Kabupaten Pelalawan. Provinsi Riau, BRG, Jakarta

Bronson B (1978) Exchange at the upstream and downstream ends: notes toward a functional model of the coastal state in Southeast Asia. In: Hutterer KL (ed) Economic exchange and social interaction in Southeast Asia: perspectives from prehistory, history, and ethnography, Michigan papers on South and Southeast Asia no 13. University of Michigan Press and Centre of South and Southeast Asian Studies, Ann Arbor, pp 39–52

Chou C (2010) The Orang Suku Laut of Riau, Indonesia: the inalienable gift of territory. Routledge, London

Dethia NS, Agustina R, Arsin FX (2020) Surat keterangan ganti rugi (SKGR) sebagai jaminan dalam perjanjian utang piutang. Indonesian Notary 2(3):425–447

Effendy HT, Hasbi M, Shomary S (eds) (2005) Lintasan sejarah Pelalawan: dari Pekantua ke kabupaten Pelalawan. Pemerintah Kabupaten Pelalawan

Effendy T (1997a) Petalangan society and changes in Riau. Bijdr Taal Land Volkenkd 153(4): 630–647

Effendy T (1997b) Bujang tan domang: sastra lisan orang Petalangan. Yayasang Obor, Jakarta

Effendy T (2002) The orang Petalangan of Riau and their forest environment. In: Benjamin G, Chou C (eds) Tribal communities in the Malay world: historical, cultural and social perspective. ISEAS Publishing, Singapore, pp 364–383. https://doi.org/10.1355/9789812306104-017

Faes J (1882) Het rijk Pelalawan. Tijdschr Indische Taal Land Volkenkd 27:489–537

Filippucci P (2016) Landscape. In: Steein F (ed) The Cambridge encyclopedia of anthropology. https://doi.org/10.29164/16landscape

Fitzpatrick D (2007) Land, custom, and the state in post-Suharto Indonesia: a foreign lawyer's perspective. In: Davidson JS, Henley D (eds) The revival of tradition in Indonesian politics: the deployment of adat from colonialism to indigenism. Routledge, Oxon, pp 130–148

Gusliana HB, Ikhsan I, Ferawati F (2019) The status of indigenous forest in Riau province. In: Proceedings of the Riau Annual Meeting on Law and Social Sciences (RAMLAS 2019), Advances in social science, education and humanities research, vol 442. Atlantis Press, Dordrecht, pp 52–56. https://doi.org/10.2991/assehr.k.200529.266

Hamidy UU (1987) Ramba kepungan sialang. Balai Pustaka, Jakarta

Hirsch E (1995) Landscape: between place and space. In: Hirsch E, O'Hanlon M (eds) The anthropology of landscape: perspectives on place and space. Clarendon Press, Oxford, pp 1–30

Ingold T (2016) Introduction. In: Janowski M, Ingold T (eds) Imagining landscapes: past, present and future. Routledge, London, pp 1–18

Jong HN (2019) Indonesia forest-clearing ban is made permanent, but labeled 'propaganda'. Mongabay, 14 Aug. https://news.mongabay.com/2019/08/indonesia-forest-clearing-ban-is-made-permanent-but-labeled-propaganda/. Accessed 14 May 2021

Jong HN (2021) Top brands failing to spot rights abuses on Indonesian oil palm plantations. Mongabay, 13 July. https://news.mongabay.com/2021/07/top-brands-failing-to-spot-rights-abuses-on-indonesian-oil-palm-plantations-rspo/. Accessed 14 Sep 2021

Kathirithamby-Wells J (1993) *Hulu-hilir* unity and conflict: Malay statecraft in East Sumatra before the mid-nineteenth century. Archipel 45:77–96. https://doi.org/10.3406/arch.1993.2894

Kawai M (2021) Indoneshia niokeru aburayashi nouen-kigyou niyoru shounou-shien-houshiki (PIR) no hensen (The transition of support to small farmers by oil palm plantation companies in Indonesia). In: Hayashida H (ed) Aburayashi nouen mondai no kenkyu (The study of oil palm plantation issues) 2. Koyo Shobou, Kyoto, pp 23–46

Li TM (2001) Masyarakat adat, difference, and the limits of recognition in Indonesia's forest zone. Modern Asian Studies 35(3):645–676

Long N (2009) Fruits of the orchard: land, space, and state in Kepulauan Riau. Sojourn 24(1):60–88

Lubis ZB (2013) Social mapping of access to peat swamp forest and peatland resources. Working paper of Kalimantan Forests and Climate Partnership (KFCP), Indonesia-Australia Forest Carbon Partnership, Jakarta

MacRae G (2003) The value of land in Bali: land tenure, landreform and commodification. In: Reuter TA (ed) Inequality, crisis and social change in Indonesia: the muted worlds of Bali. Routledge Curzon, London and New York, pp 145–168

Masuda K (2009) The reconstitution of *adat* in a dual level land conflict: a case study of a village community under forest development schemes in Sumatra. In: Afrasian Centre for Peace and Development Studies Working Paper series 61. Afrasian Centre for Peace and Development Studies, Otsu

Masuda K (2012) Indoneshia mori no kurashi to kaihatsu: tochi wo meguru 'tsunagari' to 'semegiai' no shakaishi (Livelihoods and developments in Indonesian forest: sociohistory of cooperation and conflict for land use). Akashi Shoten, Tokyo

Mujiburohman DA, Arianto T, Riyadi R (2015) Kajian Yuridis tumpeng tindih pemilikan tanah di Kabupaten Kampar Provinsi Riau. In: Pujiriyani DW, Puri WH (eds) Penataan dan pengelolaan pertanahan yang mensejahterakan masyarakat: Hasil penelitian strategis PPPM-STPN. Pusat Penelitian Pengabdian kepada Masyarakat Sekolah Tinggi Pertanahan Nasional, Yogyakarta, pp 169–195

Murdiyarso D, Dewi S, Lawrence D et al (2011) Indonesia's forest moratorium: a stepping stone to better forest governance? Working Paper no 76. CIFOR, Bogor

Nagata J (2021) Riau-shu niokeru aburayashi nouen sangyou no kakudai to kouzou-henka (The expansion and structural changes of oil palm plantation industries in Riau province). In: Hayashida H (ed) Aburayashi nouen mondai no kenkyu (The study of oil palm plantation issues) 2. Koyo Shobou, Kyoto, pp 69–91

Nagata J, Arai SW (2013) Evolutionary change in the oil palm plantation sector in Riau Province, Sumatra. In: Pye O, Bhattacharya J (eds) The palm oil controversy in Southeast Asia: a transnational perspective. ISEAS Publishing, Singapore, pp 76–96

Nieuwenhuijzen FN (1858) Het rijk Siak Sri Indrapoera. Tijdschr Indische Taal Land Volkenkd 7: 388–438

Osawa T (2022) At the edge of mangrove forest: the Suku Asli and the quest for indigeneity, ethnicity and development. Kyoto Area Studies on Asia, vol 29. Kyoto University Press, Kyoto; Trans Pacific Press, Tokyo

Porath N (2003) When the bird flies: shamanic therapy and the maintenance of worldly boundaries among an indigenous people of Riau (Sumatra). CNWS Publications, Leiden

Sloan S, Edwards DP, Laurance WF (2012) Does Indonesia's REDD+ moratorium on new concessions spare imminently threatened forests? Conserv Letters 5:222–231. https://doi.org/10.1111/j.1755-263X.2012.00233.x

Tsing A (2005) Friction: an ethnography of global connection. Princeton University Press, Princeton

Turner A (1997) Cultural survival, identity and the performing arts of Kampar's suku Petalangan. Bijdr Taal Land Volkenkd 153(4):648–671

Wilkinson RJ (1959) A Malay-English dictionary. Macmillan & Co ltd, London

Chapter 4
Inferring Recent Changes in Fish Fauna in the Middle Reaches of the Kampar River: Survey Results from the Fishing Village of Rantau Baru

Hikaru Nakagawa

Abstract Degradation of peatlands is an issue of global concern, yet ample knowledge of local conditions is lacking when it comes to determining (1) the impacts of river and floodplain development and (2) how best to conserve peat swamp ecosystems. This chapter documents the relationships between scientific and local names of fishes and recent changes in fish biodiversity in the mid-Kampar River Basin of Sumatra. A questionnaire was administered to 164 householders in the village of Rantau Baru and information on 96 species was triangulated with previous English- and Indonesian-language research. Results indicate the local extinction (defined as caught in the past but not observed during the last five years) of large predatory fishes and the invasion of several exotic species, potentially pointing to the early stage of degradation of the freshwater ecosystems. The potentiality of establishing effective freshwater protected areas in the mid-Kampar Basin is assessed by a narrative review of studies and methods from other developing countries. Local-scale ecosystem conservation that incorporates local perspectives and scientific investigation is of the highest priority to address development pressures on rivers, floodplains, and surrounding communities.

Keywords Ecosystem degradation · Ecosystem linkage · Fishery management · Indigenous knowledge · Invasive species · Local ecological knowledge

4.1 Introduction

The ecosystems of Southeast Asian rivers are known as some of the largest biodiversity hotspots in the world (Myers et al. 2000; Dudgeon 2011). The species diversity and the vast production of fish provide important food resources to the

H. Nakagawa (✉)
Aqua Restoration Research Center, Public Works Research Institute, National Research and Development Agency, Kakamigahara, Gifu, Japan

© The Author(s) 2023
M. Okamoto et al. (eds.), *Local Governance of Peatland Restoration in Riau, Indonesia*, Global Environmental Studies,
https://doi.org/10.1007/978-981-99-0902-5_4

peoples of Southeast Asian countries (FAO 2018). Despite their significance, however, Southeast Asian rivers are also known as some of the most threatened ecosystems in the world (Dudgeon 2011). This is due to human interventions and the resulting impacts, including dam construction, logging and deforestation, sand and gravel mining, sedimentation, and various types of water pollution (Allen et al. 2012). Without consideration of ecosystem conservation, the rapid development of Southeast Asian economies currently underway presents serious concerns for the food and water security of these countries at both local and national scales.

In Riau Province in eastern Sumatra, natural forest cover has decreased rapidly since the 1980s due to national and international industrial activity (in particular, the oil-palm and pulp industries), as well as agricultural activity of residents and immigrants from northern Sumatra and Java (Miettinen et al. 2016; Mizuno and Kusumaningtyas 2016; Shimamura 2016). In tropical peat swamp ecosystems, forests and grasslands grow on accumulated peatland intersected by permanent and temporal bodies of water, such as river channels, oxbow lakes, and floodplains; these ecosystems are scattered across lowland areas at 0–10° latitudes (Nofrizal et al., Chap. 5). Floodplains and their surrounding peatlands are typically seen as unsuitable for cultivation, and thus have been less developed relative to other forested areas in Indonesia. But clear-cutting in these areas has increased since the 2000s (Masuda et al. 2016). Floodplain habitats, such as riparian forests and swamps that become submerged during the rainy season, provide spawning, rearing, and foraging habitats for river fishes in tropical regions (Amoros and Bornette 2002; Correa and Winemiller 2014). Therefore, floodplain development presents a serious threat to river health, basin ecosystems, and the sustainability of inland fisheries. However, scientific knowledge about how freshwater fish use floodplain forests is lacking, and the potential risks of the loss of floodplain forests in Indonesia to future inland fisheries are largely unknown.

Local ecological knowledge (LEK) is knowledge accumulated over a lifetime through one's observations and hands-on interactions with ecological systems and natural resources, or a cumulative body of local ecosystem knowledge that transcends generations through cultural transmission (Thornton and Scheer 2012; Berkström et al. 2019). LEK includes taxon-specific information, such as preferred habitat, abundance, behavior, breeding, and these seasonal patterns. Such information greatly increases comprehension of ecosystems, and therefore can be critical to ecosystem conservation, particularly in situations where scientific data are scarce or unavailable, such as in developing countries (see for example Berkström et al. 2019). As holders of LEK, local residents are essential to environmental conservation, and including their LEK in scientific interpretations may advance the success of conservation efforts. Despite this, LEK is rarely shared with village governments or higher administrative organizations and is seldom reflected in development planning or conservation efforts (Glaser et al. 2010; Satria and Adhuri 2010). Thus, identifying how to combine science with local wisdom remains a contemporary challenge to achieving sustainable development for local communities and ecosystems (Osawa, Chap. 6).

In an effort to document LEK in an area experiencing peatland degradation, Nakagawa et al. (2021) focus on the local names of fish used in a fishing village along the middle reaches of the Kampar River in Sumatra, Indonesia. Understanding and cross-referencing local names is necessary when collecting species-specific information from local residents or from existing literature written in a local language (see Ankei 1989; Castillo et al. 2018). The taxonomic description of a species and the determination of its scientific name are guided by international codes (e.g. the International Code of Zoological Nomenclature [ICZN], the International Code of Nomenclature for Algae, Fungi, and Plants [ICN], and the International Code of Nomenclature of Prokaryotes [ICNP]) (ICZN et al. 1999; Turland et al. 2018; Parker et al. 2019). While these codes were developed in the context of conventional natural sciences, the local name for a species is typically determined by its morphological or behavioral character and is based on LEK and the historical and cultural context of the human community where the species is found (Medin and Atran 2004). Therefore, the scientific name of a species often does not directly correspond to its local name. For a speciose taxon, a local name often relates to a scientific name higher than the species level, such as the genus or family (see for example Ankei 1989; Castillo et al. 2018). In addition, a single species may have multiple local names that correspond to its various body sizes or to ontogenetic stages during which a fish is important to a community's livelihood (see Ankei 1989; Castillo et al. 2018). It is therefore critical to explicitly define the relationships between local and scientific names prior to collecting LEK, particularly when planning and implementing conservation activities at the ecological community and ecosystem levels.

This chapter presents the results of a survey that aims to (1) identify the relationships between scientific and local names of fishes and (2) understand recent changes to the fish biodiversity at the research site, a fishing village along the middle reaches of the Kampar River in Indonesia. The results of the survey are then discussed in the context of risks to ecosystem stability and the potential for ecosystem conservation in the Sumatran peatlands. In addition, the potentiality of establishing effective freshwater protected areas in the mid-Kampar Basin is assessed by a narrative review of studies and methods from other developing countries. Settlement of protected areas may be an effective strategy for biodiversity conservation from the view of the precautionary principle (Lauck et al. 1998) when ecosystem knowledge is limited, as is the case in most Southeast Asian regions.

4.2 Materials and Methods

4.2.1 Research Site

The village of Rantau Baru, Pangkalan Kerinci Sub-district, Pelalawan District, Riau Province is located along the middle reaches of the Kampar River, which flows through eastern Sumatra from west to east (Fig. 4.1a). The village is approximately

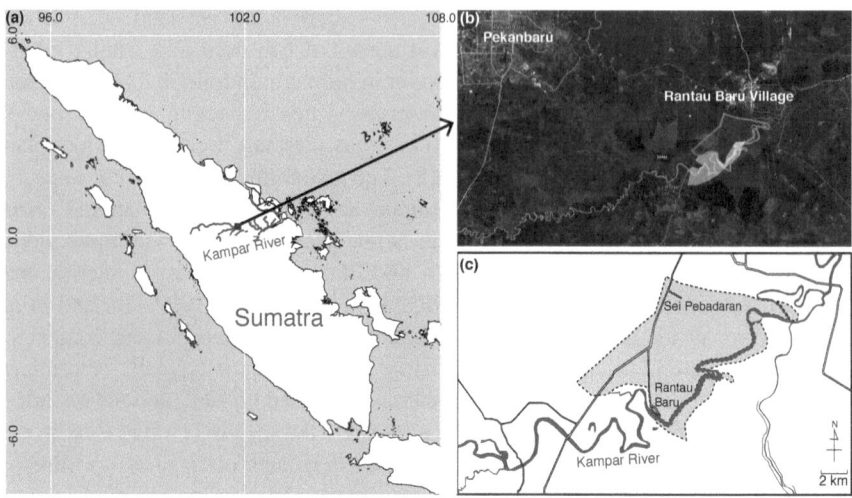

Fig. 4.1 (**a**) Location of the village of Rantau Baru on the island of Sumatra; (**b**) Satellite image of the area surrounding the village of Rantau Baru (Google Maps, https://www.google.co.jp/maps/, accessed March 19, 2020). The grey line marks the administrative boundary of the village; the white shaded area indicates the remaining floodplain forested area used by villagers as a fishing ground; (**c**) Location of the settlements of Rantau Baru and Sei Pebadaran. The shaded area indicates the administrative boundary of the village. (Modified from Nakagawa et al. 2021 with the permission of Center for Southeast Asian Studies)

200 km from the river mouth and consists of two settlements (Fig. 4.1c). The main, and older, settlement of Rantau Baru is located on the banks of the Kampar River and consists of 116 houses. The newer settlement of Sei Pebadaran was constructed on hinterland peat soil by the district government around 2005; it is 8 km north of the settlement of Rantau Baru and consists of 48 houses. The two settlements are together customarily called Rantau Baru, which is also the name of the administrative village (Osawa and Binawan, Chap. 3). The settlements of Rantau Baru are regarded as having the longest history of any in Pelalawan District. Most villagers recognize that their ancestors lived in the proto settlement upstream on the opposite shore of the Kampar River and moved to the present-day village of Rantau Baru at least a few hundred years ago. In addition, tens of immigrants from Java and northern Sumatra live in the settlement of Sei Pebadaran to work in surrounding oil palm plantations at present (Osawa and Binawan, Chap. 3).

Rantau Baru village is typical of the fishing villages of the middle reaches of the Kampar River (Nofrizal, Chap. 5). Most households, including those in Sei Pebadaran, fish commercially or for self-consumption in the Kampar River and its tributaries, as well as in the oxbow lakes, canals, and swamps near the river. Typical fishing equipment includes fixed traps, gill nets, casting nets, and long lines (Masuda 2012). The village is surrounded by a floodplain, and riparian areas are typically submerged during the rainy season. Large portions of the riverbanks are covered with peat soils.

The area surrounding the village of Rantau Baru has undergone dramatic changes during the past 20 years. In 1996, a large-scale hydroelectric dam (PLTA Koto Panjang) was constructed on the upper reaches of the Kampar River, which has changed the flood regime downstream (Fitri and Husni 2020). Since the late 1980s, peat swamps in the research area have been drained for the development of acacia and oil-palm plantations (Shimamura 2016). The drained and dried hinterlands, which were covered by forested peat swamps in the past, now experience frequent fire, and burned areas often do not recover to forests or plantations, but instead have become abandoned bush (Shimamura 2016).

4.2.2 Survey by Questionnaire

A questionnaire was administered to all 164 houses in Rantau Baru between January 27 and February 2, 2020. The questionnaire was comprised of 101 questions designed to obtain basic information about the respondents and their households, attitudes toward conservation of peat swamp forests, levels of participation in local community activities, fishery activity, land ownership status, and incomes and assets (Dewi, Chap. 7; Hasegawa, Chap. 8; Prasetyawan, Chap. 9).

Three questions regarding fish names and species sightings were asked of residents who reported fishing for consumption or commercial purposes. The questions were:

1. *Tolong tulis jenis-jenis ikan yang Anda tangkap dalam 1 tahun terakhir.* (Please write the names of fishes that you have caught during the past year.)
2. *Apakah ada jenis ikan yang ditangkap di masa lalu, tetapi dalam 5 tahun terakhir tidak ditemukan lagi? Jika ada, tolong tuliskan nama jenis ikan nya (boleh lebih dari satu).* (Is there any type of fish that was caught in the past, but has not been found during the last 5 years? If so, please write down the name of the fish (may be more than one)).
3. *Apakah ada jenis ikan yang dulu tidak ada namun sekarang ditemukan? Jika ada, tolong tuliskan nama jenis ikan nya (boleh lebih dari satu).* (Is there any type of fish that was not caught in the past but is caught now? If so, please write down the name of the fish (may be more than one)).

Before administration of the questionnaire, respondents were informed of the survey method and the objective of the research. Enumerators were careful not to show respondents fish names to avoid leading questions, and the respondents were free to provide any local name that they knew.

4.2.3 Literature Survey

Two sources from the English-language literature and three sources from the Indonesian-language literature about fish fauna in the middle reaches of the Kampar River were used to obtain a reference species list for known fishes from the area surrounding the village of Rantau Baru (Fauzi 2004; Fithra and Siregar 2010; Efizon et al. 2015; Aryani 2015; Aryani et al. 2020). Misidentifications and synonymous scientific names in the literature were corrected following Nelson et al. (2016) and FishBase (Froese and Pauly 2000). Records of fish that had been identified only to the genus level or higher were removed from the list, except for *Tor* sp., which could not be resolved to species due to taxonomic uncertainty (Pinder et al. 2019). Lists of Indonesian names of fish recorded from sites upstream and downstream of Rantau Baru, as well as from neighboring rivers (the Rokan, Siak, and Indragiri) were used supplementarily (Siregar et al. 1994; Tjakrawidjaja and Haryono 2000; Iskandar and Dahiyat 2012; Fahmi et al. 2015; Firdaus et al. 2015; Lubis et al. 2016; Purnama and Yolanda 2016; Yustina 2016).

4.2.4 Collation of Local and Scientific Names and Tabulation of Species Sightings

The scientific, English, Indonesian, and local names of fishes found in Rantau Baru were collected, along with all local alternate names. Spelling variations of local fish names that appeared to be caused by a difference in pronunciation or a listening error were first collated (for example, Kayang, Khayangan, Koloso, and Keloso were combined as the local name Kayangan/Arwana (*Scleropages formosus*)). Using the Indonesian name of fishes as a reference, the local names were then correlated to the scientific names. The resulting comprehensive list is presented in Table 4.1. Hereafter, names in quotation marks refer to the local names obtained from the questionnaires. Note that the spelling of local names provided in Table 4.1 is that as given by the respondents; thus, in several cases, the spelling of a local name does not correspond to Indonesian orthography (for example *asin*, or "salt" in standard Indonesian, is presented as *masin*, and *kucil*, or "small" in standard Indonesian, is presented as *kucir*). Table 4.1 also includes the number of respondent(s) who reported sightings of each species in response to Question 1 (fishes caught within the last year), Question 2 (fishes caught previously but not observed within the last five years), and Question 3 (fishes caught now that were not caught previously).

Finally, rarefaction curves of the number of local names relative to the number of respondents were drawn to evaluate the effect of the sampling effort on the number of local names identified both before (Fig. 4.2a) and after (Fig. 4.2b) collation of spelling variations. After accounting for spelling variation, a confidence rate of 95% was calculated using bootstrap resampling with 999 iterations.

Table 4.1 Names and sightings of fishes in the middle reaches of the Kampar River: Survey results from Rantau Baru village

No.	Order	Family	Species	English name	Indonesian name	Local name	Alternate name(s)	Q1	Q2	Q3
1	Myliobatiformes	Dasyatidae	*Fluvitrygon signifer*	White-rimmed stingray	Pari					
2	Osteoglossiformes	Osteoglossidae	*Scleropages formosus*	Asian bonytongue	Arowana/ Pejang/Taliso	Kayangan/ Arwana	Kayang; Khayangan; Koloso; Keloso		10	
3		Notopteridae	*Chitala borneensis*	Indonesian featherback	Belida	Belida	Balido	3		
4			*Chitala lopis*	Giant featherback	Belida	Belida	Balido	3		
5	Clupeiformes	Clupeidae	*Clupeichthys bleekeri*	Kapuas river sprat	Bunga air putih					
6			*Clupeichthys goniognathus*	Sumatran river sprat	Bunga air merah					
7		Chirocentridae	*Chirocentrus dorab*	Dorab wolf-herring	Parang-Parang	Ikan Parang	Ikan parang-parang; Ikan paparang; Parang-parang; Paparang; Parang; Pemparang		7	
8	Cypriniformes	Cyprinidae	*Tor* sp.	Mahseer		Gadi	Ikan gadi; Gadih		4	
9			*Barbichthys laevis*	Sucker barb	Bentulu					
10			*Barbonymus schwanenfeldii*	Tinfoil barb	Lampai/ Lampam/ Kapiah/Kapiek	Kapiek	Ikan kepetuk; Kepetok; Kepituk; Kepureh; Kapetuk; Kepetuk	26	1	
11			*Barbonymus balleroides*							

(continued)

Table 4.1 (continued)

No.	Order	Family	Species	English name	Indonesian name	Local name	Alternate name(s)	Q1	Q2	Q3
12			*Barbonymus gonionotus*	Java barb	Tawes					
13			*Albulichthys albuloides*		Dara putih	Ikan putih-putih	Ikan putitt; Pitue; Putih	4		
14			*Puntioplites waandersi*		Daro putih	Ikan putih-putih	Ikan putitt; Pitue; Putih	4		
15			*Puntioplites bulu*		Tabingalan/Bulu-Bulu/Pantau		Pantau/Tabingal	26		
16			*Puntigrus tetrazona*	Tiger barb	Pantau/Ikan Baja/Aji-aji/Sumatra		Pantau/Tabingal	26		
17			*Rasbora argyrotaenia*	Silber rasbora	Tabingalan/Bada/Pantau		Pantau/Tabingal	26		
18			*Rasbora rutteni*		Pantau/Seluang/Bada		Pantau/Tabingal	26		
19			*Rasbora tawarensis*		Pantau		Pantau/Tabingal	26		
20			*Rasbora lateristriata*	Yellow rasbora						
21			*Rasbora reticulata*							
22			*Labiobarbus leptocheilus*		Luang/Wadon-guang/Umbu-umbu/	Sisik		1		

#	Scientific name	Common name	Local name		
			Ubut-ubut/Sisik merah/Malih		
23	*Labiobarbus festivus*	Signal barb	Ingau/Mali-mali/Terpayang		
24	*Labiobarbus ocellatus*		Lamba/Mali		
25	*Labiobarbus fasciatus*		Siluang		
26	*Leptobarbus hoevenii*		Jelawat/Jelejer		
27	*Leptobarbus melanopterus*		Petulu		
28	*Osteochilus kelabau*		Kelabau	Kelabau	1
29	*Osteochilus vittatus*	Bonylip barb/Hard-lipped barb	Lelan Botiong/Paweh/Nilem/Kelabu/Puyou/Asang	Pawe	1
30	*Osteochilus borneensis*				
31	*Osteochilus microcephalus*		Lelan Kunyit/Sibruak		
32	*Osteochilus schlegelii*	Giant sharkminnow	Si buruk		
33	*Osteochilus pleurotaenia*		Lelan		
34	*Osteochilus waandersii*		Puyou		

(continued)

Table 4.1 (continued)

No.	Order	Family	Species	English name	Indonesian name	Local name	Alternate name(s)	Q1	Q2	Q3
35			*Oxygaster anomalura*		Sepimping/ Seluang ping-ping/Pimpiang	Pon-ping		1		
36			*Thynnichthys polylepis*		Motan Besar Kepala	Motan		5		
37			*Thynnichthys thynnoides*		Motan Siruncing/ Lumo	Motan		5		
38			*Crossocheilus oblongus*	Siamese flying fox/Siamese algae-eater	Selimang batu/ Lukaslawat					
39			*Crossocheilus langei*	Red algae-eater	Selimang batang					
40			*Cyclocheilichthys apogon*	Beardless barb	Lais Timah/ Kaperas/ Sipaku/ Sibahan					
41			*Cyprinus carpio*	Common carp	Ikan Mas					
42			*Epalzeorhynchos kalopterus*	Flying fox	Selimang kayu					
43			*Hampala macrolepidota*	Hampala barb	Hampal/ Barau/ Sebarau/ Sibarau					

#	Order	Family	Scientific name	English common name	Dungan		Local names		
44			*Hampala bimaculata*						
45			*Luciosoma trinema*		Seluang-juo/ Ikan jua				
46			*Lobocheilos falcifer*		Klarii				
47		Cobitidae	*Acantopsis octoactinotos*	Long-nosed loach	Awu-awu/ Anculong				
48			*Chromobotia macracanthus*	Clown loach					
49			*Syncrossus hymenophysa*	Tiger loach	Ciling-ciling/ Langli				
50	Siluriformes	Claridae	*Clarias batrachus*	Walking catfish	Lele	Lele	Lele kaleng; Lele sawit; Limbek	20	1
51			*Clarias teijsmanni*		Keli/Lele Kembang	Lele	Lele kaleng; Lele sawit; Limbek	20	1
52		Bagridae	*Hemibagrus wyckii*	Crystal-eyed catfish	Hinur/Jatisa/ Geso	Geso	Baung pilar	2	1
53			*Hemibagrus nemurus*	Asian redtail catfish	Baung rambe	Baung	Ikan baung; Anak barung; Baung kuning; Baung pisane	52	1
54			*Hemibagrus planiceps*		Baung	Baung	Ikan baung; Anak barung; Baung kuning; Baung pisane	52	1
55			*Mystus nigriceps*	Two-spot catfish	Ingir-ingir/ Baung/ Senggiring/ Tundik	Baung	Ikan baung; Anak barung; Baung kuning; Baung pisane	52	1
56			*Mystus micracanthus*	Two-spot catfish	Baung	Baung		52	1

(continued)

Table 4.1 (continued)

No.	Order	Family	Species	English name	Indonesian name	Local name	Alternate name(s)	Q1	Q2	Q3
57			*Mystus gulio*	Long whiskers catfish	Baung	Baung	Ikan baung; Anak barung; Baung kuning; Baung pisane	52	1	
58		Siluridae	*Ompok eugeneiatus*	Malay glass catfish	Silais	Selais	Ikan selais; Kapore; Selais kecir; Silais; Slais; Salais	60		1
59			*Ompok hypophthalmus*		Selais danau/ Silais	Selais	Ikan selais; Kapore; Selais kecir; Silais; Slais; Salais	60		1
60			*Kryptopterus palembangensis*		Selais	Selais	Ikan selais; Kapore; Selais kecir; Silais; Slais; Salais	60		1
61			*Kryptopterus schilbeides*		Selais/Silais	Selais	Ikan selais; Kapore; Selais kecir; Silais; Slais; Salais	60		1
62			*Kryptopterus macrocephalus*	Striped glass catfish		Selais	Ikan selais; Kapore; Selais kecir; Silais; Slais; Salais	60		1
63			*Kryptopterus limpok*	Long-barbel sheatfish	Selais janggut/ Silais	Selais	Ikan selais; Kapore; Selais kecir; Silais; Slais; Salais	60		1
64			*Wallago leeri*		Tapah/Tapak	Tapah	Tapa	6	1	
65			*Phalacronotus apogon*		Lais Ttimah					

	Order	Family	Scientific name	English name						
66			*Belodontichthys dinema*		Sengarat/ Singarek					
67		Pangasidae	*Pangasius pangasius*	Pangas catfish	Patin/Juaro/ Jambal	Juaro/ Patin	Ikan juaro; Jambal; Ikan patin jambal; Patin keramba; Patin kualo; Patin kuning; Patin kunyit	20	13	4
68			*Pangasius polyuranodon*		Juaro	Juaro/ Patin	Ikan juaro; Jambal; Ikan patin jambal; Patin keramba; Patin kualo; Patin kuning; Patin kunyit	20	13	4
69			*Pseudolais micronemus*	Short-barbel pangasius						
70			*Pangasianodon hypophthalmus*	Striped catfish						
71		Loricariidae	*Pterygoplichthys pardalis*	Amazon sailfin catfish	Sapu-sapu	Sapu-sapu/ Indosiar	Ikan sapu-sapu; Ikan indosiar; Ikan terbang;	15		
72	Beloniformes	Zenarchopteridae	*Hemirhamphodon chrysopunctatus*		Julung-julung					
73	Synbranchiformes	Synbranchidae	*Monopterus albus*	Asian swamp eel/Swamp eel/Rice eel/White ricefield eel	Belut	Belut	Bulan-bulan; Blang	22	1	
74		Mastacembelidae	*Mastacembelus maculatus*	Freckfin eel	Tilan					
75			*Mastacembelus notophthalmus*		Tilan					

(continued)

Table 4.1 (continued)

No.	Order	Family	Species	English name	Indonesian name	Local name	Alternate name(s)	Q1	Q2	Q3
76			*Mastacembelus unicolor*		Tilan					
77	Perciformes	Cichlidae	*Oreochromis niloticus*	Nile tilapia	Nila	Nila	Ikan nila		1	2
78		Pristolepididae	*Pristolepis grootii*	Indonesian leaffish	Sepatung/Katung/Katong					
79		Ambassidae	*Parambassis wolfii*	Duskyfin glassy perchlet	Sipongkah					
80	Gobiiformes	Butidae	*Oxyeleotris marmorata*	Marble goby	Betutu					
81	Anabantiformes	Helostomatidae	*Helostoma temminckii*	Kissing gourami	Tuakang/Tambakan	Tuakang				
82		Osphronemidae	*Osphronemus goramy*	Giant gourami	Gurami/Kalau	Gurami	Kalui	1	1	
83			*Belontia hasselti*	Malay combtail	Selinca	Selincah	Silinca	4		
84		Anabantidae	*Anabas testudineus*	Climbing perch	PuyuhPuyu//Betok/Puju-Puju	Betik				
85		Osphronemidae	*Trichogaster trichopterus*	Three-spot gourami	Sepat rawa	Sepat	Sepat siam	10		
86			*Trichogaster leeri*	Pearl gourami	Sepat mutiara	Sepat	Sepat siam	10		
87			*Sphaerichthys osphromenoides*	Chocolate gourami	Sepat batik/Tuwakan	Sepat	Sepat siam	10		

No.									
88		Chandidae	*Channa striata*	Striped snakehead	Gabus	Gabus	Ikan gabus; Botuik;	13	2
89			*Channa lucius*	Forest snakehead	Bujuk	Gabus	Ikan gabus; Botuik;	13	2
90			*Channa bankanensis*		Bujuk	Gabus	Ikan gabus; Botuik;	13	2
91			*Channa micropeltes*	Red snakehead/ Giant snakehead/ Indonesian snakehead	Toman	Toman		8	
92			*Channa pleurophthalma*		Serandang				
93	Pleuronectiformes	Soleidae	*Achiroides leucorhynchos*		Sebelah/Ikan lidah				
94		Cynoglossidae	*Cynoglossus microlepis*	Smallscale tonguesole	Lidah-lidah				
95	Tetraodontiformes	Tetraodontidae	*Pao palembangensis*		Buntal				
96	Decapoda	Palaemonidae	*Macrobrachium rosenbergii*	Giant river prawn	Udan	Udang	Ulang-ulang	14	1

Alternate name(s) are spelling variations of local fish names that appeared to be caused by a difference in pronunciation or a listening error. Q1, Q2, and Q3 correspond to the number of respondent(s) who reported sightings of each species in response to Question 1 (fishes caught within the last year), Question 2 (fishes caught previously but not observed within the last five years), and Question 3 (fishes caught now that were not caught previously), respectively. (Modified from Nakagawa et al. 2021 with the permission of Center for Southeast Asian Studies)

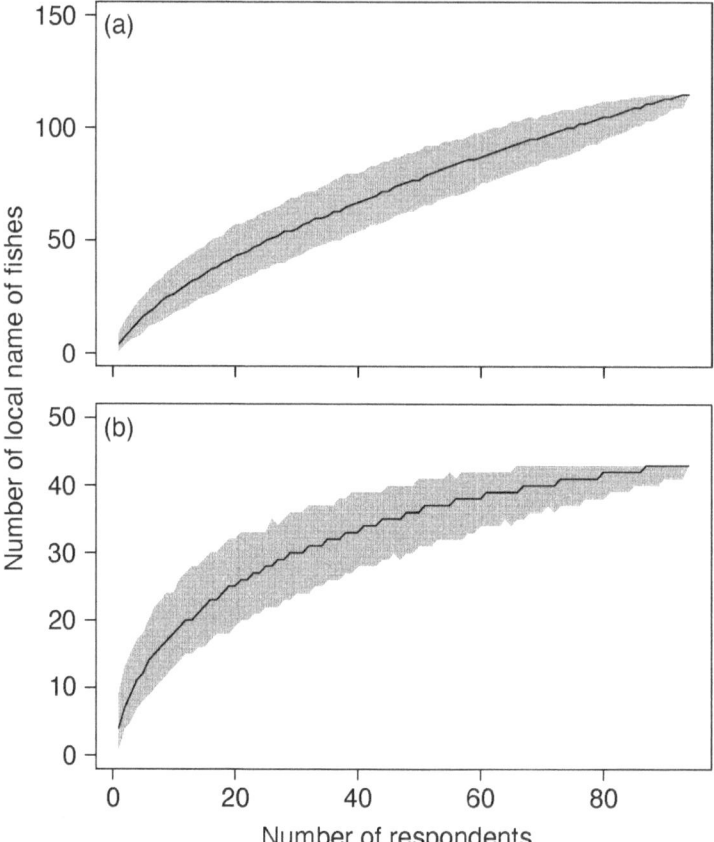

Fig. 4.2 (**a**) Rarefaction curve of the number of local names of fishes relative to the number of respondents prior to spelling collation; (**b**) Rarefaction curve of the number of local names of fishes relative to the number of respondents following spelling collation. (Reprinted from Nakagawa et al. 2021 with the permission of Center for Southeast Asian Studies)

4.3   Results and Discussion

4.3.1   Local Names of Fishes in Rantau Baru

Ninety-four of the 164 respondents provided local fish names. A total of thirty-eight local fish names were recorded following the spelling compilation and the removal of names that clearly did not relate to a specific taxon or were related to marine fish. The number of local names recorded did not reach saturation without compilation (Fig. 4.2a), but it did reach saturation at approximately 70 respondents with spelling compilation (Fig. 4.2b). The comprehensive species list in Table 4.1 was compiled using questionnaire responses and existing literature about the area. The list includes 95 species belonging to 52 genera, 27 families, and 12 orders of fishes and one prawn species. Of the total, twenty-eight local names were related to scientific names of fish taxa from the area (Table 4.1). More than half of the local names were related to a scientific name at the genus level or higher and were also related to multiple scientific names at the species level. For example, "Pantau/Tabingal" was associated with multiple species belonging to the genera *Puntioplites* or *Rasbora*. "Baung" referred to multiple species belonging to the genera *Hemibagrus* or *Mystus* and included several alternates, such as "Baung kuning" (1/52 cases in Question 1) and "Baung pisane" (2/52 cases in Question 1), although the relationship between these alternates and scientific names could not be verified. "Selais" referred to multiple species belonging to the genera *Ompok* or *Kryptopterus*. "Patin" referred to multiple species of the genus *Pangasius* or *Pangasianodon*, and respondents often gave several alternates referring to *Pangasius* spp. or *Pangasianodon* spp., such as "Juaro" (13/20 cases in Question 1) and "Patin kunyit" (1/20 cases in Question 1). However, these alternates did not correlate to scientific names at the species level.

Four local names, "Kayangan/Arwana," "Ikan Parang," "Gadi," and "Belut," directly connected to the scientific names of four fish species (*Scleropages formosus*, *Chirocentrus dora*, *Tor* sp., and *Monopterus albus*, respectively). The existence of these species has not been recorded in previous scientific research of the Rantau Baru area (Aryani 2015; Efizon et al. 2015; Fauzi 2004; Fithra and Siregar 2010). In this study, the first three of these species were reported as fish that had been caught in the past but had not been observed by respondents during the last five years (defined here as extirpated, or locally extinct). Respondents reported a total of 16 local names of fishes that "had been caught previously but have not been observed within the last five years" (Question 2), and 7 local names of fishes that were not caught previously but are caught now, or exotic species (Question 3) (Table 4.1). Excluding the previously mentioned three species ("Kayangan/Arwana," "Ikan Parang," and "Gadi"), "Patin" (13 cases), especially its alternates, "Patin kunyit" (3 cases) and "Patin juaro" (2 cases), was frequently recorded in response to Question 2. Local residents also recognized several subgroups within "Patin." The spelling of several local names compiled as "Patin" resembled the scientific name or a synonym of a species belonging to the genus *Pangasius*. We suspect that *Pangasius juaro* (a synonym of *Pangasius polyuranodon*), *Pangasius kunyit*, and *Pangasius djambal* are related to "Juaro," "Patin kunyit," and "Patin jambal," respectively. Interestingly, all three of these species are described as native to Sumatra Island, whereas local respondents defined "Patin kunyit" as the name of an extirpated species and "Patin jambal" as the name of an exotic species. This may reflect temporal changes in the composition of *Pangasius* spp. during the last several years or a cross-swapping of local and scientific names due to miscommunication among and between residents and scientists.

4.3.2 Decreased Sightings or Local Extinction of Large Predators

Among the species that were reported by respondents as unseen in the previous five years, the most frequently mentioned were *Scleropages formosus*, *Chirocentrus dora*, *Tor* sp., and *Pangasius* sp., which are all large predators that occupy the higher trophic levels of aquatic ecosystems in peat swamp areas. The absence of such species therefore may indicate that bodies of water in the sampling site are in an early stage of ecosystem degradation. Generally, top predators that have a large body size, are low in abundance, and that have a high number of home-range requirements are particularly vulnerable to habitat fragmentation or destruction (Raffaelli 2004). This vulnerability is explained by their position at the top of an ecological pyramid. Because the pyramid is formed by the constraints of prey–predator mass ratios and that of the transfer of energy from lower to upper trophic levels of the pyramid (Trebilco et al. 2012), a larger ecological pyramid is needed to maintain a larger predator population. The decreased sighting of large predators raises concerns that recent developments in peat swamp ecosystems, such as deforestation, palm plantations, and fire, may be shrinking the pyramid, as depicted in Figs. 4.3a and b. Indeed, in their investigation of trophic positions of stream fishes using stable isotope analysis in Southeastern Sabah, Malaysian Borneo, Wilkinson et al. (2021) demonstrate that while the position of meso-predators does not change in oil-palm plantations versus forests, the trophic positions of apex predators in oil-palm plantations are lower than in forests.

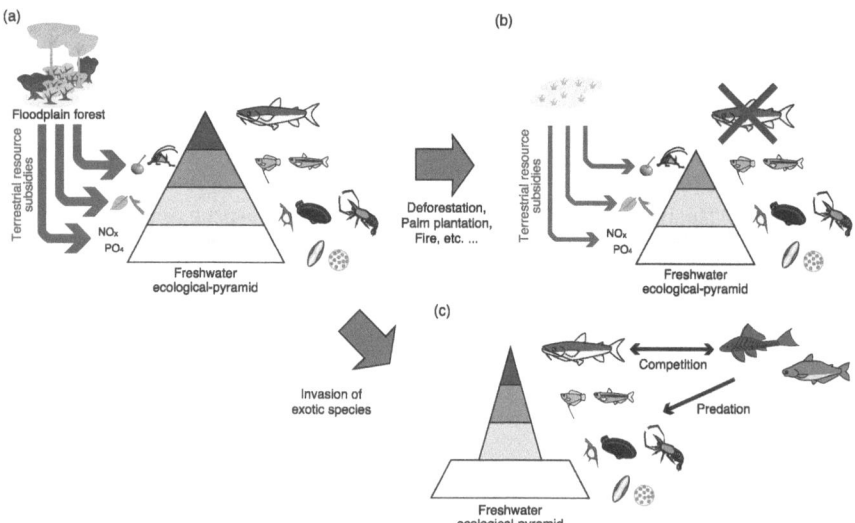

Fig. 4.3 Schematic images of the degradation process of a freshwater ecological pyramid by (**a**) land developments and (**b**) the introduction of exotic species

Local residents with LEK about river ecosystems recognize that floodplain forests are vital not only as fishing grounds, but also as spawning areas, and primary and secondary floodplain forests are relatively conserved around settlements compared to other areas around the mid-Kampar River (Nakagawa et al. 2021). However, the range of species dispersal generally correlates with the body length of the fish (Minns 1995; Radinger and Wolter 2014), and the dispersal distance of large river fish (≥ 500 mm in standard length) is often longer than 50 km (Radinger and Wolter 2014). In addition, several fish species in Southeast Asian rivers migrate distances farther than hundreds of kilometers and these fishes are often commercially important (Poulsen et al. 2002). Thus, the remaining forested area of the village, which is smaller than ten square kilometers (Fig. 4.1b), is too small to sustain populations of large fish species.

Furthermore, the loss of top predators often precipitates long-lasting impacts to natural ecosystems, including drastic changes in the species composition of lower trophic levels, ecosystem productivity and other functions, and even changes to landscape characteristics, which typically are not wanted by humans (Raffaelli 2004; Estes et al. 2011). For example, fishing down, or the negative spiral whereby the size of caught fish and the mesh in fish gear progressively decreases as larger individuals and species are successively eliminated, is typical to the collapse of aquatic ecosystems (Allan et al. 2005). Theoretical models predict that the fishing down initially leads to an increase in the weight of the total catch as the number of harvested species and individuals increases, followed by a plateau or slight decline in total catch (Welcomme 2001). If the decrease in the number of sightings and local extinction of large predatory fishes in the survey area indeed reflects the early stage of ecosystem degradation, scientific investigation to assess the effects of development in peat swamp ecosystems is urgently needed.

4.3.3 Exotic Species

In response to Question 3 about sightings of new, or exotic, species, "Sapu-sapu" was the most frequently mentioned (15 cases). "Patin" (4 cases), especially its alternate "Patin jambal" (2 cases), was the second-most frequent response. "Sapu-sapu," or *Pterygoplichthys* spp., is a major ornamental fish that has been artificially introduced into tropical and subtropical regions worldwide (Orfinger and Goodding 2018). Several species belonging to the genus *Pangasius* and *Pangasianodon* were also artificially moved beyond their native ranges, primarily for the purpose of aquaculture; among these, *Pangasianodon hypophthalmus* has been the main species introduced to the Indonesian islands (Lazard et al. 2009; Rimmer et al. 2013). Anthropogenic introduction of exotic fishes favors large species that have a high fishery or ornamental value. These species are often top predators that have never been exposed to native prey species throughout their evolutionally history and thus, become invasive (Mack et al. 2000). Therefore, contrasting with the impact of ecosystem fragmentation and destruction, which tends to affect species at higher trophic levels, the effects of exotic species tend to be more obvious at lower trophic levels, where species are directly consumed by an invader (Estes et al. 2011) (Fig. 4.3c). For example, many studies report serious decreases of native aquatic organisms in Asian and North and Central American countries following the invasion of *Pterygoplichthys* spp. (Orfinger and Goodding 2018). Most reports of the negative effects of *Pangasianodon hypophthalmus* on ecosystems come indirectly from studies of chemical and nutrient leakage from aquaculture ponds (see for example Rico et al. 2013). However, several studies directly address the negative effects of this catfish on other species via predation and/or competition (see Lazard et al. 2009; Rimmer et al. 2013).

The introduction of exotic fish species has various origins, from intensive ones with commercial purposes to unintentional ones due to deviations and abandonment of farmed or ornamental fishes. Regardless of the origin, however, assessment of the impact on the endemic ecosystem is rarely done in advance (Leprieur et al. 2009; Gozlan et al. 2010), despite these unassessed introductions often causing serious, even catastrophic, degradations of freshwater ecosystems (see for example Gophen et al. 1995; McDowall 2006; Hughes and Herlihy 2012; Matsuzaki and Kadota 2015). In villages along the middle reaches of the Kampar River, a diverse array of native species, from small to large fishes, are useful for self-consumption and commercial purposes. The invasion and increase of exotic species that potentially affect these fishery resources should therefore be observed with the utmost attention.

4.3.4 Potentiality of a Freshwater Protected Area in the Mid-Kampar River Basin

Although studies about the planning, management, and effectiveness of protected areas for sustainable fisheries and biodiversity conservation have mainly investigated marine ecosystems (Lester et al. 2009; Edgar et al. 2014), studies that report the effectiveness of protected areas in river ecosystems are increasing (Acreman et al. 2020). Peatlands, especially in floodplain areas that become seasonally submerged, are less suitable for oil-palm or acacia plantations compared to unsubmerged lands with mineral soils. At the same time, the drying of peatlands by channel construction for plantations results in the increase of fires. Therefore, the development of floodplains often does not improve the economic conditions of residents (see Kasori, Chap. 11). Indeed, if development of floodplains will decrease fishery production, then establishing freshwater protected areas that include surrounding floodplains may be a better choice than agricultural development when considering how best to sustain local economies.

Specifically, what types of planning and management of a protected area would be effective in the research site and areas like it? Acreman et al. (2020) review scientific papers on the effectiveness of protected freshwater areas and offer eight lessons on how to enhance the conservation of freshwater biodiversity. Their lessons about ecosystem monitoring (Lesson 1), size and habitat heterogeneity of the protected area (Lessons 2, 3, and 4), and trade-offs between biodiversity conservation and other human activities (Lessons 5, 7, and 8), are particularly applicable to peat swamp ecosystems in Indonesia today. Here, a perspective on the management of peat swamp freshwater ecosystems is warranted.

Ideally, to examine the effectiveness of a protected area in a freshwater system characterized by high naturally temporal variability, it is necessary to include both quantitative and comparable monitoring, such as before-after control-impact (BACI) design, which arguably requires long time frames (Adams et al. 2015). This type of monitoring also requires highly trained persons to conduct field surveys and continuous budget allocations to pay for their laborious work. Although BACI monitoring may be realized in the future, it is difficult to implement in addressing urgent issues. In this situation, local fishers sharing daily catch records with scientists may be effective. Estimation of stocks using the catch per unit effort (CPUE) is a well-established and widely used method in fishery management. Combining estimates based on the CPUE with other scientific census data can be a powerful means to grasp the conditions of the focal systems as well as their determinant factors (see for example Russ et al. 2004). However, it should be noted that, as suggested by the local fish names presented in this study, in order to gain sufficient practical data with this method, it is necessary to obtain prior coordination between scientists and fishers.

In riverine systems, many fish species use different habitats and parts of the basin at different life stages. In addition, aquatic habitats typically interface with a riparian or littoral zone, where stands of semi-aquatic and terrestrial vegetation regulate shade and water temperature, channel stability, and supplies of nutrients and organic matter to aquatic food webs (Fig. 4.3a). Furthermore, the natural flood regime is a defining feature that governs channel structure and connectivity, and substrate characteristics. Therefore, catchment-scale management is ideal for conserving biodiversity in freshwater ecosystems (Naiman et al. 2005). The Indonesian government uses the KHG (Kesatuan Hidologis Gambut), or peat hydrological unit, as the fundamental unit for managing peat swamp ecosystems (Kasori, Chap. 11). A KHG is an area of peat deposits formed between two rivers, between a river and a sea, and/or in a swamp, which is usually $0.01–10 \text{ km}^2$ in size (Ibrahim et al. 2019). A KHG does not contain a river catchment, but catchment-scale management can easily be applied to a KHG, because the centers of two neighboring KHGs usually consist of lateral slopes of a river channel. Although KHG governance at present mainly targets water management and fire prevention, it may be beneficial to incorporate a fisheries management perspective into KHG governance in the future.

Bhutan may have the most advanced case of catchment-scale ecosystem conservation in the world. In this country, Nature Needs Half (NNH), the international conservation movement that aims to protect 50% of the earth by 2050 (Pimm et al. 2018), has already protected terrestrial areas, and additional conservation plans for freshwater ecosystems that explicitly consider catchment-scale management have been suggested (see Dorji et al. 2020). While it may be difficult at present to implement an ecosystem conservation plan like Bhutan's in Indonesia, the floodplain forests intentionally left undisturbed by the villagers in the research area are clearly too small to sustain a population of large predators, as mentioned above. However, effective ecosystem conservation may be possible, even if individual protected areas are small. For example, Koning et al. (2020) demonstrate the effectiveness of protected areas created by 23 separate local residential communities in Thailand's Salween Basin. In this case, although the area of each individual protected area is small, the network of the areas works like a meta-community (Leibold et al. 2004) and has markedly increased richness, density, and biomass in fishes relative to adjacent areas. Many fisher villages similar to the research site are scattered along the Kampar River. Establishing a protected area in most of these villages, even on a small scale, could together function as one large protected area, which may in turn produce promising results for the entire basin.

4.3.5 How to Establish and Manage Freshwater Protected Areas in the Mid-Kampar River Basin

Acreman et al. (2020) indicate that biodiversity conservation in a protected area mainly fails due to external factors, such as inappropriate, illegal, or unregulated land and water management of the catchment area, or internal factors, such as human activities other than biodiversity conservation, including recreational uses, within the protected area. They emphasize how laws and regulations associated with protected areas need to be enforced and how regulatory activities should involve local communities. The challenges of local governance and politics vis-à-vis environmental issues are discussed in Chapts. 3, 8, and 11. Here, the potentiality of establishing protected areas in Rantau Baru village and the surrounding local communities is discussed. In the village, while local peoples seem to recognize the importance of floodplain forests to sustaining fishery resources as empirically based LEK (Prasetyawan, Chap. 9), floodplain development is still ongoing, even though it does not necessarily lead to an increase in the villagers' incomes (see Kasori, Chap. 11). This situation suggests that vague future concerns based on LEK do not work to disincentivize the commodification of the lands they own (see Osawa and Binawan, Chap. 3).

Conversely, the success of ecosystem conservation in this region may be achieved by the verification of local concerns via scientific investigation, sharing investigation results with local governments and peoples, and raising awareness of the importance of ecosystem conservation in the local communities. In practice, the "scenario planning method," used in the Community-Based Management of Environmental Challenges in Latin America (COMET-LA) project (Waylen et al. 2015), may be applicable in the communities of the middle Kampar. While the method is implemented by facilitators who are mix of interdisciplinary researchers and project-specific civil society organization personnel, its outputs are selected by local stakeholders via a workshop that is comprised of four steps. These steps are: (1) explore how drivers of change may influence the socio-ecosystem using morphological analysis (see Godet 2006 for detailed methods), (2) construct alternative scenarios, (3) identify 'robust' response options, and (4) discuss implications and requirements of response options (Waylen et al. 2015). These steps may be implemented in Rantau Baru as follows. First, referencing natural and sociological scientific investigations, facilitators identify the potential impacts of land developments (the drivers) and identify two opposing states, such as "floodplains will remain forested" or "floodplains will be developed to oil-palm or acacia plantations." These drivers and the contrasting states are then presented to the communities to modify or reselect. Second, alternative scenarios are created that consider balancing increases in local incomes with losses of natural resources. Community participants discuss and amend the scenarios to determine what the acceptable scenario is. Third, community participants discuss what actions, or response options, might be sufficiently relevant and robust to achieve community goals in light of possible future changes. These may include regulating the commodification of their lands and establishing protected areas. Finally, workshop participants discuss what specifically needs to be done, by when, and by whom, identifying specific actions for individuals, the community, and external actors to operationalize the robust response options. For example, the actions and responses may include local governments and stakeholders creating new rules to govern land use and fisheries and fishers recording daily catches to continuously monitor the status of natural resources.

4.4 Recreational Fishing: The Novel Commodification Activity of Fisheries in Rantau Baru

In Chap. 5, Nofrizal discusses the potential of recreational fishing to increase fisher income. Indeed, if managed appropriately, recreational fishing has numerous potential benefits. For example, recreational fishing may provide local people with alternative options for stable livelihoods and may provide national economies of developing countries an alternative revenue stream to that of extractive industries (such as logging or mining) or other activities that transform natural landscapes (such as large-scale agriculture or aquaculture) (Barnett et al. 2016). However, recreational fishing is also known to cause ecosystem conservation failure when it (1) fails to consider the local sociocultural context, (2) does not adequately distribute direct employment and knock-on benefits to local people, and (3) inappropriately regulates the ownership and tenure of natural resources and the space where recreational fishing occurs (Barnett et al. 2016; Acreman et al. 2020). It is common in developing countries, especially in cases where recreational fishing generates significant economic benefits, that very few businesses are owned or operated by indigenous people, and few indigenous people are employed by local businesses; thus, incentivizing sustainable livelihoods is necessary (Barnett et al. 2016). In any case, the implementation of recreational fishing schemes is based on the presumption that fishery production will be maintained at the same or a higher level than at present. Therefore, considering the current conditions of the mid-Kampar River Basin from the perspective of both natural resource sustainability and diversifying the economy, ecosystem conservation is the highest priority.

4.5 Conclusion

The need for ecosystem conservation in peat swamp areas has been repeatedly raised and is obvious in terms of global interests, such as achieving carbon neutrality and biodiversity conservation (Couwenberg et al. 2010; Hooijer et al. 2010; Posa et al. 2010; Miettinen et al. 2016). However, local perspectives based on scientific investigation are equally important when assessing the trade-offs of land development for income on the one hand and ecosystem conservation to sustain traditional—and future—livelihoods on the other. As we see from this chapter, this is particularly important when it comes to addressing development pressures on rivers and floodplains.

Acknowledgements We thank villagers in the Rantau Baru village for the warm understanding and the kind cooperation to our research. This research was supported by Research Institute for Humanity and Nature (RIHN: a constituent member of NIHU) Project No. 14200117. Figures 4.1 and 4.2, and Table 4.1 are reproduced from Nakagawa et al. (2021), Southeast Asian Studies Vol. 10, No. 3, under the permission of the Center for Southeast Asian Studies.

References

Acreman M, Hughes KA, Arthington AH et al (2020) Protected areas and freshwater biodiversity: a novel systematic review distils eight lessons for effective conservation. Conserv Lett 13:e12684. https://doi.org/10.1111/conl.12684

Adams VM, Setterfield SA, Douglas MM et al (2015) Measuring benefits of protected area management: trends across realms and research gaps for freshwater systems. Philos Trans R Soc B 370(1681):20140274. https://doi.org/10.1098/rstb.2014.0274

Allan JD, Abell R, Hogan Z et al (2005) Overfishing of inland waters. BioScience 55:1041–1051. https://doi.org/10.1641/0006-3568(2005)055[1041:OOIW]2.0.CO;2

Allen DJ, Darwall WRT, Smith KG (2012) The status and distribution of freshwater biodiversity in Indo-Burma. IUCN, Cambridge, UK; Gland, Switzerland

Amoros C, Bornette G (2002) Connectivity and biocomplexity in waterbodies of riverine floodplains. Freshw Biol 47:761–776. https://doi.org/10.1046/j.1365-2427.2002.00905.x

Ankei Y (1989) Folk knowledge of fish among the Songola and the Bwari: comparative ethnoichthyology of the Lualaba River and Lake Tanganyika fishermen. Afr Study Monogr, Supplementary Issue 9:1–88. https://doi.org/10.14989/68349

Aryani N (2015) Native species in Kampar Kanan River, Riau Province Indonesia. Int J Fish Aquat Stud 2:213–217

Aryani N, Suharman I, Azrita A et al (2020) Diversity and distribution of fish fauna of upstream and downstream areas at Koto Panjang Reservoir, Riau Province, Indonesia. F1000Research 8: 1435. https://doi.org/10.12688/f1000research.19679.2

Barnett A, Abrantes KG, Baker R et al (2016) Sportfisheries, conservation and sustainable liveli-hoods: a multidisciplinary guide to developing best practice. Fish Fish 17:696–713. https://doi.org/10.1111/faf.12140

Berkström C, Papadopoulos M, Jiddawi NS et al (2019) Fishers' local ecological knowledge (LEK) on connectivity and seascape management. Front Mar Sci 6(130):00130. https://doi.org/10.3389/fmars.2019.00130

Castillo TI, Brancolini F, Saigo M et al (2018) Ethnoichthyology of artisanal fisheries from the lower La Plata River Basin (Argentina). J Ethnobiol 38:406–423. https://doi.org/10.2993/0278-0771-38.3.406

Correa SB, Winemiller KO (2014) Niche partitioning among frugivorous fishes in response to fluctuating resources in the Amazonian floodplain forest. Ecol 95:210–224. https://doi.org/10.1890/13-0393.1

Couwenberg J, Dommain R, Joosten H (2010) Greenhouse gas fluxes from tropical peatlands in Southeast Asia. Glob Change Biol 16:1715–1732. https://doi.org/10.1111/j.1365-2486.2009.02016.x

Dorji T, Linke S, Sheldon F (2020) Freshwater conservation planning in the context of nature needs half and protected area dynamism in Bhutan. Biol Conserv 251:108785. https://doi.org/10.1016/j.biocon.2020.108785

Dudgeon D (2011) Asian river fishes in the Anthropocene: threats and conservation challenges in an era of rapid environmental change. J Freshw Fish Biol 79:1487–1524. https://doi.org/10.1111/j.1095-8649.2011.03086.x

Edgar GJ, Stuart-Smith RD, Willis TJ et al (2014) Global conservation outcomes depend on marine protected areas with five key features. Nature 506:216–220. https://doi.org/10.1038/nature13022

Efizon D, Putra RM, Kurnia F et al (2015) Keanekarangaman jenis-jenis ikan di oxbow pinang dalam desa buluh cina Kabupaten Kampar, Riau. In: Ramli Z, Razman MR, Efizon D (eds) Prosiding seminar antarabangsa ke 8: ekologi, habitat manusia dan perubahan persekitaran. Universiti Kebangsaan Malaysia, Langkawi, pp 23–46

Estes JA, Terborgh J, Brashares JS et al (2011) Trophic downgrading of planet Earth. Science 333:301–306. https://doi.org/10.1126/science.1205106

Fahmi MR, Ginanjar R, Kusumah RV (2015) Diversity of ornamental fish in peatlands biosphere reserve Bukit-Batu, Riau Province. Pros Sem Nas Masy Biodiv Indon 1:51–58. https://doi.org/10.13057/psnmbi/m010108

FAO (Food and Agriculture Organization of the United Nations) (2018) The state of fisheries and aquaculture 2018: meeting the sustainable development goals. FAO, Rome

Fauzi M (2004) Struktur komunitas ikan sungai Kampar yang dipengaruhi perubahan massa air akibat bendungan PLTA Koto panjang. J Perikan Kelaut 9:47–60

Firdaus F, Pulungan CP, Efawni E (2015) A study on fish composition in the Air Hitam River, Pekanbaru, Riau Province. JOM FAPERIKA UNRI 2:1–14

Fithra RY, Siregar YI (2010) Keanekaragaman ikan Sungai Kampar inventarisasi dari Sungai Kampar Kanan. J Ilmu Lingkungan 4:139–147. https://doi.org/10.31258/jil.4.02.p.139-147

Fitri AR, Husni D (2020) Disaster education to increase family resilience (community based participatory action research on post flood reconstruction phase). In: Agung S, Nanto D, Adrefiza A et al (eds) ICEMS 2019. Proceedings of the 5th international conference on education in Muslim society, Jakarta, September–October 2019. EAI, Gent. https://doi.org/10.4108/eai.30-9-2019.2291125

Froese R, Pauly D (eds) (2000) FishBase 2000: concepts, design and data sources. ICLARM, Los Baños

Glaser M, Baitoningsih W, Ferse SCA et al (2010) Whose sustainability? Top-down participation and emergent rules in marine protected area management in Indonesia. Mar Policy 34:1215–1225. https://doi.org/10.1016/j.marpol.2010.04.006

Godet M (2006) Creating futures: scenario planning as a strategic management tool, 2nd edn. Economica, Paris

Gophen M, Ochumba PBO, Kaufman LS (1995) Some aspects of perturbation in the structure and biodiversity of the ecosystem of Lake Victoria (East Africa). Aquat Living Resour 8:27–41. https://doi.org/10.1051/alr:1995003

Gozlan RE, Britton JR, Cowx I et al (2010) Current knowledge on non-native freshwater fish introductions. J Fish Biol 76:751–786. https://doi.org/10.1111/j.1095-8649.2010.02566.x

Hooijer A, Page S, Canadell JG et al (2010) Current and future CO_2 emissions from drained peatlands in Southeast Asia. Biogeosciences 7:1505–1514. https://doi.org/10.5194/bg-7-1505-2010

Hughes RM, Herlihy AT (2012) Patterns in catch per unit effort of native prey fish and alien piscivorous fish in 7 Pacific Northwest USA rivers. Fisheries 37:201–211. https://doi.org/10.1080/03632415.2012.676833

Ibrahim FK, Suryadi Y, Soekarno I et al (2019) Water management system of peatlands for palawija plants on KHG Pulang Pisau, Central of Kalimantan. MATEC Web Conf 270:04006. https://doi.org/10.1051/matecconf/201927004006

ICZN (International Commission on Zoological Nomenclature) (1999) International code of zoological nomenclature, 4th edn. International Trust for Zoological Nomenclature, London

Iskandar J, Dahiyat Y (2012) Keaneka ragaman ikan di sungai Siak Riau. Bionatura 14:51–58

Koning AA, Pelales KM, Fluet-Chouinard E et al (2020) A network of grassroots reserves protects tropical river fish diversity. Nature 588:631–635. https://doi.org/10.1038/s41586-020-2944-y

Lauck T, Clark CW, Mangel M et al (1998) Implementing the precautionary principle in fisheries management through marine reserves. Ecol Appl 8:S72–S78. https://doi.org/10.1890/1051-0761(1998)8[S72:ITPPIF]2.0.CO;2

Lazard J, Cacot P, Slembrouck J et al (2009) Fish farming of Pangasiids. Cah Agric 18:164–173

Leibold MA, Holyoak M, Mouquet N et al (2004) The metacommunity concept: a framework for multi-scale community ecology. Ecol Lett 7:601–613. https://doi.org/10.1111/j.1461-0248.2004.00608.x

Leprieur F, Brosse S, García-Berthou E et al (2009) Scientific uncertainty and the assessment of risks posed by non-native freshwater fishes. Fish Fish 10:88–97. https://doi.org/10.1111/j.1467-2979.2008.00314.x

Lester SE, Halpern BS, Grorud-Colvert K et al (2009) Biological effects within no-take marine reserves: a global synthesis. Mar Ecol Prog Ser 384:33–46. https://doi.org/10.3354/meps08029

Lubis AY, Efawani F, Windarti W (2016) Re-inventarisation of fish in the Sali River, Pekanbaru Regency, Riau province. JOM Faperi 3:1–10

Mack RN, Simberloff D, Lonsdale WM et al (2000) Biotic invasions: causes, epidemiology, global consequences, and control. Ecol Appl 10:689–710. https://doi.org/10.1890/1051-0761(2000)010[0689:BICEGC]2.0.CO;2

Masuda K (2012) Indoneshia mori no kurashi to kaihatsu: tochi wo meguru 'tsunagari' to 'semegiai' no shakaishi (Livelihoods and developments in Indonesian forest: sociohistory of cooperation and conflict for land use). Akashi Shoten, Tokyo

Masuda K, Mizuno K, Sugihar K (2016) A socioeconomic history of the peatland region: From trade to land development, and then to conservation. In: Mizuno K, Fujita MS, Kawai S (eds) Catastrophe and regeneration in Indonesia's peatlands: ecology, economy and society, Kyoto CSEAS series on Asian studies, vol 15. NUS Press; Kyoto University Press, Singapore; Kyoto, pp 148–184

Matsuzaki SS, Kadota T (2015) Trends and stability of inland fishery resources in Japanese lakes: introduction of exotic piscivores as a driver. Ecol Appl 25:1420–1432. https://doi.org/10.1890/13-2182.1

McDowall RM (2006) Crying wolf, crying foul, or crying shame: alien salmonids and a biodiversity crisis in the southern cool-temperate galaxioid fishes? Rev Fish Biol Fish 16:233–422. https://doi.org/10.1007/s11160-006-9017-7

Medin DL, Atran S (2004) The native mind: biological categorization and reasoning in development and across cultures. Psychol Rev 111:960–983. https://doi.org/10.1037/0033-295x.111.4.960

Miettinen J, Shi C, Liew SC (2016) Land cover distribution in the peatlands of Peninsular Malaysia, Sumatra and Borneo in 2015 with changes since 1990. Glob Ecol Conserv 6:67–78. https://doi. org/10.1016/j.gecco.2016.02.004

Minns CK (1995) Allometry of home range size in lake and river fishes. Can J Fish Aquat Sci 52: 1499–1508. https://doi.org/10.1139/f95-144

Mizuno K, Kusumaningtyas R (2016) Land and forest policy in Southeast Asia. In: Mizuno K, Fujita MS, Kawai S (eds) Catastrophe and regeneration in Indonesia's peatlands: ecology, economy and society, Kyoto CSEAS series on Asian studies, vol 15. NUS Press; Kyoto University Press, Singapore; Kyoto, pp 19–68

Myers N, Mittermeier RA, Mittermeier CG et al (2000) Biodiversity hotspots for conservation priorities. Nature 403:853–858. https://doi.org/10.1038/35002501

Naiman RJ, Décamps H, McClain ME et al (2005) Riparia: ecology, conservation, and management of streamside communities. Academic Press, Cambridge. https://doi.org/10.1016/B978-0-12-663315-3.X5000-X

Nakagawa H, Osawa T, Binawan A et al (2021) Local names of fishes in a fishing village on the bank of the middle reaches of the Kampar River, Riau, Sumatra Island, Indonesia. Southeast Asian Stud 10:435–454. https://doi.org/10.20495/seas.10.3_435

Nelson JS, Grande TC, Wilson MCH (2016) Fishes of the world, 5th edn. Wiley, Hoboken

Orfinger AB, Goodding DD (2018) The global invasion of the suckermouth armored catfish genus *Pterygolichthys* (Siluriformes: Loricariidae): annotated list of species, distributional summary, and assessment of impacts. Zool Stud 57:7. https://doi.org/10.6620/ZS.2018.57-07

Parker CT, Tindall BJ, Garrity GM (2019) International code of nomenclature of prokaryotes: prokaryotes code (2008 revision). Int J Syst Evol Microbiol 69:S1–S111. https://doi.org/10. 1099/ijsem.0.000778

Pimm SL, Jenkins CN, Li BV (2018) How to protect half of Earth to ensure it protects sufficient biodiversity. Sci Adv 4:eaat2616. https://doi.org/10.1126/sciadv.aat2616

Pinder AC, Britton JR, Harrison AJ et al (2019) Mahseer (*Tor* spp.) fishes of the world: status, challenges and opportunities for conservation. Rev Fish Biol Fish 29:417–452. https://doi.org/ 10.1007/s11160-019-09566-y

Posa MRC, Wijedasa LS, Corlett RT (2010) Biodiversity and conservation of tropical peat swamp forests. BioScience 61:49–57. https://doi.org/10.1525/bio.2011.61.1.10

Poulsen AF, Poeu O, Viravong S et al (2002) Fish migrations of the lower Mekong River Basin: implications for development, planning and environmental management. MRC Technical Paper no. 8, Mekong River Commission, Phnom Penh, p 62

Purnama AA, Yolanda R (2016) Diversity of freshwater fish (Pisces) in Kumu River, Rokan Hulu Distinct, Riau Province, Indonesia. Aquac Aquar Conserv Legis 9:785–789

Radinger J, Wolter C (2014) Patterns and predictors of fish dispersal in rivers. Fish Fish 15:456–473. https://doi.org/10.1111/faf.12028

Raffaelli D (2004) How extinction patterns affect ecosystems. Science 306:1141–1142. https://doi. org/10.1126/science.1106365

Rico A, Phu TM, Satapornvanit K et al (2013) Use of veterinary medicines, feed additives and probiotics in four major internationally traded aquaculture species farmed in Asia. Aquaculture 412–413:231–243

Rimmer MA, Sugama K, Rakhmawati D et al (2013) A review and SWOT analysis of aquaculture development in Indonesia. Rev Aquac 5:255–279. https://doi.org/10.1111/raq.12017

Russ GR, Alcala AC, Maypa AP et al (2004) Marine reserve benefits local fisheries. Ecol Appl 14: 597–606. https://doi.org/10.1890/03-5076

Satria A, Adhuri DS (2010) Pre-existing fisheries management systems in Indonesia, focusing on Lombok and Maluku. In: Ruddle K, Satria A (eds) Managing coastal and inland waters: pre-existing aquatic management systems in Southeast Asia. Springer, Dordrecht, pp 31–55. https://doi.org/10.1007/978-90-481-9555-8_2

Shimamura T (2016) An overview of tropical peat swamps. In: Mizuno K, Fujita MS, Kawai S (eds) Catastrophe and regeneration in Indonesia's peatlands: ecology, economy and society, Kyoto

CSEAS series on Asian studies, vol 15. NUS Press; Kyoto University Press, Singapore; Kyoto, pp 123–147

Siregar S, Putra RM, Sukendi (1994) Fauna ikan di perairan sekitar bukit Tigapuluh Seberida, Sumatera. In: Sandbukt O, Wiriadinata H (eds) Rain forest and resource management: proceeding of the NORINDRA seminar, Jakarta, May 1993. Indonesian Institute of Science, Jakarta, pp 67–70

Thornton TF, Scheer AM (2012) Collaborative engagement of local and traditional knowledge and science in marine environments: a review. Ecol Soc 17:8. https://doi.org/10.5751/ES-04714-170308

Tjakrawidjaja AH, Haryono (2000) Keanekaragaman jenis ikan di areal penambangan gambut Perawang dan sekitarnya, kabupaten Bengkalis-Riau. In: Sjafei DS et al (eds) Prosiding Seminar Nasional Keanekaragaman Hayati Ikan, Bogor, June 2000. Pusat Studi Ilmu Hayat Institut Pertanian Bogor, Bogor, pp 55–60

Trebilco R, Baum JK, Salomon AK et al (2012) Ecosystem ecology: size-based constraints on the pyramids of life. Trends Ecol Evol 28:423–431. https://doi.org/10.1016/j.tree.2013.03.008

Turland NJ, Wiersema JH, Barrie FR et al (eds) (2018) International Code of Nomenclature for algae, fungi, and plants (Shenzhen Code) adopted by the Nineteenth International Botanical Congress Shenzhen, China, July 2017. Regnum Vegetabile 159 [Online]. Koeltz Botanical Books, Glashütten. https://doi.org/10.12705/Code.2018

Waylen KA, Martin-Ortega J, Blackstock KL et al (2015) Can scenario-planning support community-based natural resource management? Experiences from three countries in Latin America. Ecol Soc 20:28. https://doi.org/10.5751/ES-07926-200428

Welcomme RL (2001) Inland fisheries: ecology and management. Fishing News Books, Oxford

Wilkinson CL, Chua KWJ, Fiala R et al (2021) Forest conservation to oil palm compresses food chain length in tropical streams. Ecology 102:e03199. https://doi.org/10.1002/ecy.3199

Yustina (2016) The impact of forest and peatland exploitation towards decreasing biodiversity of fishes in Rangau River, Riau-Indonesia. Int J Appl Bus Econ Res 14:1043–1055

Chapter 5
Fisheries of the Rantau Baru and Kampar Rivers, Sumatra, Indonesia

Nofrizal, Romie Jhonnerie, Thamrin, Tengku Said Raza'i, Zulfan Sa'am, and Hikaru Nakagawa

Abstract Rivers and peat swamps provide fishing grounds that can support the people living in Rantau Baru. Survey activities were conducted to describe the capture fisheries business carried out by fishers. The survey results show that 109 of 623 residents work as fishers, including women. Small-scale traditional fishing gear, such as traps, gillnets, mini long lines, set nets, pole and line, and cash nets are used, but traps are the dominant gear. Transportation to catch fish relies on boats, outboard motorboats, and fishing vessels. The outboard motorboat is widely used by fishers because of its small size and ability to navigate shallow and narrow waters. At least 44 species of fish from 10 families are caught and sold by fishers. Catches fluctuate according to the seasons, with increases during the flood season and decreases during the dry season. The fishing grounds also have potential for recreational fishing activities, as fishers earned US$37,242.67 from boat rental services for fishing-related tourism activities in 2019. This chapter provides an overview of the fishing activities and the economic value generated from fishing activities in the rivers and peat swamps of Rantau Baru.

Nofrizal (✉) · Thamrin
Marine Science Post Graduate Study, Faculty of Fisheries and Marine Science, Riau University, Pekanbaru, Indonesia

Environmental Science, Post Graduate Studies, Riau University, Pekanbaru, Indonesia

R. Jhonnerie
Marine Science Post Graduate Study, Faculty of Fisheries and Marine Science, Riau University, Pekanbaru, Indonesia

T. S. Raza'i
Marine Science and Fisheries Faculty, Raja Ali Haji Maritime University, Tanjung Pinang, Riau, Indonesia

Z. Sa'am
Environmental Science, Post Graduate Studies, Riau University, Pekanbaru, Indonesia

H. Nakagawa
Aqua Restoration Research Center, Public Works Research Institute, Gifu, Japan

M. Okamoto et al. (eds.), *Local Governance of Peatland Restoration in Riau, Indonesia*, Global Environmental Studies,
https://doi.org/10.1007/978-981-99-0902-5_5

Keywords Capture fisheries · Peat swamps · Recreational fishing

5.1   Introduction

Indonesian peat swamp ecosystems, which include forests, bush, swamps, ponds, and rivers located in and around peatlands are mainly scattered across lowland areas in Indonesia, covering a total of 206,000 km,[2] with 35% located in Sumatra, 32% in Kalimantan, and 33% in Papua (Muchlisin et al. 2015; Miettinen et al. 2016). Until a few decades ago, central and local governments in Indonesia regarded peat swamp forests as economically useless land and tried to develop them for capitalistic gains (Humphreys 2013; FAO 2016). For example, the central government encouraged transmigration from Java and other densely populated areas to peat swamp forests (Whitten 1987; Fearnside 1997). The government initiated an unsuccessful mega rice project in the peat areas of Central Kalimantan relocating transmigrants there (Boehm and Siegert 2000; Giesen 2008). The government divided peat areas for concessions and granted logging permissions to the companies without proper monitoring (Yolamalinda 2013; Enrici and Hubacek 2018), and overlooked extensive illegal logging (Lambin et al. 2018). These legal and illegal activities led to the construction of numerous drainage channels for agricultural development and the extraction of logs in peatlands (Hergoualc'h et al. 2018). However, more recently this condition has been changing because of the increase of knowledge about the ecosystem functions of tropical peatlands (Hergoualc'h et al. 2018).

The ecosystems of Southeast Asian rivers are known as one of the largest biodiversity hotspots in the world (Dudgeon 2011). The species diversity and vast fish supplies are important food resources to the people in Southeast Asian countries (FAO 2018). Temporal and permanent inland water bodies such as rivers, oxbow lakes, and swamps in peatlands provide critical spawning, rearing, and foraging habitat for river fishes in tropical regions, and support a high secondary production of diverse fish species (Amoros and Bornette 2002; Correa and Winemiller 2014; Hergoualc'h et al. 2018). In lowland areas of Indonesia, such water bodies mostly consist of peat swamp ecosystems and have functioned as valuable fishing grounds to sustain local fishery catches (Haryono 2007; Posa et al. 2011). The recent development of peatlands presents a serious threat to river health, basin ecosystems, and the sustainability of inland fisheries (e.g., Yustina 2016).

Engaging local stakeholders is key to ecosystem conservation and natural resource management (Sterling et al. 2017). Most inland fishery catches in Southeast Asia are provided by local residents from small-scale fisheries (Salayo et al. 2008; Cooke et al. 2016). As local fishers not only sell their catches to obtain income, but often self-consume their catches, the anthropogenic degradation of freshwater ecosystems not only leads to decreases in their income, but also in daily food consumption. Therefore, local residents engaging in small-scale inland fisheries should be the key stakeholders in decision making around the development and conservation of peat swamp ecosystems. However, at present, the central and local governments undervalue and overlook the potential of inland fisheries and do not prioritize the empowerment of inland small-scale fishers (Cooke et al. 2016). In Indonesia, information about inland small-scale fishery activities by local fishers is extremely limited except for a few qualitative reports (Allison and Ellis 2001; Masuda 2012; Stacey et al. 2019; Stacey et al. 2021). There is almost no comprehensive analysis of the fishing gear and techniques employed in the Kampar River area of Sumatra. The first steps toward achieving viable peat swamp ecosystem management to secure the sustainability of local fisheries are (1) assessing the actual and concrete situation of small-scale fisheries and (2) identifying strategies to motivate local fishers to protect the peat swamp ecosystem. Therefore, this study examines the livelihoods of people who depend on the fisheries of submerged forests around peat swamps. Awareness of the community's dependence on these natural resources provides a basis to promote the preservation of them among community, private, and government actors.

In this chapter, we introduce the local fisheries and their commodification by fishers in Rantau Baru, a typical fishing village in the mid-Kampar Basin. We specifically focus on the utilization of peat swamp ecosystems. We illustrate how the new industry of recreational fishing in the village and has the potential to improve both the incomes of local households and peatland ecosystem conservation.

5.2 Research Site and Methodology

Rantau Baru is located in the administrative area of Pangkalan Kerinci Sub-district in Pelalawan District, Riau Province, Indonesia. The village, which covers an area of approximately 10 km^2, is in a lowland area, and includes swamps, peatlands, typical peat swamp forests, the Kampar River and its tributary Bokol-Bokol, and several oxbow lakes (Kiyap, Awareness, Seluk Kuras, Badagu, and Sepunjung). Oxbow lakes and rivers are important fishing grounds for the peoples working as a fishers in Rantau Baru.

The original data used in this chapter were obtained mainly through field research in Rantau Baru that was conducted during 55 short-term trips to the villages from January 2020 to October 2021. These trips included observations of fishing and fish processing activities and unstructured interviews with villagers. We also conducted a questionnaire survey of 51 households.

Based on the survey results, at least 109 of the total 623 residents in Rantau Baru (17.5%) fish as a permanent job (Table 5.1). In addition, many people in the village have side jobs as fishers. Civil servants, government contract employees, and entrepreneurs also catch fish. Women, housewives, and widows are also engaged in fishery activity, making salted, dried, and smoked fish to augment income.

Fishing grounds in Rantau Baru are centered on the mainstream of the Kampar River. The Kampar is the main transportation route used to access other fishing grounds as well as being an important fishery ground itself. Global Positioning System (GPS) loggers were used to track fishers' trajectories, and showed that most fishing activity was conducted in the mainstream of the Kampar River,

Table 5.1 Total population of Rantau Baru by occupation in 2020

No.	Backwoods	Housewife	Fisher	Entrepreneur	Student	Unemployed	Farmer	Company employee	Civil Servant	Government contract employee
1.	Sepujung	17	14	9	3	6	2	1	-	5
		10	10	2	10	5	-	2	-	3
		11	5	9	15	3	1	-	-	-
		15	9	8	19	3	1	2	1	5
	Subtotal	**53**	**38**	**28**	**47**	**17**	**4**	**5**	**1**	**13**
2.	Malako Kecil	20	16	9	15	8	1	2	-	5
		15	10	5	21	3	1	-	1	1
		17	18	3	26	1	-	-	-	3
		13	12	11	4	3	-	-	-	2
	Subtotal	**65**	**56**	**28**	**66**	**15**	**2**	**2**	**1**	**11**
3.	SeiPebadara	13	5	4	21	2	4	1	-	1
		14	3	10	18	1	-	5	-	2
		7	3	4	16	-	4	-	-	-
		9	4	5	10	-	1	2	-	2
	Subtotal	**43**	**15**	**23**	**65**	**3**	**9**	**8**	**0**	**5**
	Total	**161**	**109**	**79**	**178**	**35**	**15**	**15**	**2**	**29**

Sometimes people who work as housewives, entrepreneurs, civil servants, and other occupations also work as fishers as part-time jobs

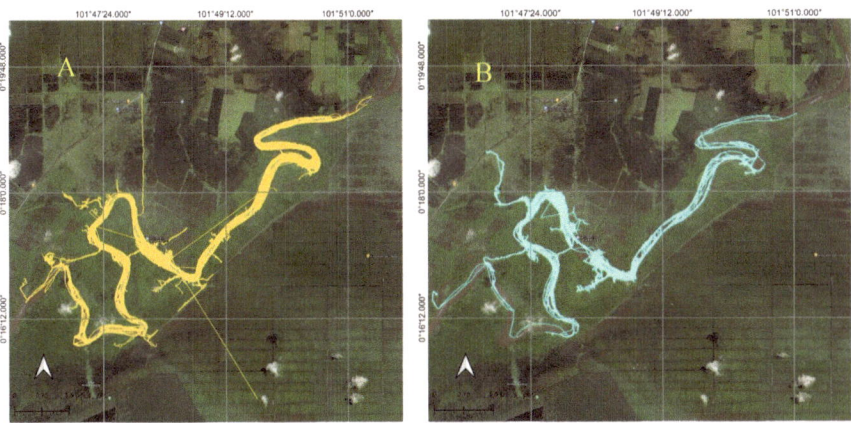

Fig. 5.1 GPS tracking handle for observation of the fishing grounds of (**a**) fishermen and (**b**) fisherwomen in Rantau Baru

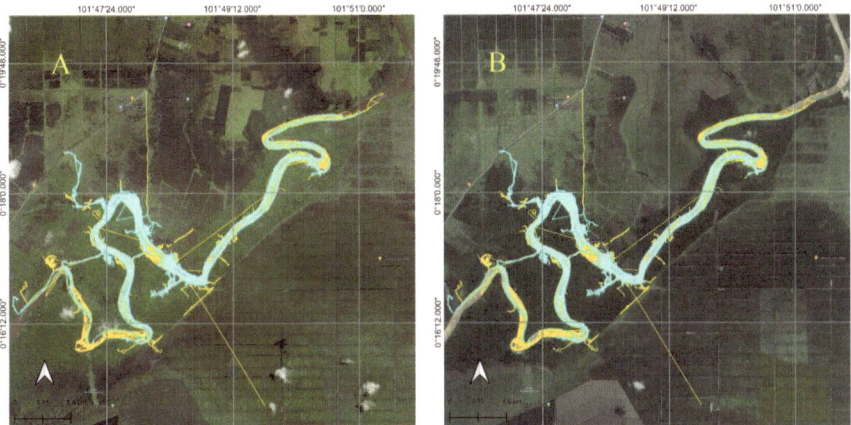

Fig. 5.2 Fishing grounds of fishermen and fisherwomen in (**a**) the dry and (**b**) the rainy (flood) seasons

followed by activity in swamps, tributaries, secondary flows, and artificial canals in plantation areas (Fig. 5.1).

As Chap. 6 explains, the adat (communal law) committee of Rantau Baru village has a regulation to annually auction certain fishing areas of tributaries, lakes, and swamps and only those who win the auction have the right to fish there. The regulation also allows fisherwomen, especially widows, to catch fish in all the available fishing grounds in Rantau Baru so that they can continue to earn some income (Fig. 5.2).

5.3   Fishing Activities in Rantau Baru

5.3.1   Fishing Gear

Fisheries in Rantau Baru are artisanal; fishers use low-tech fishing gear with simple tools that do not require special knowledge or skills to operate. Such gear makes it possible for anyone to work as a permanent or part-time fisher. The fishing gear used by fishers include traps, gill nets, mini long lines, poles, lines, lift nets, and cast nets (Figs. 5.3, 5.4, 5.5, 5.6, and 5.7). Each type of fishing gear has a different target fish, and is employed according to the conditions of the fishing location. We can find many similarities between traditional fishers in Rantau Baru and those in villages worldwide, especially in their choice of fishing gear depending on the similarity of fishing ground conditions and the types of primary target fishes (e.g. Rahman et al. 2017). Most fishing gear material, such as nets, ropes, yarn, sinkers, and buoys, are bought at local markets in the district capital and the provincial capital. Other materials such as wood, bamboo, rattan and cork can be found in and near Rantau Baru.[1] Traditional fishing gear is generally made using materials that are readily available around residences. Therefore, traditional fishers depend on the peat swamp forest around the village not only for fishing grounds, but also for fishing gear.

The most commonly used fishing gear by fishers in Rantau Baru are large and small traps. The large traps (pengilars) catch almost all species of fish and shrimp both in rivers and swamp areas (Fig. 5.4a). Small traps are used to catch swamp eel (*Monopterus albus*) (Fig. 5.4b) and are constructed using plastic pipes, woven bamboo, or rattan. A funnel at the front of the trap prevents the swamp eel from escaping once they enter the trap. Sometimes, the fishers construct two funnels, at the front and center of the trap. The end of the trap is covered with plastic or a coconut shell. This cover can be opened and closed, and the bait is placed in it to entice the fish into the trap (Fig. 5.4b).

Two types of line fishing are used in Rantau Baru: pole and line, and long line. Pole and line are used as substitutes while operating the main fishing gear. Once fishers set up their main fishing gear, such as a trap or stationery gill net, they use the pole and line while waiting for the hauling time. The construction of the mini long line in Rantau Baru is quite simple (Fig. 5.5). This fishing gear consists of only three main parts: the main line made of polyester with a diameter of 3 mm, a branch line made of 1 mm monofilament, and a number 7 hook. The branch line is not equipped with a snapper, so fishers tie the branch line to the main line using a double English knot during the setup. The branch line is not equipped with a swivel, and the distance between each branch line depends on the fishing ground conditions, so it does not have a fixed range. The main target of line fishing is predatory fish from the Bagridae, Claridae, Siluridae, Pangasidae, and Chandidae families. Line fishing is

---

[1] Ali et al. (2015) state that traditional fishermen along the Ramnadad River, Southern Bangladesh, also use traps as their main fishing gear. The traps used in the Ramnadad River are constructed with a bamboo frame and iron wire.

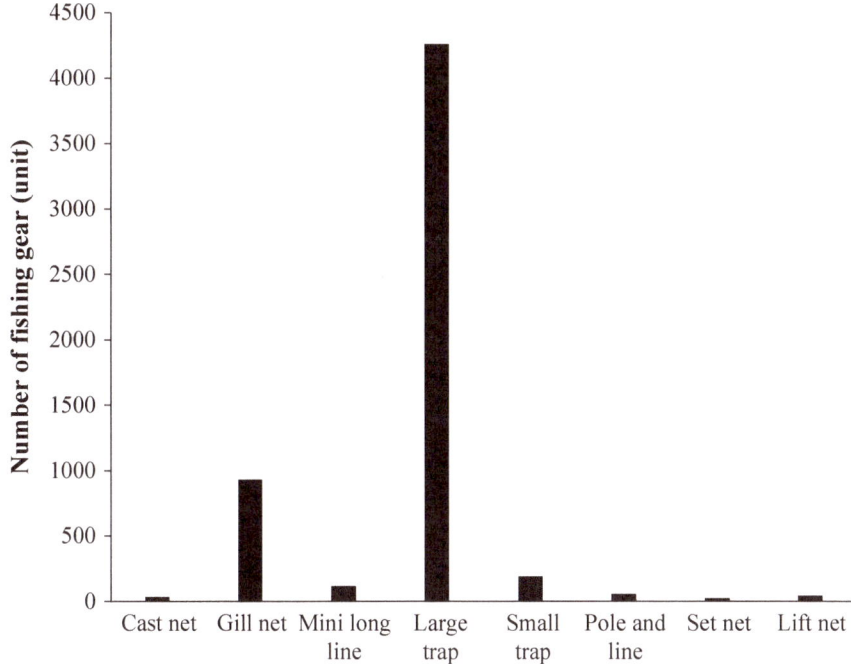

Fig. 5.3 Number and types of fishing gear operated by fishers in Rantau Baru

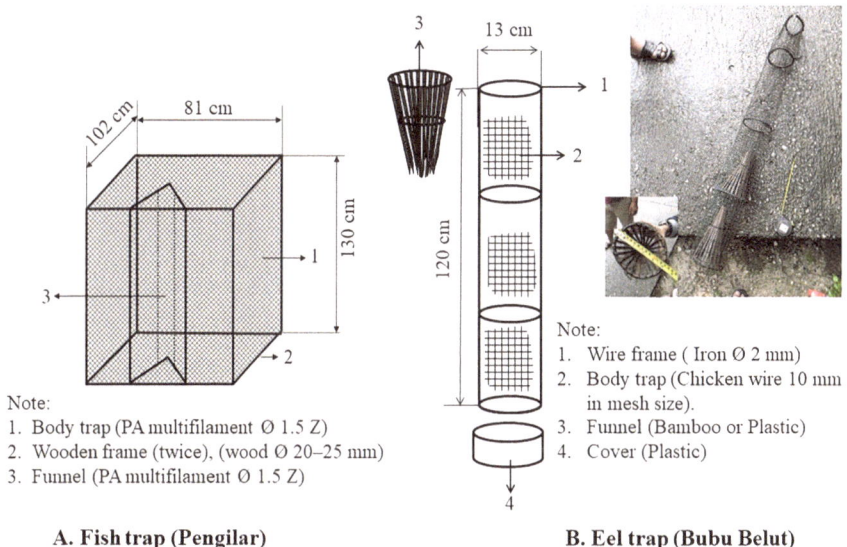

Note:
1. Body trap (PA multifilament Ø 1.5 Z)
2. Wooden frame (twice), (wood Ø 20–25 mm)
3. Funnel (PA multifilament Ø 1.5 Z)

Note:
1. Wire frame (Iron Ø 2 mm)
2. Body trap (Chicken wire 10 mm in mesh size).
3. Funnel (Bamboo or Plastic)
4. Cover (Plastic)

A. Fish trap (Pengilar)   **B. Eel trap (Bubu Belut)**

Fig. 5.4 Design and construction of (**a**) large traps for fish and (**b**) small traps for eels

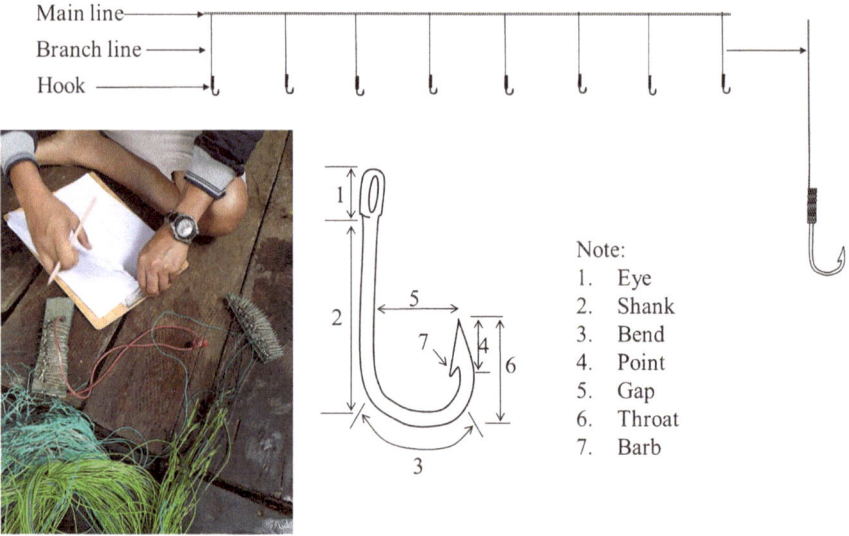

Fig. 5.5 Illustration of the mini long line construction in Rantau Baru

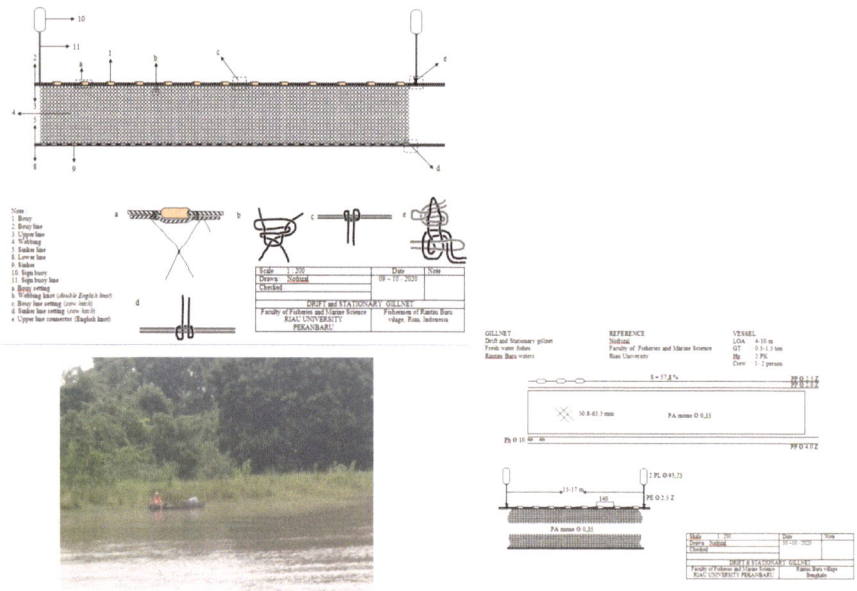

Fig. 5.6 Design and construction of the gill net used in Rantau Baru

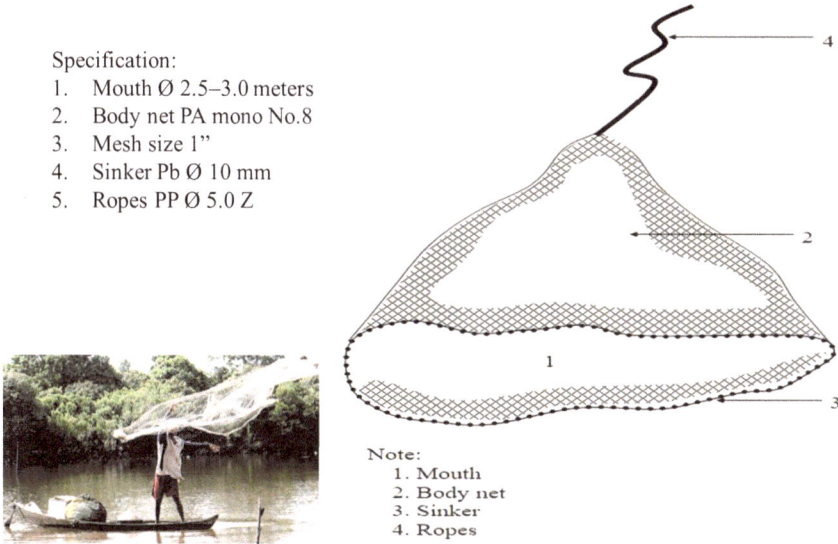

Specification:
1. Mouth Ø 2.5–3.0 meters
2. Body net PA mono No.8
3. Mesh size 1"
4. Sinker Pb Ø 10 mm
5. Ropes PP Ø 5.0 Z

Note:
1. Mouth
2. Body net
3. Sinker
4. Ropes

Fig. 5.7 Illustration of the cast net used by local fishermen in Rantau Baru

equipped with live fish bait when used, so the size of the catch is relatively bigger than that of other fishing methods. Line fishing is used in almost all the fishing areas in Rantau Baru, such as the Kampar River mainstream, tributaries, and oxbow lakes. In the flood season, the mini long line is used in floodplain areas. During the dry season, mini long lines are used from the riverbanks. The ideal mini-long line fishing area is a submerged area. The mini-long line is used from a stationary position and is used periodically for hauling.

A gill net is widely used by both men and women fishers to catch large fishes, such as those in the Notopteridae and Osphronemidae families (Fig. 5.6). It is possible to operate a gill net using a small boat. The cast net is a type of fishing gear widely used in oxbow lakes and rivers with sloping coastal contours. Almost all species can be caught by a cast net. The diameter of the cast net used in Rantau Baru varies from 2.5 to 5.0 m, with webbing made of polyamide monofilament No. 8 and a mesh size of 2.5 cm. The bottom of the cast net is equipped with a chain sinker made of lead with a diameter of 10 mm, while the upper part is made of polypropylene rope with a diameter of 5 mm (Fig. 5.7). Generally, the cast net is used by fishers during the dry season when the water discharge decreases, and the waters become shallow in the oxbow lakes, swamps, and small rivers and the fish are confined in these areas. On the other hand, the cast net is not for use during the rainy season when the waters of the Kampar River merge with the oxbows and the swamp by flooding and the fish spread and migrate all over the oxbow lakes, swamps, and small rivers.

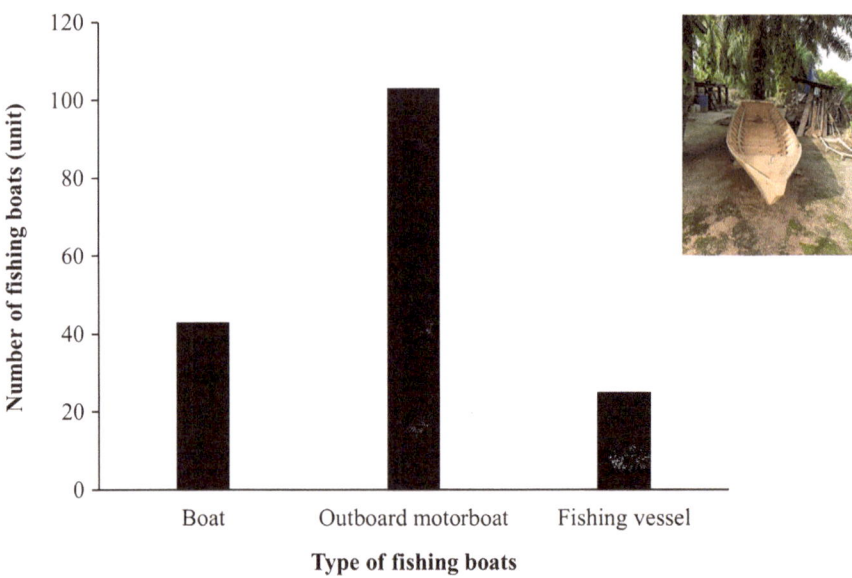

Fig. 5.8 The number of fishing vessels in Rantau Baru in 2020

5.3.2 Boats (pompong)

Rantau Baru experiences severe flooding periodically every year. The depth of the floods can reach 1.5–2.0 m for a period of 2–4 months. This means that boats and motorboats are the main transportation for the Rantau Baru community. Three types of boats are used by fishers, namely, boats or canoes, outboard motorboats, and fishing boats. The boat capacity is only 0.1–0.2 gross tonnage (GT), while the maximum outboard motorboat capacity is 0.5 GT. Some fishers own a fishing boat with a capacity of 2–3 GT (Fig. 5.8). Fishers mainly use outboard motorboats, as they are inexpensive and equipped with a small, fuel-efficient engine. In addition, outboard motorboats are small and have a shallow draft, which gives them good maneuverability to reach shallow and narrow waters, optimizing their fishing capacity.

Small fishing boats with the capacity of 1.5–2 GT are efficient in terms of operation, maintenance, and variable costs when carrying out fishing operations (Ahmad and Nofrizal 2015). Since the fishing grounds in local river and swamp waters are limited in range and depth and continuous fishing activity does not exceed one day, it is rational for traditional fishers with limited capital to use a small boat.

Construction of fishing boats is done in Rantau Baru or a nearby village. The primary material used is wood obtained from the forest in and around the peat swamp area. Boat engines and other equipment, such as propeller axles and propellers, are imported from Pekanbaru City. Knowledge about the manufacture of fishing boats has been passed down from generation to generation (Nofrizal and

Ahmad 2013). The fundamental obstacle in manufacturing fishing boats is the scarcity of wood and planks to build them. Deforestation and the Forestry Ministry's logging moratorium policy area causing a reduction in the construction of wooden fishing boats. Under these conditions, shipbuilders look for other jobs as fishers, farmers, builders, and others (Nofrizal et al. 2014a, 2014b). In turn, this deindustrialization makes it challenging to find fishing vessels, creating an obstacle for fishers in Rantau Baru trying to develop their fishing operations. The same problem also arises in the wooden shipyard industry of Samut Sakhon, Thailand (Kanoksilapatham 2016) and Bagan Siapiapi, Indonesia (Nofrizal et al. 2014a, 2014b).

5.4 Commodification of Fishery Catches

According to secondary data from the Pelalawan District Fisheries and Marine Service Department, fish production in Pangkalan Kerinci Sub-district reached 245.39 tons in 2020, the fifth largest in the region, with most of the catch coming from Rantau Baru village. Kuala Kampar and Teluk Meranti sub-districts are the largest fish producers in Pelalawan District along the Malacca Straits (Fig. 5.9), but

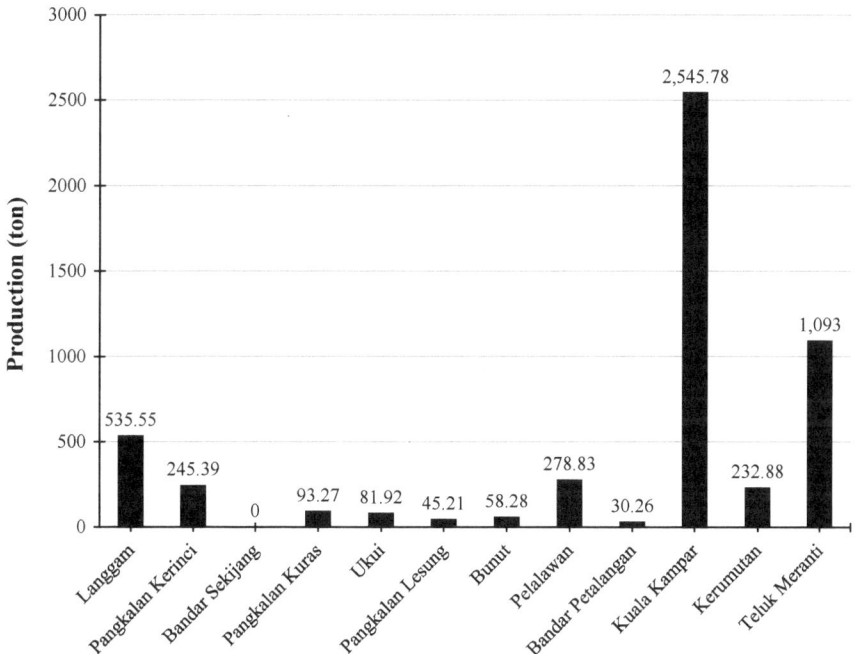

Fig. 5.9 Fisheries production in the villages of Pelalawan District, Riau Province, Indonesia in 2020

their production comes not only from freshwater fish, but also from marine fisheries. So, the freshwater production from Pangkalan Kerinci, especially from Rantau Baru village, is quite significant in Pelalawan District.

Table 5.2 shows the survey results on the retail price of fishery products from fishers and traders in the Pelalawan market. We observed at least 44 fish species belonging to ten families that had been caught and were being sold by fishers in Rantau Baru. Based on interviews with 155 respondents who work as fishers as well as in fish processing, local wholesale prices of fresh fish and shrimp range from US$0.69 to US$9.00 per kilogram. Giant prawn (*Macrobrachium rosenbergii*) are the most expensive fishery commodity, making it the highest priority target of local fishers. Giant prawn is a seasonal catch, occurring when the river water discharge increases.

Fishery catches in Pangkalan Kerinci, especially in Rantau Baru, change according to the seasons, as depicted in Fig. 5.10. During the monsoon season (October to April), the land floods can last for 3–4 months. Fishery production in this season is larger than in the dry season (May to September). The fishery catch trends were almost constant from 2015 to 2019.

According to interviews with middlemen traders, local market traders, and fishers regarding the price of fishery commodities from Rantau Baru, local wholesale prices in Rantau Baru are significantly lower than market prices. For example, the price of shrimp commodities for fishers in Rantau Baru is US$9.00 per kilogram. The price can increase by as much as 27.77%, with price of US$12.46 per kilogram at the consumer level.

Fishers in Rantau Baru process their fishery products into both smoked and dried fish to increase the selling price. The processing technology is very simple, using only a furnace, chicken wire netting, and nets to dry and smoke fish. Furthermore, the fuel for smoking fish is wood obtained from the forest around the village. One advantage of processing fish is that the products can become more durable. We observed the increase in the price of fresh fish products processed into smoked and dried fish can reach US$7.6 per kilogram, especially for fish commodities made from Siluridae (Table 5.2). Several types of fish are not processed by the fishers of Rantau Baru. These include *Chitala* sp., *Albulichthys albuloides*, *Puntioplites waandersi*, *Puntioplites feathers*, *Puntigrus tetrazona*, *Rasbora argyrotaenia*, *Rasbora rutteni*, *Rasboraawarensis*, *Monopterus albus*, *Trichogaster trichopterus*, *Trichogaster leerii*, *Sphaerichthys osphromenoides*, *Channa striata*, *Channa Lucius*, *Channa bankanensis*, and *Macrobrachium rosenbergii*. This is because neither the texture nor the taste of these species are good when the fish is smoked or dried. Consequently, these processed fish are not accepted in the market or by consumers.

According to a survey of the distribution and sale of catches from the Central Kampar watershed, such catches are sold both at local markets in the district and in Pekanbaru City, the provincial capital. Pelalawan District is famous for producing fresh and processed fish, such as smoked, dried, and salted fish. The most famous freshwater fish-producing areas in Pelalawan District are Langgam, Rantau Baru, and Teluk Meranti villages. According to the results of surveys and interviews with middlemen traders, fishery products from Pelalawan District are also well known on

Table 5.2 Survey results of local fish wholesale and retail prices in Rantau Baru and in Pelalawan and Pekanbaru markets

No.	Family	Species	Local name	Fishing gear	Local market price (US$/kg)		Consumer level price (US$/kg)	
					Fresh	Smoked	Fresh	Smoked
1	Notopteridae	Chitala borneensis	Belida	Pole and line, mini long line, gill net, large trap	4.16	–	6.9	–
2		Chitala lopis	Belida	Pole and line, mini long line, gill net and large trap	4.16	–	6.9	–
3	Cyprinidae	Barbonymus schwanenfeldii	Kapiek	Pole and line, large trap, gill net and cast net	0.83	1.38	2.41	–
4		Albulichthys albuloides	Ikan putih-putih	Gill net, large trap, cast net, pole and line	–	–	–	–
5		Puntioplites waandersi	Ikan putih-putih	Gill net, large trap, cast net, pole and line	–	–	–	–
6		Puntioplites bulu	Pantau/Tabingal	Cast net, scop net, small trap, gill net	1.73	–	4.14	–
7		Puntigrus tetrazona	Pantau/Tabingal	Cast net, scop net, small trap, gill net	1.73	–	4.14	–
8		Rasbora argyrotaenia	Pantau/Tabingal	Cast net, scop net, small trap, gill net	1.73	–	4.14	–
9		Rasbora rutteni	Pantau/Tabingal	Cast net, scop net, small trap, gill net	1.73	–	4.14	–
10		Rasbora tawarensis	Pantau/Tabingal	Cast net, scop net, small trap, gill net	1.73	–	4.14	–
11		Labiobarbus leptocheilus	Sisik	Gill net, large trap, cast net	2.08	3.11	3.45	–
12		Osteochilus kelabau	Kelabau	Gill net, large trap, cast net, mini long lie, pole and line	2.42	3.45	2.76	–
13		Osteochilus vittatus	Pawe	Gill net, large trap, cast net, mini long lie, pole and line	1.04	2.07	1.72	–

(continued)

Table 5.2 (continued)

No.	Family	Species	Local name	Fishing gear	Local market price (US$/kg) Fresh	Smoked	Consumer level price (US$/kg) Fresh	Smoked
14		Oxygaster anomalura	Pin-ping	Gill net, large trap, cast net, mini long lie, pole and line	0.69	1.72	1.72	6.90
15		Thynnichthys polylepis	Motan	Scop net, gill net, cast net	1.39	5.53	1.72	6.90
16		Thynnichthys thynnoides	Motan	Scop net, gill net, cast net	1.39	5.53	1.72	6.90
17	Clariidae	Clarias batrachus	Lele dumbo	Large trap, small trap, pole and line, mini long line	1.04	4.15	2.76	4.83
18		Clarias teijsmanni	Lele akar	Large trap, small trap, pole and line, mini long line	1.04	4.15	4.83	–
19	Bagridae	Hemibagrus wyckii	Geso	Mini long line, pole and line, large trap	3.81	–	8.97	
20		Hemibagrus nemurus	Baung	Mini long line, pole and line, large trap, gill net, cast net	3.46	8.30	6.90	17.25
21		Hemibagrus planiceps	Baung	Mini long line, pole and line, large trap, gill net, cast net	3.46	8.30	6.90	17.25
22		Mystus nigriceps	Baung	Mini long line, pole and line, large trap, gill net, cast net	3.46	8.30	6.90	17.25
23		Mystus micracanthus	Baung	Mini long line, pole and line, large trap, gill net, cast net	3.46	8.30	6.90	17.25
24		Mystus gulio	Baung	Mini long line, pole and line, large trap, gill net, cast net	3.46	8.30	6.90	17.25
25	Siluridae	Ompok eugeneiatus	Selais	Mini long line, pole and line, large trap, gill net, cast net	2.77	10.37	6.90	17.25
26		Ompok hypophthalmus	Selais	Mini long line, pole and line, large trap, gill net, cast net	2.77	10.37	6.90	17.25
27		Kryptopterus palembangensis	Selais	Mini long line, pole and line, large trap, gill net, cast net	2.77	10.37	6.90	17.25

28		*Kryptopterus schilbeides*	Selais	Mini long line, pole and line, large trap, gill net, cast net	2.77	10.37	6.90	17.25
29		*Kryptopterus macrocephalus*	Selais	Mini long line, pole and line, large trap, gill net, cast net	2.77	10.37	6.90	17.25
30		*Kryptopterus limpok*	Selais	Mini long line, pole and line, large trap, gill net, cast net	2.77	10.37	6.90	17.25
31		*Wallago leeri*	Tapah	Large trap, mini long line, pole and line, gill net	4.16	8.30	8.97	–
32	Pangasiidae	*Pangasius sutchi*	Patin	Gill net, cast net, pole and line, large trap, gill net	3.46	–	10.35	–
33		*Pangasius polyuranodon*	Juaro	Gill net, cast net, pole and line, large trap, gill net	3.46	10.37	4.21	14,03
34	Synbranchidae	*Monopterus albus*	Belut	Small trap	2.77	–	4.14	–
35	Osphronemidae	*Osphronemus goramy*	Gurami	Pole and line, gill net, large trap	2.42	4.84	4.14	–
36		*Belontia hasselti*	Selincah	Large trap, gill net, cast net, pole and line	1.73	2.76	2.41	5.52[a]
37		*Trichogaster trichopterus*	Sepat	Large trap, gill net, cast net	2.08	–	4.14	5.52[a]
38		*Trichogaster leerii*	Sepat	Large trap, gill net, cast net	2.08	–	4.14	5.52[a]
39		*Sphaerichthys osphromenoides*	Sepat	Large trap, gill net, cast net	2.08	–	4.14	5.52[a]
40	Chandidae	*Channa striata*	Gabus	Large trap, pole and line, mini long line	1.04	–	3.45	6.90
41		*Channa lucius*	Gabus	Large trap, pole and line, mini long line	1.04	–	3.45	6.90
42		*Channa bankanensis*	Gabus	Large trap, pole and line, mini long line	1.04	–	3.45	6.90
43		*Channa micropeltes*	Toman	Large trap, pole and line, mini long line	1.73	4.15	4.14	8.28
44	Palaemonidae	*Macrobrachium rosenbergii*	Udang	Large trap, pole and line, cast net	9.00	–	12.46	–

[a]Salted fish product

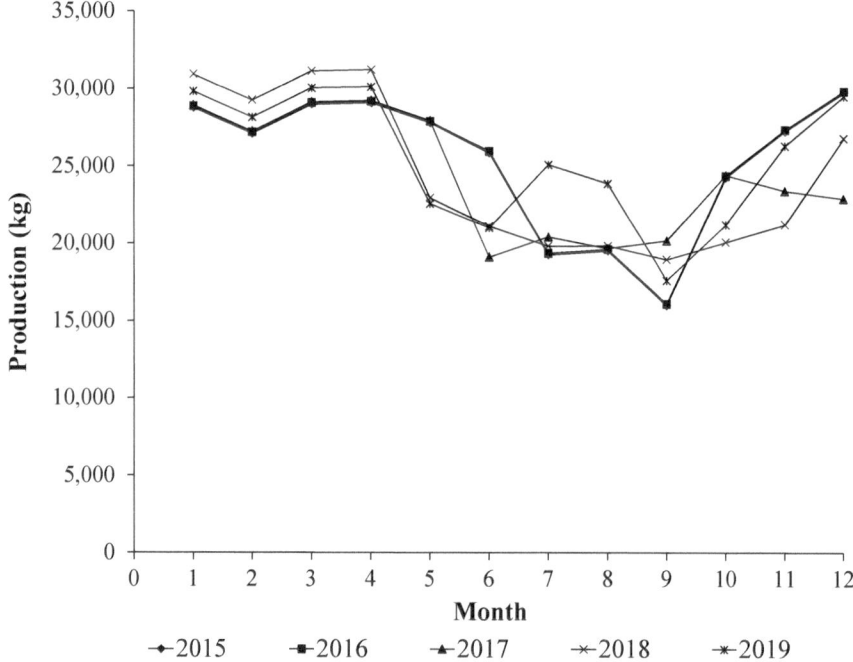

Fig. 5.10 Seasonal changes in fishing catches in Pangkalan Kerinci Sub-district during 2015-2019

Java Island, especially in Jakarta, because of the provincial highway in the east that crosses Pelalawan District. This eastern route is the shortest and is mainly used by vehicles driving to Java, Jambi, South Sumatra, and Lampung. Some travelers on this route buy fish, especially the smoked fish products, as souvenirs.

The decline and fluctuation of the catch from year to year is a problem for the long term prospects for traditional fisheries in Rantau Baru. One of the viable strategies to achieve both sustainable fisheries and sustainable peat swamp ecosystem management is to empower the already popular recreational fishing in Rantau Baru.

5.5 Recreational Fishing in Rantau Baru

Recreational fishing in Rantau Baru has the potential to reduce the number of unemployed, which was recorded as high as 35 persons (or 5.62% of the labor force) in 2020. The unemployed can work as fishers or fishing tour guides. Freshwater recreational fishing has a long tradition and is now enjoyed by millions worldwide (Cowx 2001). Recreational fishing is an activity carried out by individuals for sport as well as domestic consumption, not for commerce (Cowx 2001). In European countries and the United States, recreational fisheries are essential sources of income and employment in regional and national economies, providing

Table 5.3 The number of boat rentals and fishing excursions to Rantau Baru in 2019

Month	Type of boat for rent			Number of fishing tourists	Income from recreational fishing (US$)
	Boat (Unit)	Outboard motorboat (Unit)	Fishing vessel (Unit)		
January	0	0	16	48	336.97
February	0	0	14	42	294.85
March	0	0	2	6	42.12
April	0	0	16	56	336.97
May	8	138	212	975	5461.79
June	9	138	210	976	5423.18
July	5	134	196	905	5086.21
August	9	130	194	903	5030.04
September	7	134	197	923	5114.29
October	6	126	199	908	5096.74
November	6	119	193	865	4921.23
December	2	1	4	16	98.28
Total	**52**	**920**	**1,453**	**6,623**	**37,242.67**

practitioners with social, cultural, physiological, and physical benefits, food security, and exerting biological impacts on fish stocks (Hickley and Tompkins 1998; Cowx and Arlinghaus 2008). Successful recreational fishing can provide human and financial resources for the sustainable management of fishing sites and their surrounding environments. In some developed countries, recreational fishing is the main form of inland water and ocean use (Cowx 2001). Approximately one-tenth of the population of the entire country regularly undertakes recreational fishing in Europe, the USA, and Canada. However, national and international policies for managing and developing the conservation of resources and ecosystems for recreational fisheries have been largely ignored, perhaps due to perceptions of the sector as less profitable than commercial fisheries (Cooke and Cowx 2006).

Rantau Baru is known by the people in Riau as one of the best recreational fishing destinations in the province. The popularity and condition of fishing spots in Rantau Baru can be seen through social media and YouTube.[2] The peak times for tourist visits to Rantau Baru are Saturdays and Sundays; however, tourists still visit to fish throughout the week. Recreational fishing contributes to the increase of incomes for the people in Rantau Baru. According to interviews with boat owners and fishing tourist guides, fishing tourism activity provided an income of US$37,242.67 from boats and boat rentals in 2019. In that year, 6,623 tourists came to Rantau Baru for fishing (Table 5.3). Floods in the rainy season cut off the road to Rantau Baru, making it difficult for tourists to visit by car for recreational fishing. Subsequently,

---

[2]YouTube video that describes the conditions of fishing grounds and recreational fishing spots in Rantau Baru, Pelalawan District, Riau Province, Indonesia. (https://www.youtube.com/watch?v=nwzrNrDOm4w)

the number of visitors for fishing fluctuates significantly between the rainy and dry season, rising from just 6 in March 2019 to 976 in June, causing incomes to fluctuate from only US\$42.12 to US\$5,461.79.

If infrastructure, such as the road to the village, restaurants, guesthouses, and angler shops, is developed, more recreational fishing tourists will visit Rantau Baru in any season and that will increase the villager income significantly. Even now, there is an unwritten environmentally friendly regulation on the prohibition of fishing methods and gear destructive to the environment, including the use of drag nets, electric fishing rods, intoxicants. The rising popularity of recreational fishing will further motivate villagers to recognize the importance of peat forest and swamp conservation and develop more sustainable and environmentally friendly management practices for the peat forests and swamps in Rantau Baru.

5.6   Conclusion

Submerged forests in peatlands have fishery potential due to the biodiversity of the fish and aquatic animals living in them. This fishery resource has potential as both a source of livelihood for fishers in Rantau Baru and as a fishing tourism destination for people in Riau Province. The fishing gear of local fishermen makes use of materials from the surrounding environment. Meanwhile, fishing tourism activities have increased the family economy of fishers through boat rentals and tourist guide services. The preservation of submerged forests in the environment must be maintained to sustain local community life and the biodiversity of inland waters.

Acknowledgments We extend our gratitude to the Research Institute for Humanity and Nature, which provided financial support for our research and survey activities in Rantau Baru. We would also like to thank Riau University for providing administrative support during the implementation of this research. Finally, we thank all the students in the doctoral and master's programs in Environmental Science, Post Graduate Studies, Riau University, who were significantly involved in helping to collect the data.

References

Ahmad M, Nofrizal (2015) Efisiensi ekonomi dan produktivitas kapal perikanan tangkap ikan kurau (*Eletheronema tetradactylum*). J Ilmu Teknol Kelaut Trop 7(1):39–47. https://doi.org/10.29244/jitkt.v7i1.9770

Ali MM, Hossain MB, Al Masud M et al (2015) Fish species availability and fishing gears used in the Ramnabad River, Southern Bangladesh. Asian J Agric Res 9(1):12–22. https://doi.org/10.3923/ajar.2015.12.22

Allison EH, Ellis F (2001) The livelihoods approach and management of small-scale fisheries. Mar Policy 25(5):377–388. https://doi.org/10.1016/S0308-597X(01)00023-9

Amoros C, Bornette G (2002) Connectivity and biocomplexity in waterbodies of riverine floodplains. Freshw Biol 47(4):761–776. https://doi.org/10.1046/j.1365-2427.2002.00905.x

Boehm HDV, Siegert F (2000) Application of remote sensing and GIS to monitor peatland multi-temporal in Central Kalimantan. In: International symposium on Tropical Peat Swamps, Safeguarding a Global Natural Resource, Bogor, 27–29 July 1999

Cooke SJ, Allison EH, Beard TD Jr et al (2016) On the sustainability of inland fisheries: finding a future for the forgotten. Ambio 45(7):753–764. https://doi.org/10.1007/s13280-016-0787-4

Cooke SJ, Cowx IG (2006) Contrasting recreational and commercial fishing: searching for common issues to promote unified conservation of fisheries resources and aquatic environments. Biol Conserv 128(1):93–108. https://doi.org/10.1016/j.biocon.2005.09.019

Correa SB, Winemiller KO (2014) Niche partitioning among frugivorous fishes in response to fluctuating resources in the Amazonian floodplain forest. Ecology 95(1):210–224. https://doi.org/10.1890/13-0393.1

Cowx IG (2001) Recreational sport fishing in fresh waters. In: Safran P (ed) Fisheries and aquaculture, vol 3. Encyclopedia of Life Support Systems, Oxford, pp 355–374

Cowx IG, Arlinghaus R (2008) Recreational fisheries in the twenty-first century: towards a code of conduct. In: Aas Ø (ed) Global challenges in recreational fisheries. Blackwell, Oxford, pp 338–352. https://doi.org/10.1002/9780470697597.ch17

Dudgeon D (2011) Asian river fishes in the Anthropocene: threats and conservation challenges in an era of rapid environmental change. J Fish Biol 79(6):1487–1524. https://doi.org/10.1111/j.1095-8649.2011.03086.x

Enrici AM, Hubacek K (2018) Challenges for REDD+ in Indonesia: a case study of three project sites. Ecol Soc 23(2):art 7. https://doi.org/10.5751/ES-09805-230207

FAO (2018) The state of world fisheries and aquaculture 2018: meeting the sustainable development goals. FAO, Rome

FAO (Food and Agriculture Organization of the United Nations) (2016) State of the world's forests. Forests and agriculture: land-use challenges and opportunities. FAO, Rome

Fearnside PM (1997) Transmigration in Indonesia: lessons from its environmental and social impacts. Environ Manage 21(4):553–570. https://doi.org/10.1007/s002679900049

Giesen W (2008) Biodiversity and the ex-mega rice project area in Central Kalimantan. Master plan for the rehabilitation and revitalisation of the ex-mega rice project area in Central Kalimantan. Technical report no 8. Euroconsult Mott MacDonald, Arnhem and Deltares, Delft. https://doi.org/10.13140/RG.2.2.33951.00168

Haryono (2007) Komposisi dan kelimpahan jenis ikan air tawar pada lahan gambut di wilayah Propinsi Riau. Berita Biologi 8(4):231–239

Hergoualc'h K, Carmenta R, Atmadja S et al (2018) Managing peatlands in Indonesia challenges and opportunities for local and global communities. CIFOR Infobrief no. 205, CIFOR, Bogor. https://doi.org/10.17528/cifor/006449

Hickley P, Tompkins H (eds) (1998) Recreational fisheries: social, economic and management aspects. Blackwell, Oxford

Humphreys D (2013) Deforestation. In: Falkner R (ed) The handbook of global climate and environment policy. Wiley, Chichester, pp 72–88. https://doi.org/10.1002/9781118326213.ch5

Kanoksilapatham B (2016) Wooden fishing vessel building at Samut Sakhon: its last breath. Int J Cult Hist 2(1):25–34. https://doi.org/10.18178/ijch.2016.2.1.032

Lambin FE, Gibbs KH, Heilmayr R et al (2018) The role of supply-chain initiatives in reducing deforestation. Nat Clim Change 8:109–116. https://doi.org/10.1038/s41558-017-0061-1

Masuda K (2012) Indoneshia mori no kurashi to kaihatsu: tochi wo meguru 'tsunagari' to 'semegiai' no shakaishi (Livelihoods and developments in Indonesian forest: sociohistory of cooperation and conflict for land use). Akashi Shoten, Tokyo

Miettinen J, Shi C, Liew SC (2016) Land cover distribution in the peatlands of Peninsular Malaysia, Sumatra and Borneo in 2015 with changes since 1990. Glob Ecol Conserv 6:67–78. https://doi.org/10.1016/j.gecco.2016.02.004

Muchlisin ZA, Akyun Q, Rizka S et al (2015) Ichthyofauna of Tripa peat swamp forest, Aceh province, Indonesia. Check List 11(2):art 1560. https://www.biotaxa.org/cl/article/view/11.2.1560

Nofrizal, Ahmad M (2013) Pengembangan galangan kapal kayu tradisional di Bagansiapiapi. Universitas Riau Press, Pekanbaru

Nofrizal, Ahmad M, Syaifuddin (2014a) Industri galangan kapal tradisional di Bagansiapiapi. J Perikan Kelaut 19(2):9–21. https://doi.org/10.31258/jpk.19.2.9-21

Nofrizal, Ahmad M, Syaifuddin (2014b) Pengembangan galangan kapal kayu tradisional di Bagansiapiapi: melalui pelatihan pembuatan kapal FRP untuk peningkatan sumberdaya manusia galangan kapal. Universitas Riau Press, Pekanbaru

Posa MRC, Wijedasa LS, Corlett RT (2011) Biodiversity and conservation of tropical peat swamp forests. BioScience 61(1):49–57. https://doi.org/10.1525/bio.2011.61.1.10

Rahman MB, Hoque MSZ, Rahman MM et al (2017) Exploration of fishing gear and fisheries diversity of Agunmukha River at Galachipa Upazila in Patuakhali District of Bangladesh. Iran J Fish Sci 16(1):108–126. http://dorl.net/dor/20.1001.1.15622916.2017.16.1.9.0

Salayo N, Garces L, Pido M et al (2008) Managing excess capacity in small-scale fisheries: perspectives from stakeholders in three Southeast Asian countries. Mar Policy 32(4):692–700. https://doi.org/10.1016/j.marpol.2007.12.001

Stacey N, Gibson E, Loneragan NR et al (2019) Enhancing coastal livelihoods in Indonesia: an evaluation of recent initiatives on gender, women and sustainable livelihoods in small-scale fisheries. Marit Stud 18(3):359–371. https://doi.org/10.1007/s40152-019-00142-5

Stacey N, Gibson E, Loneragan NR et al (2021) Developing sustainable small-scale fisheries livelihoods in Indonesia: trends, enabling and constraining factors, and future opportunities. Mar Policy 132:art 104654. https://doi.org/10.1016/j.marpol.2021.104654

Sterling EJ, Betley E, Sigouin A et al (2017) Assessing the evidence for stakeholder engagement in biodiversity conservation. Biol Conserv 209:159–171. https://doi.org/10.1016/j.biocon.2017.02.008

Whitten AJ (1987) Indonesia's transmigration program and its role in the loss of tropical rain forests. Conserv Biol 1(3):239–246. https://doi.org/10.1111/j.1523-1739.1987.tb00038.x

Yolamalinda (2013) Decentralizations effect to Indonesia forest: illegal logging practice. Economica 1(2):214–222. https://doi.org/10.22202/economica.2013.v1.i2.122

Yustina (2016) The impact of forest and peatland exploitation towards decreasing biodiversity of fishes in Rangau River, Riau-Indonesia. Int J Appl Bus Econ Res 14:10343–10355

Chapter 6
Rethinking the Local Wisdom Approach in Peatland Restoration through the Case of Rantau Baru: A Critical Inquiry to the Present-Day Concept of *Kearifan Lokal*

Takamasa Osawa

Abstract In recent years, studies have promoted the efficacy of "local wisdom" in contributing to the prevention of peatland degradation and its fires in Indonesia. However, during the past quarter of a century, the related concept of indigenous knowledge (IK) has been criticized by various scholars for its deficiencies. The same deficiencies are found in the present-day use of the concept of local wisdom in academic papers. The ideology and idealism surrounding the concept narrows researchers' epistemic perspectives to local lives, and the designation of knowledge as local wisdom confines the trans-regional problem to local areas. Through observation of the situation in Rantau Baru, this chapter examines the validity of IK, questions the present-day application of local wisdom to tropical peatland problems in academic research, and suggests the need to investigate the dynamism, interaction, and transformation of knowledge beyond the framework of local areas in order to better understand local realities and build a broader network of cooperation.

Keywords Local wisdom (*kearifan lokal*) · Indigenous knowledge (IK) · Participation · Cooperation · Idealism · Rantau Baru

6.1 Introduction

Let me begin with a short description of my research in Rantau Baru. When I joined the Tropical Peatland Society Project at the Research Institute for Humanity and Nature in October 2017 as a social anthropologist, I was expected to explore the relationship between people's lives and the peat environment at the village level, focusing on livelihoods and other cultural practices. This focus concurred with the

T. Osawa (✉)
Institute of Liberal Arts and Science, Kanazawa University, Kanazawa, Ishikawa, Japan
e-mail: tosawa@staff.kanazawa-u.ac.jp

© The Author(s) 2023
M. Okamoto et al. (eds.), *Local Governance of Peatland Restoration in Riau, Indonesia*, Global Environmental Studies,
https://doi.org/10.1007/978-981-99-0902-5_6

research and policies developed by the Indonesian Peatland Restoration Agency (*Badan Restrasi Gambut*, BRG) and its academic partners, which seek to identify local knowledge and practices that might contribute to the 3R restoration strategies (rewetting, revegetation, and revitalization) for degraded peatland. I initially thought this was a relevant approach to seek solutions to the peatland problem. Rantau Baru was a good research site for accessing local wisdom, as the villagers there have lived in peat environments for several generations.

However, as soon as I began fieldwork in the village, I encountered a challenge to finding the local wisdom we had expected. As described in Chap. 3, locals have used peat hinterlands very minimally throughout their peat environment history. Although many villagers responded in the questionnaire survey that they inherited knowledge of peatland cultivation from their ancestors (see Chap. 7, Fig. 7.3), according to villager narratives, actual use of peatland has been quite limited. Indeed, traditional peatland use consisted almost exclusively of maintaining a few *sialang* tree areas owned by the village,[1] and occasional logging of timber in the flooded forests during the rainy season (with the use of canoes). Beginning in the mid-1990s, the peat swamp forests around the village were not drained by villagers, but rather by oil palm companies, and since the same period, peatland fires have been burning remaining forests (Binawan and Osawa, Chap. 10).[2] Villagers have either sold usage rights of sections of bare peatland to companies and urban residents or attempted to plant oil palm trees themselves by adopting the companies' agricultural techniques. Repeated peatland fires and the haze they generate have threatened villagers' lives, and the locals today recognize the urgent necessity of preventing future fires (see Chaps. 3 and 5). They understand that the repeated fires are not caused by their traditional methods of using the land, but mainly by "the carelessness of fires among anglers from cities" (see Osawa and Binawan, Chap. 3) and illegal land clearing by fire in oil palm plantations owned by companies or urban residents (Binawan and Osawa, Chap. 10). Given the villagers' negligible economic reliance on the peat hinterlands, their knowledge of the peat environment is limited.

Where, then, should we have found local wisdom, knowledge, and practices? Traditional peatland knowledge and practice is inherited from ancestors, and for generations, the peat swamp forests were hard to access or use for cultivation, and forest products were collected only during the flood season. Indeed, the village elders often justified their limited peatland knowledge by explaining how difficult it was to access and use the forests. Therefore, traditional knowledge and practices relating to the peatlands represented a small part of their life.

The prevailing view is that local wisdom can contribute to peatland restoration strategies—rewetting and revegetating—and revitalizing the livelihoods of local

---

[1] Bees build their hives in *sialang* trees, which are located on the riverbanks (see Chap. 3). The (peatland) forested area around the *sialang* trees, or the *sialang* areas, are protected by customary law or *adat*.

[2] In other areas of Riau, villagers employed by oil palm and acacia companies logged timber in the peat swamp forests, then the companies drained the forests (Lubis 2013; Masuda et al. 2016). However, such logging did not happen in Rantau Baru (see Chap. 3).

communities. In other words, local wisdom is knowledge and practice that may include innovative techniques for restoring the peatland or effective methods to mobilize and engage local communities in restoration activities. What could be an example of such "local wisdom"? For instance, it may be possible to claim that (1) the local practice of protecting *sialang* areas can be useful for peatland conservation, (2) traditional social institutions and networks can help organize local fire-prevention groups, or (3) knowledge of palm oil cultivation gained through working in company plantations can be applied to the peatland in Rantau Baru to develop more effective methods of peatland usage. As we can see from these three examples alone, what could be considered "local wisdom," in terms of its origin or how long it has been held or practiced, is ambiguous and arbitrary. Furthermore, the conceptual framework used to extract and formulate local wisdom is not in villagers' hands, but rather in those of researchers, and thus risks ignoring the village's rich social and cultural contexts. For the past several decades, anthropologists and development sociologists have debated and critiqued these shortcomings and other aspects of local wisdom, and its "parent concept" of Indigenous Knowledge.

Indigenous Knowledge (or IK) is often invoked in international development aid programs to support improvements in Global South standards of living and in environmental conservation programs to promote the sustainable use of ecosystems and resources. Proponents of an IK approach argue that it contributes to both the practical and ethical goals of such programs by not only providing scientists and development practitioners with previously unknown facts and new "know-how," but also by empowering locals who have been marginalized by centralized policies. Anthropologists and development sociologists, however, have questioned the value and validity of IK, arguing that the concept is too ambiguous and may cause adverse effects. The variant concept of local wisdom, which is most often used in Indonesia, involves a similar level of ambiguity and risk as IK. While many studies on peatland restoration in recent years claim the value of local wisdom, deeper analysis of the concept and its application reveals that despite its ethical ideals, its contribution to restoration goals is at best uncertain, and at worst, burdensome for locals.

This chapter presents a critical inquiry of the relevance and validity of the concept of local wisdom and associated academic approaches to the peatland problem. First, I examine the emergence of the IK concept in the fields of international development and environmental conservation and provide an overview of the current debates surrounding the concept among anthropologists and development sociologists. Second, I outline the history and use of the local wisdom concept in Indonesia and examine its application in recent academic research on peatland problems. In doing so, I demonstrate how it has evolved into its ideology. Finally, I highlight the limitations of the local wisdom approach given the realities of Rantau Baru and provide some suggestions on how to better study peatland problems in local areas.

Although there are many approaches to peatland restoration, I focus on the local wisdom approach in this chapter. The utopian idealism surrounding the local wisdom concept narrows researchers' epistemic perspectives to local lives. In seeking local wisdom, the approach also confines the problem to local areas, thereby invariably attributing responsibility to locals. Instead of relying on the IK or local wisdom concept, which can suspend knowledge in a static state, researchers should consider knowledge as fluid and dynamic, and enhance communication, interactions, and transformations among different knowledge sources. In particular, strengthening cooperation between locals, urban residents, companies, and government entities may provide more opportunities to build a broad network and achieve peatland restoration goals.

In this chapter, I use the terms "IK" and "local wisdom" to refer to the concepts that have been formulated in the transnational and Indonesian national discourses respectively. However, when referring to local ways of life and intellections emerging from everyday life, I use expressions such as "local knowledge" and "traditional knowledge and practice."

6.2   IK and the Hegemony of Knowledge

6.2.1   Development of the IK Concept and Its Application

The concept of IK can be understood, first and foremost, by its relationship with "Western," "scientific," or "centralized" knowledge. There is no doubt that information transfer has occurred across various regions of the world since ancient times. Typically, when one region gains a dominant position as a center, its knowledge spreads out and becomes predominant, often even in the margins. By the middle of the seventeenth century, European countries had gained the predominant position through colonialization in several regions in the world. While introducing European technologies and knowledge to their colonies, colonial powers extracted knowledge from the colonized areas to develop scientific knowledge in home countries. Between the sixteenth and seventeenth centuries, particular attention was paid to local plants and herbs that were useful for medical care (Ellen and Harris 2000, pp. 8–12). Although this process depended on local knowledge, it ultimately incorporated that knowledge into existing Western, scientific, and centralized knowledge systems, muting how it had been embedded in local society and culture. Ellen and Harris (2000, p. 11) claim that "the European relationship with local Asian knowledge was [. . .] to acknowledge it through scholarly and technical appropriation and yet somehow to deny it by re-ordering it in cultural schema which link it to an explanatory system which is proclaimed western." As colonialism expanded, the local knowledge of the colonies came to be regarded as primitive and marginal vis-à-vis Western scientific knowledge, and this view prevailed until the latter half of the twentieth century.

The relationship between Western scientific knowledge and local knowledge began to be reconsidered in the mid-1960s. This shift was motivated by both practical and romantic reasons (Ellen and Harris 2000, pp. 12–14). On the one hand, modern scientific knowledge recognized that local knowledge in non-Western worlds held practical potential, and thus could contribute to the development of scientific knowledge (Warren et al. 1995, p. xvii). In particular, academic and industrial researchers have documented local knowledge of flora and fauna with the expectation that it serves the development of modern biotechnology and medicine (Posey 2000, p. 35; Slikkerveer and Slikkerveer 1995). On the other hand, the poetry and aesthetics that emerge from the relationship between locals and their natural surroundings were idealized and praised (Ellen and Harris 2000, p. 13).

Since the 1970s, studies have repeatedly pointed to the limits of top-down or centralized modernization approaches planned in laboratories and offices by academic scholars and central government officials (e.g., Ferguson 1994). To overcome these limits, "bottom-up" approaches associated with keywords such as "participation," "grassroots," and "empowerment" have been developed. These approaches stress the importance of local agencies, and thus involve local populations in planning and decision making. At the same time, they relativize the supremacy of development plans that are formulated based on Western, scientific, modern, or centralized knowledge (Hobart 1993). As part of this process, the valuation of local knowledge has been gradually incorporated into international development aid to the Global South. It was in this context that the term "indigenous knowledge" was first explored as a concept in the edited volume *Indigenous Knowledge Systems and Development* (Brokensha et al. 1980).

An appreciable number of social and cultural anthropologists, development sociologists, and human geographers have acknowledged the positive role of IK, emphasizing the importance of documenting and employing "indigenous knowledge systems" (Warren et al. 1995; Slikkerveer and Dechering 1995). Warren et al. (1995) highlight IK's significance and practicability in development programs, as it "forms the information base for a society which facilitates communication and decision-making" (Warren et al. 1995, p. xv). In explaining their interdisciplinary approach to IK systems, they note that "Once the methodologies for documenting these [indigenous knowledge] systems are introduced into training institutes in a given country, the recorded systems can be systematically deposited and stored for use by development practitioners" (Warren et al. 1995, p. xviii). During the 1980s, the concept of IK was also incorporated into ecological conservation efforts in reaction to failed programs that excluded locals (Berkes 2004, p. 623; Lanzano 2013, p. 3; Slikkerveer and Dechering 1995). In the decades that followed, IK has been idealized as something accumulated and developed by locals living harmoniously and symbiotically with their natural surroundings for generations.

Since the 1990s, various international institutions have announced their positive engagement with IK systems. For example, in 1996, the World Bank declared its commitment to IK as a "Knowledge Bank." In 2002, the United Nations Educational, Scientific and Cultural Organization (UNESCO) launched Local and Indigenous Knowledge Systems Programmes (LINKS) to preserve and promote IK. Since 2010, several international conferences related to development and environmental issues have confirmed the importance of IK (Slikkerveer 2019, pp. 35–39). Many related projects are operating all over the world as part of the Sustainable Development Goals scheme. UNESCO's website (n.d.) defines IK and its role as follows:

Local and indigenous knowledge refers to the understandings, skills and philosophies developed by societies with long histories of interaction with their natural surroundings. [...] This knowledge is integral to a cultural complex that also encompasses language, systems of classification, resource use practices, social interactions, ritual and spirituality. These unique ways of knowing are important facets of the world's cultural diversity, and provide a foundation for locally-appropriate sustainable development.

In this definition, UNESCO (1) asserts IK's practical efficacy as "a foundation for locally-appropriate sustainable development," and (2) assigns it an ethical value as "important facets of cultural diversity" (including language, ritual, spirituality, and so on).

6.2.2 Debates Surrounding Indigenous and Local Knowledge

Both the conceptualization and application of IK have been the subject of criticism in anthropology and development sociology. First, the meaning of "indigenous" has been problematized. Indigenous knowledge is necessarily linked to "indigenous peoples," who can be defined as those having historical continuity in a place, cultural distinctiveness, social marginalization, self-identification, and self-governance (Dove 2006; Osawa 2022). However, this definition is controversial, and its interpretation varies depending on the state or area. Because the term, regardless of its interpretation, marks a clear distinction between non-indigenous and "indigenous" or "native" peoples, it may cause tensions and discrimination between the two (Dove 2006; Ellen and Harris 2000, pp. 2–3).

The fluid interpretations of IK variations, such as "traditional ecological knowledge" and "local knowledge," are similarly criticized. Lanzano (2013, p. 4), for example, notes that "tradition" can be constructed for a specific political purpose and "local" denotes a marginalized position in relation to a larger, centralized power. In short, labeling knowledge as "indigenous," "traditional," or "local" establishes a contrast with other kinds of knowledge, such as "universal," "Western," "scientific," "modern," or "centralized," thus resulting in "othering elements or systems of knowledge that do not fit in the corpus of" Western, modern, and centralized knowledge (Lanzano 2013, p. 4; see also Ellen and Harris 2000, p. 26).

Along the same lines, the meaning of "knowledge" has also been problematized. In his pioneering work, Agrawal (1995) questions the dichotomy of Western scientific knowledge versus IK, and he deconstructs the universality and assumed supremacy of the former. He further argues that applying IK to development and environmental programs causes IK to be appropriated by the more power-laden scientific knowledge, resulting in the "strangulation [of IK] by centralized control and management" (Agrawal 1995, p. 428). In their critical examination of IK from an anthropological perspective, Ellen and Harris (2000) point out that the epistemic origin of both scientific knowledge and IK is most often unknown, and this anonymity sustains the distinction between them. When the origin of knowledge is revealed, the validity of and emphasis on IK may be put into question (see also Dove

2000). Taking a philosophical approach, Horsthemke (2021) notes that knowledge in the IK concept is presumed as the three types of knowledge, i.e. "promotional, theoretical or factual; practical or skill-type; and finally, acquaintance- or familiarity-type" (Horsthemke 2021, p. 6), but in reality, knowledge is composed of "beliefs," "truths," and "appropriate justifications," which are sustained by experience, evidence, and testimony, and such knowledge cannot be divided into "local" or "indigenous" forms (Horsthemke 2021, pp. 43–96).

Scholars have also criticized the methodology of IK systems. Ellen and Harris (2000) remind us that although IK is essentially negotiable, fluid, and embedded in a specific social and cultural context, the IK approach involves a process of "collecting," "codifying," and "decontextualizing" IK in order to incorporate it into Western scientific knowledge. This process does not provide an adequate understanding of locals' knowledge. Worse still, the knowledge thus extracted may then be imposed on local communities as "top-down" policies instead of the "bottom-up" ideal. According to Olivier de Sardan (2005), the need to demonstrate the "participation" of locals, a main tenet of the IK approach, contributes to an "ideological populism," which "paints reality in the colours of its dreams, and has a romantic vision of popular knowledge." He then draws a contrast between ideological and *methodological* populism, noting that the latter "considers that 'grassroots' groups and social actors have knowledge and strategies that should be explored, without commenting on their value or validity" (Olivier de Sardan 2005, p. 9).

Upon reading "recent works constructed around local knowledge or agency of 'grassroots' actors," de Sardan notes:

> [. . .] we observe that one can simultaneously succumb to ideological populism, through a systematic idealization of competences of the people, in terms either of autonomy or of resistance, while obtaining innovative results thanks to methodological populism, which sets itself the task of describing the agency and the pragmatic and cognitive resources that all actors have, regardless of the degree of domination or deprivation in which they live. (Olivier de Sardan 2005, p. 9)

In other words, the IK approach involves the systematic idealization of knowledge, obscuring the relationships between locals and the states or capitals that have intervened in their lives.

In terms of the conservation of ecosystems and resources, Berkes (2004) points out that although community-based approaches, including the IK approach, may be effective in conservation programs when locals and conservation practitioners share the same objectives, this is not always the case. He underlines that typically "local rules are about use, allocation, and conflict management and not about preservation per se" (Berkes 2004, p. 625), and that local people make resource use decisions in the context of larger, international, and capital systems (Berkes 2004). Describing historical and present-day forest preservation practices in Burkina Faso, Lanzano summarizes the problems with IK-inspired research:

> [. . .] IK-inspired research bears some ambiguities, such as the risk of proposing reductionist and effectiveness-oriented explanations of complex social and cultural phenomena. Here is where the reflection over indigenous knowledge connects with the 'ecologically noble savage' debate, raising doubts over the possibility of clearly identifying conservationists'

attitudes among indigenous people when conservation is defined as any action purposely intended to preserve resources (Lanzano 2013, p. 14).

On the other hand, some scholars have defended the value of the IK concept and approaches. The development anthropologist Sillitoe (1998, 2002) recognizes the ambiguity of the concept and the risks of IK approaches, but emphasizes its usefulness to development projects. He draws a distinction between IK research as an applied method and anthropological research as an intellectual pursuit. He insists that IK can "introduce a locally informed perspective into development to promote an appreciation of indigenous power structures and know-how" (Sillitoe 1998, p. 224). Sillitoe (2002) also underlines that the application of IK can be an effective countermeasure when responding to the imposition of Western, scientific, and centralized knowledge occurring in the context of the rapid globalization and modernization penetrating local lives. In his later work, Sillitoe (2007) uses the term "local science" (as opposed to "global science") instead of IK. Slikkerveer et al. (2019) also describes IK as "neo-ethnoscience" and applies it to poverty reduction efforts in Indonesia, proposing a method of "integrated community managed development."

Indeed, in the field of environmental conservation, studies have increasingly tried to integrate IK systems and natural science (e.g., Alexander et al. 2011; Geleta 2015). For example, Berkes and Berkes (2009) demonstrate the need to incorporate local knowledge in scientific approaches, characterizing the former as involving a holistic perspective that assesses complex ecosystems qualitatively with many variables. Nakagawa (Chap. 4) emphasizes the contributions of local environmental knowledge in assessing the conditions of the freshwater ecosystem of the Kampar River.

Ultimately, the crux of the debate around the validity of IK can be located in the relationship between knowledge and power. On the one hand, IK has emancipatory potential as a means for locals to break free from the "top-down" imposition of scientific knowledge and centralized policies. It also enables scientists to openly turn their attention toward local knowledge and practices that may contain phenomena and principles unknown to them. On the other hand, the IK concept and approach may alienate locals and allow their knowledge, dislodged from rich local contexts, to be further dominated by Western scientific knowledge. Despite its idealized features, IK, as an ambiguous framework created by observers, conceals the inequality of power between Western scientific and local knowledge. Yet, this dilemma does not stop countless IK projects from continuing to operate. In this situation, the inclusion of an IK approach should be assessed according to the balance of advantages and disadvantages that emerge depending on the goals and specific context of each program (Lanzano 2013, pp. 5–6).

6.3   Local Wisdom in Indonesia and Peatland Restoration Policy

6.3.1   Local Wisdom: From Concept to Policy to Ideology

In studies of Indonesia, there is less consistency in the usage of the term IK than in studies of other countries. Several variant expressions are used to indicate various aspects of locals' knowledge. For example, in their analysis of archetypal anthropology research topics in Indonesia, such as ritual, social change, and identity, Puspitorini and Hunter (2020) employ the English term "indigenous knowledge," combining UNESCO's definition and the classic definition of "local knowledge" by Clifford Geertz relating to "cultural patterns." Slikkerveer et al. (2019) also use the term "indigenous knowledge," but specifically to refer to social institutions and networks that can contribute to poverty reduction. Nugroho et al. (2018) adopt the term "local knowledge" in contrast to "scientific" and "professional" knowledge and explore the potential of "local knowledge" to contribute to policymaking. Recent studies have increasingly used the term "local wisdom" in English, or *kearifan lokal* in Indonesian. This is often the case in studies that focus on relationships between local communities and their natural environments and on resource management practices inherited over generations.

The term "local wisdom" became widely used during the era of decentralization and democratization in Indonesia. Several factors explain this. As mentioned above, the term "indigenous" is necessarily linked to the concept of "indigenous peoples," which is transnationally defined. However, the Indonesian government does not use the term "indigenous" in national policies to refer to specific groups of people, declaring instead that *all* ethnic groups in Indonesia are "indigenous" or "native" (Ellen and Harris 2000, p. 5; Osawa 2022, p. 13). In addition, since the end of Suharto's centralized regime, decentralization and democratization, including the empowerment of locals, have been prioritized. In this context, then, it is not surprising that the term "local" has been adopted rather than "indigenous": not only is it more appropriate, it also can mitigate, to a certain extent, the risk of being drawn into debates surrounding indigeneity.

On the other hand, use of the term "wisdom," or *kearifan*, has a clear history of use in Indonesian studies that pre-dates the use of "local wisdom." For example, we can find similar terms, such as "environmental wisdom" (*kearifan lingkungan*) (Zakaria 1994) and "traditional wisdom" (*kearifan tradisional*) (Nababan 1995), in works published during the mid-1990s. These studies assert the importance of employing traditional knowledge and practice in environmental conservation and sustainable resource management efforts. In the context of studies on agriculture in swampy lands, Hidayat (2000) used the term "cultural wisdom" (*kearifan budaya*) to refer to tidal irrigation without extensive drainage, which has been practiced among the farmers of Banjar Malays in Kalimantan.

The term "local wisdom" became more frequently used during the subsequent decade. For example, Lubis (2005, p. 239) directly associates "local wisdom" with "indigenous knowledge" and discusses its value in resource management. During the early 2000s in general, "local wisdom" was seen as something that (1) was accumulated by traditional (or *adat*) communities living close to natural surroundings, especially forests, and (2) was about to disappear in a coming wave of modernization (Lubis 2005; Santoso 2006; Nurjaya 2007). At the same time, anthropologists linked the term to the concept of "local genius," regarding *kearifan lokal* as a long-enduring knowledge formed to sustain a people's existence in a specific locale (Sartini 2004). In this view, locals incorporate external cultural knowledge and practices with internal ones (see Ruastiti 2011). Noor and Jumberi (2007), agricultural scientists who studied the use of peat and swampy lands in Kalimantan, adopted the terms "local cultural wisdom" (*kearifan budaya lokal*) and "local wisdom," noting that local knowledge can be transformed in communication.[3] Some scholars during this period therefore considered *kearifan lokal* not solely as existing traditional knowledge and practice, but as something gained from interaction with external cultural knowledge.

However, the usage of the concept of local wisdom has changed in subsequent years. The present-day concept adopts a hybrid of the above views: while it refers to locals' cultural knowledge and practice accumulated through interaction with their natural surroundings over generations that continues to exist today, acknowledgment of the influences of external cultural knowledge is muted. This hybrid form of the concept is included in Indonesia's Law Number 32 of 2009 on Environmental Protection and Management (Maria 2018). In the law, "local wisdom" is defined as a value formed in an idealized relationship with the natural surroundings, or an embodiment of "the noble values prevailing in the society's life to protect and manage sustainable living environment"[4] (Maria 2018, p. 2). In this definition, the historical continuity of local wisdom is not emphasized. The law obligates the government to implement environmental policies with "recognition of and respect for the local wisdom and environmental wisdom" and to provide local communities with "social, cultural and economic benefit."[5]

In 2017, the Ministry of Environment and Forestry introduced Ministerial Regulation Number P.34, which also declares the need to recognize and respect local wisdom.[6] While adopting the same definition of local wisdom as that in the environmental law of 2009, this regulation also uses the term "traditional knowledge" (*pengetahuan traditional*). According to the regulation, traditional knowledge is

---

[3] Noor and Jumberi (2007, pp. 4–5) emphasize that local cultural wisdom includes "belief systems, norms and culture, and what is expressed as traditions and myths" (author's translation).

[4] *Undang-Undang Republik Indonesia Nomor 32 Tahun 2009 tentang Perlindungan dan Pengelolaan Lingkungan Hidup*, Chapter I, Article 1, Paragraph 30.

[5] Elucidation of Law Number 32 of 2009 on Environmental Protection and Management Section I, Paragraph 2.

[6] *Peraturan Lingkungan Hidup dan Kehutanan Republik Indonesia Nomor P.34/ MENLHK/ SETJEN/ KUM.1/ 5/ 2017 tentang Pengakuan dan Perlindungan Kearifan Lokal dalam Sumber Daya Alam dan Lingkungan Hidup* [Ministerial Regulation Number P. 34 on Recognition of and Respect for Local Wisdom in Natural Resources and Environment].

"part of local wisdom," has strong links with *adat* law community (*Masyarakat Hukum Adat*, explained later), exhibits historical continuity over generations, and contributes to the sustainable management of natural resources and the environment.[7]

Concurring with Regulation Number P.34, the Indonesian jurist Kristiyanto (2017) claims that local wisdom is broader in scope than "traditional knowledge," because it is something implemented, articulated, and manifested under environmental laws in Indonesia (Kristiyanto 2017, p. 161). He explains that:

> Local wisdom is perceived by individuals or communities who interact with the natural surroundings and is therefore cultural knowledge owned by certain groups. It includes models of sustainable natural resource management and how to maintain relationships with nature through wise and responsible use. Thus, local wisdom is a system that integrates knowledge, culture, and institutions, as well as the practice of managing natural resources (Kristiyanto 2017, p. 161 [author's translation]).

In using the terms "implementation," "articulation," and "manifestation," Kristiyanto seems to see local wisdom not just as locals' knowledge, but also as something manifested in communication between locals and the government. He deems that the "models" of sustainable resource management can be found in local wisdom, and throughout his argument, stresses the efficacy of local wisdom to promote local "participation" (*partisipasi*) in policy implementation (Kristiyanto 2017).

When IK is expressed as "local wisdom," it is almost always done so in the context of human-environment relations. Local wisdom is seen as a connection between people and their natural surroundings that has been developed over many generations. It therefore denotes a competency on the part of locals to manage the surrounding environment and resources sustainably. Recognizing and respecting local wisdom, then, (1) provides locals with social, cultural, and economic benefits, (2) contributes to forming "models" of sustainable management, and (3) facilitates "participation" in policymaking. The concept of local wisdom allows us to create an image of an idealized harmony between people and their natural surroundings based on long enduring and static features of the way locals live. The concept also assumes a homogeneity and internal cohesion within each local community. In the process of its inclusion in environmental law and national policy, this narrative of local wisdom, with its dual focus on practical efficacy and harmony with natural surroundings, seems to have transformed into a kind of ideology.

It should be noted here that local wisdom in this context does not emphasize the values of language, rituals, and spirituality that are stressed in UNESCO's definition of IK. In addition, although some scholars, such as Sartini (2004) and Kristiyanto (2017), imply that elements of local wisdom are dynamically generated and transformed through communication with external cultures, laws, and policies, this

---

[7] *Peraturan Lingkungan Hidup dan Kehutanan Republik Indonesia Nomor P.34/ MENLHK/ SETJEN/ KUM.1/ 5/ 2017 tentang Pengakuan dan Perlindungan Kearifan Lokal dalam Sumber Daya Alam dan Lingkungan Hidup*, Chapter I, Article 1, Paragraph 3.

is a relatively minor view, and this point is usually muted in recent studies of the peatland problem, which emphasize the "local" in "local wisdom," as mentioned later.

6.3.2 BRG's Approach to Local Wisdom

Similar to the above interpretations, BRG uses the term local wisdom to refer to locals' traditional use of natural resources and cultivation methods and asserts its efficacy in peatland restoration.[8] The homepage of BRG's website (n.d.) shows their respect for local wisdom by sharing a link to an article on the *Mongabay* news platform (Arumingtyas 2017). While the article does not provide a definition, it underlines the need to form a "model" of peatland restoration based on local wisdom and simultaneously suggests its efficacy to facilitate "participation" for the purpose of improving locals' economic situation. The guidelines of BRG's Peatland Care Village (*Desa Peduli Gambut*, DPG) program (BRG 2017, p. 10, 66; Hasegawa, Chap. 8) quotes passages from Law Number 32 of 2009 on Environmental Protection and Management when defining local wisdom. The reports compiled by the DPG program, which describe the situation of administrative villages in the peat environment, include a section titled "local wisdom and knowledge" (*kearifan dan pengtahuan lokal*) or "local wisdom in natural resource management" (*kearifan lokal dalam pengelolaan sumber daya alam*). These sections outline the local agricultural products commonly cultivated in peatlands, traditional institutions of environmental management, and so forth (e.g., BRG 2019).

In sum, BRG views local wisdom as locals' relationship with the environment and seeks to harness its potential to achieve improvements in the local economy, which is related to the strategy of "revitalization." However, it should be stressed here that BRG does not emphasize the historical continuity of local wisdom and its potential to restore the ecosystem of degraded peatland and prevent peatland fires, which is particularly related to the strategies of "rewetting" and "revegetation." Quoting the 2017 Ministerial Regulation of the Ministry of Environment and Forestry, the DPG program guidelines recognize the limit of local wisdom, as it is applicable to only specific areas where the locals live (BRG 2017, p. 66).[9] Therefore, the DPG reports treat local wisdom as part of basic village information, but do not evaluate it as directly related to restoring degraded ecosystems and preventing fires. In arguing for the need to promote social forestry programs in local areas, Haris Gunawan, one of BRG's four deputies during 2016–2021, and Afriyanti (Gunawan

---

[8] Perhaps BRG's emphasis on local wisdom is rooted in the studies of peatland management among Banjar Malays in South Kalimantan (mentioned later), in which local wisdom and similar expressions have been used since the beginning of the 2000s (Hidayat 2000; Noor and Jumberi 2007).

[9] While the BRG acknowledges in the text that local wisdom is rooted in a particular region, many government agencies, including the BRG, often use the term in the sense of "neo-ethnoscience" (Slikkerveer 2019), which is not tied to a particular region.

and Afriyanti 2019) seem to use the term "local wisdom" as a synonym of "local practices" (*praktik-praktik lokal*), which can be seen as traditional, but changeable. Indeed, they conclude that "the conventional local wisdom still needs to be upgraded (*ditingkatkan*)" (Gunawan and Afriyanti 2019, p. 236 [author's translation]) through the introduction of wetland cultivation and forestry methods that BRG promotes. Myrna Safitri (2020), who has also been one of the four deputies since 2016, explores how environmental laws treat the legitimacy of land clearing by fire. She examines the definition of local wisdom, which she sees as transforming in accordance with changes in the environment and suggests the need to create a "new local wisdom" (*kearifan lokal baru*) by introducing the new technology of peatland clearing without fire (*Pengelolaan Lahan Gambut Tampa Bakar*: PLTB). The BRG leaders therefore view local wisdom less as something that is always harmonious with the environment or historically continuous, and more as dynamic and flexible one.

This view of local wisdom is consistent with BRG's approaches, in which they have implemented a variety of programs that "renew" or "upgrade" local knowledge and practices. For example, through its DPG program, BRG selects, trains, and dispatches a facilitator to each administrative village for a certain period to promote restoration programs in the village (BRG 2017, pp. 30–31; Ramdhan and Siregar 2018, p. 155). BRG also promotes social forestry programs in peatlands that are managed by the locals themselves (Gunawan and Afriyanti 2019). These programs involve adapting local knowledge and practice to realize restoration goals. For BRG, local wisdom is something that (1) should be respected as required by the environmental law, (2) can contribute to formulating models of peatland management, and (3) can facilitate local participation. However, to achieve peatland restoration, it is essential to improve upon such wisdom through communicating with locals, exchanging knowledge among all stakeholders, and introducing suitable technologies and methods. Based on this perspective, BRG views local wisdom as changeable through communication and interaction, and indeed has tried to change it.

6.3.3 Academic Approaches to Local Wisdom in Relation to the Peatland Problem

Even before the conceptualization of local wisdom in the environmental law of 2009 and the establishment of BRG in 2016, many studies had examined the potential of traditional knowledge and practices to mitigate the drying and degradation of peatlands (Hidayat 2000; Noor and Jumberi 2007; Noorginayuwati et al. 2007). This is especially evident in the studies on rice and vegetable cultivation using tidal irrigation in peat swamps among Banjar Malays in Kalimantan. In this tidal irrigation system, shallow drainage ditches are dug in tidally-influenced wetlands to create rice paddies and vegetable gardens. Thus, groundwater levels in peatlands are not lowered excessively, preventing peatland drying and fires. Since peatland degradation and fires became a problem in the 1990s, researchers have studied not only these agricultural techniques, but also traditional customs, institutions, and beliefs in the area, which have been comprehensively described as "cultural wisdom" (Hidayat 2000) or "local cultural wisdom" (Noor and Jumberi 2007). In recent years, however, quite a few researchers identify a particular knowledge or practice as "local wisdom." These studies assume the historical continuity and static nature of local wisdom (in contrast to BRG's view). Employing various methodologies from a range of disciplines, they attempt to discover innovative technologies to mitigate the degradation of peatland and peatland fires and to formulate effective models to facilitate local participation in peatland restoration activities.

Examples of such knowledge include traditional hydrological technology using shallow and narrow trenches (Astiani et al. 2019), tidal irrigation technology in peatlands (Hairani and Noor 2020), local construction and management of canal blockages (Utami and Salim 2021), traditional cultivation of agricultural products such as sago, coconut, coffee, and rubber (Jalil and Sulistyani 2020; Jufri et al. 2018; Lestani et al. 2019), and techniques to improve the quality of peatland timber (Supriyati et al. 2016). This knowledge, thus identified as local wisdom, is deemed essential for peatland restoration in various case studies. However, we can find a tendency in these studies to presuppose historical continuity and sustainability, thus revealing the influence of an idealized conceptualization and discourse of local wisdom, or indeed an ideology, which restricts researchers' perspectives.

How does this happen? First, this research approach focuses on or extracts traditional knowledge or practice only as a prescription that can contribute to peatland restoration in a results-oriented way and fails to consider how and why such knowledge or practices are adopted within the local context. For example, although clearing of peatland by fire is commonly practiced to improve yields (Murniati and Suharti 2018, pp. 1396–1397; Lestani et al. 2019), studies fail to mention it in terms of local wisdom at all.[10] The historical context of locals' knowledge and its significance to daily practices is also muted. For example, Astiani et al. (2019) demonstrate the efficacy of traditionally constructed shallow and narrow trenches (*parit cacing*) in preventing fires and CO_2 emissions and the need to spread the principles of the technique to other areas. However, by failing to associate the technique with its historical context (i.e., limited available construction tools), they neglect the present-day likelihood that farmers may prefer to dig deeper trenches using backhoes. Locals' knowledge and practice are embedded in their lives, which are sustained by their institutions, materiality, environmental characteristics, aspirations, relational reciprocity, cosmology and so on. By reducing this complex interconnectivity to a single concept of "local wisdom," the researchers simplify this knowledge and practice, only highlighting its efficacy vis-à-vis a particular external goal.

---

[10] As an exceptional case, Lestani et al. (2019) consider past land clearing by fire in Siak District, Riau, as local wisdom. This practice has been banned since 2014 to prevent peatland fires (Murniati and Suharti 2018, p. 1399).

Second, although the studies more or less mention local wisdom as historically accumulated in a harmonious relationship with the environment, they rarely examine the origin of the knowledge and practice they identify as local wisdom. For example, some studies see traditional peatland use for general livelihood in Riau (such as the cultivation of coconut, sago palm, rubber and areca nuts) as local wisdom and suggest that it can be effective in controlling peatland degradation and fires (Jalil and Sulistyani 2020; Utami and Salim 2021). However, as Dove (2000) points out, the planting method used for rubber trees was introduced to smallholders a century ago, and since then, the cultivation area has repeatedly expanded and reduced depending on the international market price of the product. The history and nature of the cash crops of coconut, coffee, and areca nuts are similar (see also Furukawa 1994, p. 152). The origins of this practice (the cultivation of livelihood crops on peatland), then, do not square with the image of local wisdom as something developed in a harmonious relationship with nature over many generations.

In a similar vein, Utami and Salim (2021) describe canal blockings constructed and maintained by locals in a village of Sungai Tohor in the Meranti Islands District, Riau, as local wisdom. Although they describe the villager who created the method (to make the land suitable for sago cultivation), they do not mention when this was, how the villager got the idea, or the recent phenomenon of national NGOs and the government providing significant financial support for the construction.[11] By failing to do so, they obscure any external influences that may have played a role in the origin and evolution of this "local wisdom." While additional knowledge could have quickly been gained through communication with external sources of knowledge, the land use and cultivation is unilaterally labeled as "local wisdom" with no detailed examination that might uncover how and why the practice developed over time. Although it is possible to refer to local knowledge and practices that were introduced to the local community in the recent past as "local wisdom," such labeling relies on an idealized and static definition that is subject to ambiguity. This labeling of "local wisdom" also draws a sharp contrast with Western interaction with IK before the mid-twentieth century, which involved Western science habitually co-opting IK and in the process concealing its origins in local knowledge and practices (Ellen and Harris 2000, pp. 12–14). However, the similarity remains in that such identification of "local wisdom" is entirely dependent on the intent and interests of researchers or non-locals.

Finally, the studies single out a small number of traditional peatland uses, failing to acknowledge or value the heterogeneity of knowledge, or the dynamism of economic activity and agency within each local community. Dewi (Chap. 7) and Hesegawa (Chap. 8) respectively point out a significant gap in the peatland knowledge of men and women and the difficulty of integrating opinions in Rantau Baru, demonstrating the diversity of knowledge and attitudes within a community.

---

[11] The village of Sungai Tohor is a center of peatland restoration activity in Riau Province, and the government and national NGOs have supported the construction and maintenance of canal blockings (Hutagaol et al. 2017, p. 20, 50).

Additionally, people living in peat environments choose multiple livelihoods in response to market demand and productivity (Lubis 2013, p. 66; Masuda et al. 2016, pp. 207–208). For locals, selling peatland to companies and urban residents, or planting oil palms and constructing large and deep ditches, are economically rational choices to improve their economic standing (Lubis 2013, pp. 49–50; Osawa and Binawan, Chap. 3). Reducing the variety of local knowledge and practice to "local wisdom" ignores such dynamism of economic activity. Moreover, local aspirations for peatland use are also often heterogeneous, even within one community, thus the adoption of a "representative method" as local knowledge may not guarantee the participation of all, or even a majority of, villagers. Even if one could formulate a peatland management model as local wisdom, it may be challenging to apply such a model in communities that have used peatlands in different ways.

In short, the prevailing academic approach to local wisdom in peatland studies does not adequately consider the contexts, dynamism, and diversity of local knowledge and practices. Designating one specific practice or piece of knowledge as local wisdom thus runs the risk of simplifying its history, misunderstanding its institutions, overlooking its heterogeneous use within a local community, and disregarding the diverse ways it is communicated and adapted in relationships with non-locals. In arbitrarily selecting an aspect of local knowledge and practice as local wisdom, researchers codify it into something of use for addressing the peatland problem through scientific procedures or by connecting it to a disciplinary paradigm. Through this process, local wisdom is removed from the messy, dynamic context of everyday lives in a locality and becomes instead a researcher's perspective of the world. In this way, the local wisdom approach does not relativize the relationship between Western, scientific, or centralized knowledge and local or indigenous knowledge. Instead, it reinforces the boundaries between both sets of knowledge and validates the supremacy of the former. The concept of local wisdom may have merit in shifting scientists' (especially, natural scientists') attention to the life of locals. However, it simultaneously carries the risk of confining scientists' research perspective to a particular aspect of knowledge or one practice among many in local life at a specific time, which is deemed suitable to label as "local wisdom."

6.3.4   Ideological Idealism: Adat and Local Wisdom

This narrow research perspective is justified by the ideology that promotes an idealized narrative of local wisdom. According to that narrative, local wisdom has accumulated among locals through lengthy experiences living together and harmoniously with their natural surroundings. Therefore, the locals can competently manage the environment and its resources sustainably (see also Li 2001, p. 657). More specifically, in the context of peatland problems, it is assumed that people have lived in the peat environment since before the dramatic increase in fires and thus, (1) their past peatland use is sustainable, and (2) their knowledge can be applied to solving the peatland problem.

We can trace this idealized narrative and today's usage of local wisdom back to the colonial era conceptualization of *adat*. The image of a community's local wisdom coincides, or indeed could be a descendant of, "*adat* law community" (*Masyarakat Hukum Adat*), which was conceived and articulated by Cornelis van Vollenhoven and Dutch Leiden scholars around the turn of the twentieth century (Burns 1989; Li 2001, p. 659; Henley and Davidson 2007, pp. 19–25). The concept of *adat* (tradition, custom, or customary law) was developed to govern the islands outside Java and was regarded as an all-inclusive world view. The scholars viewed *adat* community as an organic whole in which people were well-organized and connected with the natural world through spiritual beliefs and practices. According to this view, "*adat* law" (*adatrecht*; *hukum adat*) in a community is seen as able to restore and maintain balance in the world (Burns 1989, pp. 56–57). Although the concepts of *adat* and local wisdom are based on a similar imagining, *adat* is used to demonstrate local philosophy, religions, and laws, and local wisdom is used to demonstrate local technology and science. Both concepts are conceived and applied by outsiders to govern or "manage" local areas. Both have also been firmly incorporated into Indonesian law: Law Number 32 of 2009 on Environmental Protection and Management obligates the government to recognize and respect "*adat* law" and "*adat* law community" together with local wisdom.[12]

Following independence, the concept of "*adat* law community" was clearly included in the Republic of Indonesia's 1945 constitution (see also Binawan and Osawa, Chap. 10). However, as various national laws were imposed across the archipelago during Suharto's era, the legal and religious aspects of *adat* law were gradually muted, and *adat* was reduced to forms of art, such as song, dance, dress, and architecture (Acciaioli 1985; Osawa 2022, pp. 175–178). Since the beginning of the decentralization and democratization era in 1998, the concept of "*adat* law community" has been revitalized and is regaining its legal position in terms of land rights and religious beliefs (Henley and Davidson 2007; Warman 2014). The concept of "local wisdom" also emerged in this context, and its enshrinement in law is a powerful symbol of the local voices that were oppressed during Suharto's centralized regime; thus, in some ways, it embodies the zeitgeist of today (Li 2001). Local wisdom should not be applied as an analytical tool devoid of this context, however. To do so would not only fail to withstand the validity and value of academic research, but it would also fail to accurately comprehend and document locals' voices and their world. We can see academic approaches that ignore such context as ideological idealism and a version of "ideological populism" (Olivier de Sardan 2005, p. 9).

I do not intend to claim that all applications of local wisdom are inadequate. Use of it by specific development and conservation programs might empower locals who have had to struggle with the "top-down" imposition of centralized or global knowledge (especially during Suharto's centralized regime), and thus relativize the supremacy of centralized or global knowledge. In specific contexts, it might also

---

[12]Law Number 32 of 2009 on Environmental Protection and Management, Chapter IX Article 63.

promote the participation of locals. It has proven to be particularly valuable when deployed by locals themselves, or their agents (such as NGOs), or as a legal and ethical concept to correct social inequality and resource exploitation. It should also be positively evaluated that, through the use of local wisdom in the environmental law, the Indonesian government has attempted to recognize local diversities and contexts.

However, adopting "local wisdom" as an analytical concept is unsuitable in academic research. Without using the ambiguous buzzword, academic studies can analyze the problems in peatland areas, the traditional practices and knowledge, and locals' contributions, which can serve to enhance understanding and mitigate those problems. The powerful ideological idealism of the concept risks narrowing researchers' attention and preventing adequate consideration of local realities, leading researchers to misunderstand local situations and miscommunicate with local people. This results in formulating less effective development plans and environmental programs.

6.4 Practical Limitations of Local Wisdom in Solving the Peatland Problem

Here I return to analysis based on the case of Rantau Baru, through which I would like to elucidate the peatland problem in Riau and specify the difficulties in applying a local wisdom approach to it. I was initially hesitant to seek local wisdom in Rantau Baru because of the limitations of the local wisdom approach at a theoretical level and its disconnect with the realities of the field situation. Again, the local wisdom approach in the majority of academic studies is adopted based on an uncertain assumption that (1) locals have lived in and around peat environments continuously since before the increase in peatland fires, (2) their traditional peatland use is sustainable, and (3) their knowledge can be applied to finding a solution to the peatland problem. The deficiencies in these assumptions are easily observable in the field. I would like to discuss these in turn by focusing on the limits of tradition, the externality of peatland, and the trans-locality of the problem.

6.4.1 Limits of Tradition

I have described the limited peatland use in Rantau Baru in Chap. 3 (Osawa and Binawan) and at the beginning of this chapter. Although villagers have lived at the edges of peatlands at least for several centuries and peatland has been recognized as part of village territory, it was difficult to access and use. It was less an area used for livelihoods and more of a hinterland, the boundaries of which are designated by the *Adat Melayu Petalangan* (see Chap. 3). Villagers have used peatland for oil palm plantations for only two decades. Rantau Baru is not an exceptional case, and this pattern of peatland use is relatively standard in Riau. Limited use of peatland can also be attributed to the rich sediment available on the riverbanks where people live. According to our questionnaire survey in 2020, of the 107 respondents who lived in the main settlement of Rantau Baru (on the riverbank), 89 people (or 83%) could distinguish peat from sediment soils. Among those 89 people, 27 (or 30%) had worked in the peatland during the previous year. While the number of the people working in peatland is rather limited even today, the number must have been much smaller several decades ago, when villagers mainly worked in swidden fields on the riverbanks and the hinterlands were covered by thick forests (Osawa and Binawan, Chap. 3).

Several field workers who investigated peatlands in Riau before or just after the expansion of oil palm and acacia plantations note that peatland is not suitable for producing crops or living, causing people to depend on river trade that links downstream and upstream areas, gathering forest products (predominantly, timber logging) in hinterland forest, and fishing in the rivers (Abe 1993; Furukawa 1994; Masuda et al. 2016; Momose 2002).[13] Furukawa (1994) characterizes land use among Riau Malays living in lowland areas as a "culture of transit," in which people do not accumulate or maintain a base for life in a fixed location, but rather use resources in transient ways. In Kapuas District, Central Kalimantan, the Ngaju Dayak use hinterland peat forest only for collecting forest products such as timber and rattan, while living on the riverbanks and cultivating agricultural crops on the sediment soils (Lubis 2013, pp. 8–17).

It is noteworthy here that the term "*gambut*," which means "peat" in Indonesian, originally comes from the language of Banjar Malays in South Kalimantan Province and was introduced to Indonesian in the 1970s (Noor 2010). Banjar Malays, who live in the coastal areas of the province, traditionally used some of their low marshland as rice paddies using tidal irrigation and since the 1920s they have greatly expanded the use of the land for rice cultivation (Noor and Jumberi 2007). The Banjar Malays employ traditional techniques to use peatland without extensive drainage of the swampy peatlands (Hidayat 2000; Noor 2010; Noor and Jumberi 2007). However, such techniques are not found everywhere. Indeed, in Riau, peatlands, for the most part, have not been actively used and local knowledge of peatland is rather limited.

These facts cast doubt on the assumption that local people have used peatland sustainably for generations and exhibit the limit of attempts to treat traditional land use as a prescription for the peatland problem. The degradation and great fires were primarily caused by the logging and drainage of peatlands by national and

---

[13] As an exception, Furukawa (1994, pp. 148–154) reports that paddy cultivation using tidal irrigation was conducted around the Tembilahan area in the estuary of the Indragiri River. But he believes that this method was brought by Banjar Malays from Kalimatan around the turn of the twentieth century. In the same region, Abe (1993) states that while peatland was used for the cultivation of rice and coconut, during the 1980s, cultivators moved from place to place every several years.

international companies supported by government policies (Mizuno et al. 2016). What is essential to achieve peatland restoration, then, is not adopting or applying local traditional peatland uses based on the assumption that they are sustainable or include innovative methods. Such an overestimation of local wisdom could bias researchers and hinder understanding and communication with the locals. Rather, it is essential to create new knowledge through communication among all the stakeholders based on a detailed understanding of local knowledge, practice, and history, which BRG's DPG program has tried to document, as mentioned above.

6.4.2 Recognitions of Space and the Externality of Peatland

Related to the history of peatland use in Rantau Baru, the assumption of local wisdom also reveals differences between locals and non-locals in the epistemology of the landscape and the use of space. Before the area began suffering from frequent fires during the mid-1990s, the landscape of Rantau Baru consisted of the rivers, riverbanks as living space, and the hinterland forests covered by thick peat soil (Osawa and Binawan, Chap. 3). These spaces were recognized as different and each had their own distinct uses.

According to Griven H. Putera, a novelist born and raised in Rantau Baru who writes about Malay personality and landscape, and UU. Hamidy, a local anthropologist in Riau with extensive knowledge of Malay cultures, the word "peat" (*gambut* or *gambui*)[14] does not appear in any of the old Malay poetry (*pantun*) in Riau that they have read or heard. Instead, they frequently see the terms "forest" (*utan*) and "swamp" (*rawang; awang*) referring to the peat environment. Although they did not know that the term "*gambut*" came from the language of Banjar Malays, they agree with the assessment that the expressions *gambut* or *gambui* were introduced to Riau Malays during the last several decades, and that prior to this, the words "forest" and "swamp" were used to refer to the peat environment (personal communication). The limited use of the peatlands among Riau Malays throughout their history supports this view. Given that Riau Malays in rural areas often contrast "settlement" (*kampung*) with "forests" (*utan*), it is clear that they recognize the peat environment as a distinct geographical space outside their settlements. That is, in the local worldview, peatland is external to their territory, in contrast with riverbank areas, which villagers recognize as living space. While they have occasionally accessed peatlands to obtain resources, peatlands were only opened by modern hydrological technology and repeated fires during the last few decades. Now that the peatlands are accessible, local people hope the land can contribute to raising their standard of living. To that end, some Rantau Baru villagers have planted oil palms on the land and some have sold or will sell the land, while others maintain the space as ancestral

---

[14] According to Putera, "*gambui*" was used to refer to "peat" around thirty years ago in Rantau Baru. Today, *gambut* is more generally used.

land (see Osawa and Binawan, Chap. 3). Their recognition of peatland space can thus be characterized by its recent economic potential and externality from settlement areas, not as an area for sustaining the essential part of traditional livelihoods.

This externality makes it difficult to expect solutions to the peatland problem to come from villagers' autonomous efforts, which the application of local wisdom is expected to facilitate. For the villagers, peatland is not an essential part of their life, but rather an additional space, to which they cannot invest much capital and labor. Through the statistical analysis of Willing to Pay (WTP), Prasetyawan (Chap. 9) reveals that the villagers pay more attention to the environment of the riverine space than that of the peatland. In addition, while those who have a higher education level can comprehend peat conservation policies and are relatively interested in peatland, around 54% of villagers only have a primary level education and have a limited interest in peatland (Prasetyawan, Chap. 9). Hasegawa (Chap. 8) finds that although the village office provides it with financial support, the village fire prevention group (*Masyarakat Peduli Api*, MPA) was inactive due to a lack of social cooperation. Behind this rejection of cooperative efforts to prevent fires, we see a lack of motivation to participate in activities to protect the hinterland peatlands, which do not directly contribute to livelihoods. If peatlands were significant to their livelihoods, villagers would likely try to protect the area regardless of individual interest.

Expecting community-directed efforts of villagers or extracting a part of knowledge and practice from them is not an effective way to solve peatland problems. Instead, it is necessary for researchers to interrogate villagers' aspirations and exchange epistemologies of environment and landscape over extended periods of time. This process can be summarized as a *sharing* of knowledge. Knowledge sharing is not just an action to "educate" locals in accordance with scientific knowledge and government policy (see also Safitri 2020, p. 208). It is equally essential that researchers learn from locals about local realities and perspectives (see Ingold 2018, pp. 1–25; Lubis 2005). This mutual education process naturally takes a long time and requires close communication, but it allows for knowledge to be shared effectively, together. Ultimately, it can also promote villager cooperation with and participation in any prevention and restoration activities that are established.

To date, BRG has tried to communicate and share knowledge with villagers by dispatching a facilitator to a village for a long period through the DPG program. However, this has not worked well in Rantau Baru. According to the villagers, a DPG program facilitator visited the village in 2019, but only occasionally to ask them questions and investigate the village situation, not to socialize or educate, objectives emphasized in the program.[15] While a few villagers communicated with the facilitator, most villagers did not know about the program at all. The results of our questionnaire survey in 2020 confirms this, as almost all respondents had little or

---

[15] According to the DPR program report on Rantau Baru, based on data obtained in a survey conducted between June 20 and July 11, 2019 (BRG 2019, p. 3).

no experience of peatland restoration education or socialization activities.[16] As the DPG program is still ongoing, it is difficult to fully assess its efficacy (see Hasegawa, Chap. 8). However, studies to date imply that the facilitation of communication has not been progressing as planned in other villages as well (Ananti 2020; Susanto 2020).

6.4.3 The Trans-Locality of the Problem

Finally, the most significant deficiency of applying a local wisdom approach to the peatland problem is that it is not a problem that is caused or that can be solved by the local communities alone. Interaction with and action by various stakeholders beyond the local community are required. The lands that are degraded and host frequent fires are not lands that Rantau Baru villagers have used. They are lands on the margins of their village, downstream of industrial oil palm plantations, and that have been or will be sold to companies and urban residents (Banawan and Osawa, Chap. 10). This means that the villagers have played, at most, only a small role in their degradation.

Identifying and highlighting local wisdom in academic papers and media comes down to attributing the solution (and, at worst, the cause) of this trans-local problem to the local communities. This evokes the impression among urban residents that because the peatland problem is happening in rural areas, it should be solved by the locals. If the main issue was improving the local standard of living, peatland degradation may well be recognized as a problem of the local areas. However, the peatland problem is an environmental problem with which many stakeholders are essentially concerned. The companies and private owners in the cities are significantly related to the causes, and peatland fires most often happen outside the zone of everyday life for villagers. Focusing on local wisdom may distract us from unearthing the roots of the problem and impose primary responsibility for solving the problems back on the locals.

Therefore, to solve the peatland problem, it is not necessary to highlight local wisdom, which may ostracize locals from non-local sectors, but rather to promote cooperation and the sharing of knowledge among the various sectors involved in and impacted by the problem. It is particularly necessary to link the problem with non-locals. Ultimately, establishing and stimulating a network of knowledge and, if possible, cooperation among local and non-local sectors must be paramount in seeking solutions.

---

[16] Of a total of 152 respondents, 126 people (or 81%) answered that they had not participated in any peatland restoration training or socialization activity, and 12 people (or 8%) answered that they had participated in such activity, but at the frequency of less than once in 2 years.

6.5 Concluding Remarks

Investigating local knowledge and practice is undoubtedly essential for mitigating and solving peatland degradation and fires. Such investigation provides researchers with a richer understanding of locals' realities, which in turn enhances meaningful communication with them and the formulation of effective restoration plans.

However, many academic studies in recent years have adopted a local wisdom approach to such investigation and to the peatland problem in general. By labeling specific knowledge and practices as "local wisdom," they seek to discover innovative or "grassroots" methods to solve the problem. This approach, like the IK approach, has many deficiencies. "Local wisdom" leaves questions about historical continuity and interaction with outsiders unanswered, and it allows outsiders to identify something as local wisdom without adequate consideration of local contexts. The approach can therefore lead to misunderstanding locals' knowledge and practice, reinforce the boundary between scientific and local knowledge, and ostracize locals from non-local sectors. In terms of addressing the peatland problem, it restricts understanding of the societies living near the peatlands and the peatland problem itself, which risks limiting responsibility for the problem to local people and agencies. These deficiencies are manifested in the ideological idealism that the concept embodies. Local wisdom gives researchers the illusion of a static existence among locals, despite the dynamism of local knowledge and practice that is a result of interaction and communication with others.

To overcome these deficiencies, instead of local wisdom, researchers should focus on the interaction and transfer of knowledge among various stakeholders related to the problem. On a more practical level, what is needed is an approach that makes it possible for stakeholders to negotiate among themselves over a long period and that simultaneously facilitates continuous communication and cooperation among them, mainly across local and non-local divides. At this practical level, improved understanding of the locals' situation will facilitate continuous, interactive communication and cooperation.

Finally, I would like to reaffirm the significance of knowledge sharing. In the process of knowledge sharing, researchers not only transmit their knowledge based on scientific procedures to the locals, but they also must be educated by local people about the complex contexts and diverse realities of each case. These interactions have the potential to transform researchers' perspectives and achieve fruitful communication and cooperation. The potential of knowledge sharing clearly illuminates the invalidity of circumscribing knowledge or wisdom to locals alone in addressing the problem of peatland restoration.

References

Abe K (1993) Sumatora deitan shicchirin no kindai: shiron (Peat swamp forest in Sumatra: a perspective). Jpn J Southeast Asian Stud 31(3):191–205. https://doi.org/10.20495/tak.31.3_191

Acciaioli G (1985) Culture as art: from practice to spectacle in Indonesia. Canberra Anthropol 8(1–2):148–172. https://doi.org/10.1080/03149098509508575

Agrawal A (1995) Dismantling the divide between indigenous and scientific knowledge. Dev Change 26(3):413–439. https://doi.org/10.1111/j.1467-7660.1995.tb00560.x

Alexander C, Bynum N, Johnson E et al (2011) Linking indigenous and scientific knowledge of climate change. BioScience 61(6):477–484. https://doi.org/10.1525/bio.2011.61.6.10

Ananti R (2020) Evaluasi program Badan Restorasi Gambut dalam merestorasi hutan dan lahan gambut di kepenghuluan Teluk Nilap Kecamatan Kubu Babussalam Kabupaten Rokan Hilir tahun 2018. JOM FISIP 7(2):1–15. https://jom.unri.ac.id/index.php/JOMFSIP/article/view/2 8567. Accessed 28 Sep 2021

Arumingtyas L (2017) Negeri ini kaya kearifan lokal kelola gambut. Mongabay, 14 Feb. https://www.mongabay.co.id/2017/02/14/negeri-ini-kaya-kearifan-lokal-kelola-gambut/. Accessed 6 Jul 2017

Astiani D, Taherzadeh MJ, Gusmayanti E et al (2019) Local knowledge on landscape sustainable-hydrological management reduces soil CO_2 emission, fire risk and biomass loss in West Kalimantan Peatland, Indonesia. Biodiversitas 20(3):725–731. https://doi.org/10.13057/biodiv/d200316

Berkes F (2004) Rethinking community-based conservation. Conserv Biol 18(3):621–630. https://doi.org/10.1111/j.1523-1739.2004.00077.x

Berkes F, Berkes MK (2009) Ecological complexity, fuzzy logic, and holism in indigenous knowledge. Futures 41(1):6–12. https://doi.org/10.1016/j.futures.2008.07.003

BRG (2019) Profil desa peduli gambut: Desa Rantau Baru, Kecamatan Pangkalan Kerinci, Kabupaten Pelalawan, Provinsi Riau. BRG, Jakarta

BRG (n.d.) Negeri ini kaya kearifan local kelola gambut. https://brg.go.id/negeri-ini-kaya-kearifan-lokal-kelola-gambut/. Accessed 6 Jul 2021

BRG (Badan Restorasi Gambut) (2017) Pedoman pelaksanaan: program desa peduli gambut. BRG, Jakarta

Brokensha D, Warren DM, Werner O (eds) (1980) Indigenous knowledge systems and development. University Press of America, Lanham

Burns P (1989) The myth of adat. J Leg Plur 28:1–127. https://doi.org/10.1080/07329113.1989.10756409

Dove MR (2000) The life-cycle of indigenous knowledge, and the case of natural rubber production. In: Ellen R, Parkes P, Bicker A (eds) Indigenous environmental knowledge and its transformations: critical anthropological perspectives. Harwood Academic Publishers, Amsterdam, pp 213–251

Dove MR (2006) Indigenous people and environmental politics. Annu Rev Anthropol 35:191–208. https://doi.org/10.1146/annurev.anthro.35.081705.123235

Ellen R, Harris H (2000) Introduction. In: Ellen R, Parkes P, Bicker A (eds) Indigenous environmental knowledge and its transformations: critical anthropological perspectives. Harwood Academic Publishers, Amsterdam, pp 1–33

Ferguson J (1994) The anti-politics machine: development, depoliticization, and bureaucratic power in Lesotho. University of Minnesota Press, Minneapolis

Furukawa H (1994) Coastal wetlands of Indonesia: environment, subsistence and exploitation (trans: Hawkes P). Kyoto University Press, Kyoto

Geleta M (2015) Conversation links and gaps between scientific knowledge and indigenous people. Sch J Sci Res Essay 4(9):162–168

Gunawan H, Afriyanti D (2019) Potensi perhutanan sosial dalam meningkatkan partisipasi masyarakat dalam restorasi gambut. J Ilmu Kehutanan 13(2):227–236. https://doi.org/10.22146/jik.52442

Hairani A, Noor M (2020) Water management on peatland for food crop and horticulture production: research review in Kalimantan. International symposium on wetlands environmental management, Banjarbaru, November 2019. IOP conference series: earth and environmental science, vol 499. IOP Publishing, Bristol, art 012006. https://doi.org/10.1088/1755-1315/499/1/012006

Henley D, Davidson JS (2007) Introduction: radical conservatism – the protean politics of adat. In: Davidson JS, Henley D (eds) The revival of tradition in Indonesian politics: the deployment of adat from colonialism to indigenism. Routledge, Oxon, pp 1–49

Hidayat T (2000) Studi kearifan budaya petani Banjar dalam pengelolaan lahan rawa pasang surut. J Kalimantan Agrikultura 7(3):105–111

Hobart M (ed) (1993) An anthropological critique of development: the growth of ignorance. Routledge, London

Horsthemke K (2021) Indigenous knowledge: Philosophical and educational consideration. Lexington Books, Lanham

Hutagaol J, Elizal, Kamali A (eds) (2017) Laporan pemetaan sosial Desa Sungai Tohor Kecamatan Tebing Tinggi Timur Kabupaten Meranti tahun 2017. BRG, Jakarta

Ingold G (2018) Anthropology: why it matters. Polity Press, Cambridge

Jalil A, Sulistyani A (2020) Lukun villager's local wisdom on managing fire disaster impact in Kepulauan Meranti Regency of Riau Province. Int J Adv Sci Technol 29(4):2622–2631

Jufri S, Amin B et al (2018) Local wisdom of the community in conserving forests and land in Meranti Islands regency, Riau province. Int J Appl Environ Sci 13(9):801–810

Kristiyanto EN (2017) Kedudukan kearifan lokal dan peranan masyarakat dalam penataan ruang di daerah. J Rechts Vinding 6(2):151–169

Lanzano C (2013) What kind of knowledge is 'indigenous knowledge'? Critical insights from a case study in Burkina Faso. Transcience 4(2):3–18

Lestani MM, Diana L, Erdiansyah E (2019) Local wisdom of land cleaning by the society of Siak Malay in past. Adv Soc Sci Educ Humanit Res 42:101–103. https://doi.org/10.2991/assehr.k.200529.278

Li TM (2001) Masyarakat adat, difference, and the limits of recognition in Indonesia's forest zone. Mod Asian Stud 35(3):645–676

Lubis ZB (2005) Menumbuhkan (kembali) kearifan lokal dalam pengelolaan sumberdaya alam di Tapanuli Slatan. Antropol Indones 29(3):239–254. https://doi.org/10.7454/ai.v29i3.3544

Lubis ZB (2013) Social mapping of access to peat swamp forest and peatland resources. Working paper of Kalimantan Forests and Climate Partnership (KFCP)

Maria (2018) Local wisdom of indigenous society in managing their customary land: a comparative study on tribes in Indonesia. E3S Web Conf 52:art 00023. https://doi.org/10.1051/e3sconf/20185200023

Masuda K, Kusumaningtyas R, Mizuno K (2016) Local communities in the peatland region: demographic composition and land use. In: Mizuno K, Fujita MS, Kawai S (eds) Catastrophe and regeneration in Indonesia's peatlands: ecology, economy and society, Kyoto CSEAS series on Asian studies, vol 15. NUS Press; Kyoto University Press, Singapore; Kyoto, pp 185–210

Mizuno K, Fujita MS, Kawai S (eds) (2016) Catastrophe and regeneration in Indonesia's peatlands: ecology, economy and society. NUS Press; Kyoto University Press, Singapore; Kyoto

Momose K (2002) Environments and people of Sumatran peat swamp forests II: distribution of villages and interactions between people and forests. Southeast Asian Stud 40(1):87–108. https://doi.org/10.20495/tak.40.1_87

Murniati, Suharti S (2018) Towards zero burning peatland preparation: incentive scheme and stakeholders role. Biodiversitas 19(4):1396–1405. https://doi.org/10.13057/biodiv/d190428

Nababan A (1995) Kearifan tradisional dan pelestarian lingkungan hidup di Indonesia. Analisis CSIS 24(6):421–435

Noor M (2010) Lahan gambut: pengembangan, konservasi, dan perubahan iklim. Gadjah Mada University Press, Yogyakarta

Noor M and Jumberi A (2007) Kearifan lokal dalam perspektif pengembangan pertanian di lahan rawa. http://repository.pertanian.go.id/handle/123456789/6293. Accessed 27 Mar 2022

Noorginayuwati, Rafieq A, Noor M et al (2007) Kearifan lokal dalam pemanfaatan lahan gambut untuk pertanian di Kalimantan. http://repository.pertanian.go.id/handle/123456789/6294. Accessed 27 Aug 2022

Nugroho K, Carden F, Antlov H (2018) Local knowledge matters: power, context and policy making in Indonesia. Policy Press, Bristol. https://doi.org/10.51952/9781447348085

Nurjaya IN (2007) Kearifan lokal dan perngelolan sumberdaya alam. J Ilmiah 8(40). https://blogmanifest.wordpress.com/2008/01/03/kearifan-lokal-dan-pengelolaan-sumberdaya-alam/. Accessed 3 Aug 2021

Olivier de Sardan JP (2005) Anthropology and development: understanding contemporary social change (trans: Alou AT). Zed Books, London; New York

Osawa T (2022) At the edge of mangrove forest: the Suku Asli and the quest for indigeneity, ethnicity and development, Kyoto area studies on Asia, vol 29. Kyoto University Press; Trans Pacific Press, Kyoto; Tokyo

Posey DA (2000) Ethnobiology and ethnoecology in the context of national laws and international agreements affecting indigenous and local knowledge, traditional resources and intellectual property rights. In: Ellen R, Parkes P, Bicker A (eds) Indigenous environmental knowledge and its transformations: critical anthropological perspectives. Harwood Academic Publishers, Amsterdam, pp 35–54

Puspitorini D, Hunter TM (2020) Introduction. In: Puspitorini D, Hunter TM (eds) Nusantara's indigenous knowledge. Focus on civilizations and cultures. Nova Science Publishers, New York, pp vii–xxiii

Ramdhan M, Siregar ZA (2018) Pengelolaan wilayah gambut melalui pemberdayaan masyarakat desa pesisir di kawasan hidrologis gambut Sungai Katingan dan Sungai Mentaya provinsi Kalimantan Tengah. J Segara 14(3):145–157

Ruastiti NM (2011) The concept of local genius in Balinese performing arts. Mudra 26(3):241–245

Safitri MA (2020) Sinergi adaptasi kearifan lokal dan pemberdayaan hukum dalam penanggulangan kebakara. Bina Hukum Lingkungan 4(2):198–215. https://doi.org/10.24970/bhl.v4i2.99

Santoso I (2006) Eksistensi kearifan lokal pada petani tepian hutan dalam memelihara kelestarian ekosistem sumber daya hutan. Wawasan 11(3):11–20

Sartini (2004) Menggali kearifan lokal Nusantara: sebuah kajian filsafati. J Filsafat 37(2):111–120. https://doi.org/10.22146/jf.33910

Sillitoe P (1998) Development of indigenous knowledge: a new applied anthropology. Curr Anthropol 39(2):223–252. https://doi.org/10.1086/204722

Sillitoe P (2002) Participant observation to participatory development: making anthropological work. In: Sillitoe P, Bicker A, Pottier J (eds) Participating in development: approaches to indigenous knowledge, ASA Monographs, vol 39. Routledge, London, pp 1–23

Sillitoe P (2007) Local science vs. global science: an overview. In: Sillitoe P (ed) Local science vs global science: approaches to indigenous knowledge in international development, Environmental anthropology and ethnobiology, vol 4. Berghahn Books, New York; Oxford, pp 1–22

Slikkerveer LJ (2019) The indigenous knowledge systems' perspective on sustainable development. In: Slikkerveer LJ, Baourakis G, Saefullah K (eds) Integrated community-managed development: Strategizing indigenous knowledge and institutions for poverty reduction and sustainable community development in Indonesia. Cooperative management. Springer, Cham, pp 33–66. https://doi.org/10.1007/978-3-030-05423-6_2

Slikkerveer LJ, Baourakis G, Saefullah K (eds) (2019) Integrated community-managed development: Strategizing indigenous knowledge and institutions for poverty reduction and sustainable community development in Indonesia. Cooperative management. Springer, Cham. https://doi.org/10.1007/978-3-030-05423-6

Slikkerveer LJ, Dechering WHJC (1995) LEAD: the Leiden ethnosystems and development programme. In: Warren DM, Slikkerveer LJ, Brokensha D (eds) The cultural dimension of development: indigenous knowledge systems. Intermediate Technology Publications, London, pp 435–440

Slikkerveer LJ, Slikkerveer MKL (1995) Teman obat keluarga (TOGA): indigenous Indonesian medicine for self-reliance. In: Warren DM, Slikkerveer LJ, Brokensha D (eds) The cultural dimension of development: indigenous knowledge systems. Intermediate Technology Publications, London, pp 13–34

Supriyati W, Alpian A, Prayitno TA et al (2016) Local wisdom in utilizing peat swamp soil and water to improve quality of gelam wood. Trop Wetland J 2(2):27–37. https://doi.org/10.20527/twj.v2i2.29

Susanto D, Sanusi, Widyanti R (2020) Implementasi kebijakan restorasi gambut di Kalimantan Selatan dari persfektif komunikasi kebijakan: studi kasus di Kecamatan Candi Laras Utara Kabupaten Tapin. Dissertation, Islamic University of Muhammad Arsyad Al Banjari Kalimantan

UNESCO (n.d.) Local and Indigenous Knowledge Systems: what is local and Indigenous knowledge? http://www.unesco.org/new/en/natural-sciences/priority-areas/links/related-information/what-is-local-and-indigenous-knowledge. Accessed 3 Jul 2021

Utami W, Salim MN (2021) Local wisdom as a peatland management strategy of land fire mitigation in Meranti regency, Indonesia. Ecol Environ Conserv 27:s127–s137

Warman K (2014) Peta perundang-undangan tentang pengakuan hak masyarakat hukum adat. https://procurement-notices.undp.org/view_file.cfm?doc_id=39284. Accessed 26 Apr 2021

Warren DM, Slikkerveer LJ, Brokensha D (1995) Introduction. In: Warren DM, Slikkerveer LJ, Brokensha D (eds) The cultural dimension of development: indigenous knowledge systems. Intermediate Technology Publications, London, pp xv–xviii

Zakaria YR (1994) Hutan dan kesejahteraan masyarakat. WALHI, Jakarta

Chapter 7
The Dimension of Gender in Peatland Management in Rantau Baru Village

Kurniawati Hastuti Dewi

Abstract Although harnessing full community participation in natural resource management produces positive ecological and economic outcomes, the specific roles of men and women in peatland communities are often overlooked. This study investigates the differentiated knowledge and roles of both men and women in peatland management in Rantau Baru, a fishing and farming Peat Care Village (*Desa Peduli Gambut*) in Riau Province, Indonesia. Primary data were collected through a survey of 152 households conducted from January–February 2020 and subsequent follow up interviews with community members. Modifying the Harvard Analytical Framework, the study examines knowledge levels of men and women as well as productive (peatland cultivation and fishery) activity, reproductive or domestic (childcare and household finance) activity, and sociopolitical (community meetings) activity. It finds that men are significantly more knowledgeable about peatlands than women and that peatland agricultural activities are dominated by men, but that gender roles are more evenly distributed in fishery activities. Women and men play complementary roles in "reproductive activities" of the household, but women do not participate nearly as much as men in the public sphere of "sociopolitical activities," such as attending community, association, and village meetings. The study provides new insight into the community's knowledge of peatland according to gender, and the potential role of both male and female community members in peatland restoration. Any project or program on peatland restoration should recognize the basic features and differences of gender roles and the specific needs of men and women to ensure the optimal contribution of all community members to peatland management and restoration.

Keywords Gender roles · Peatland management · Agriculture · Fishing · Peatland restoration · Rantau Baru · Riau

K. H. Dewi (✉)
Research Center for Politics, National Research and Innovation Agency (BRIN), Jakarta Selatan, Indonesia
e-mail: kurniawati.hastuti.dewi@brin.go.id

147

M. Okamoto et al. (eds.), *Local Governance of Peatland Restoration in Riau, Indonesia*, Global Environmental Studies,
https://doi.org/10.1007/978-981-99-0902-5_7

7.1 Introduction

Peatland restoration aims not only to rehabilitate the ecological functions of peatlands, but also, increasingly, to improve the welfare of the communities surrounding peatlands. Community involvement in peatland restoration is expected to create sustainable peatlands by increasing welfare and ecological function (Safitri 2020), and research has shown that improving local livelihoods and involving community members in restoration efforts results in better outcomes for sustainable peatland restoration. Ensuring the involvement of all community members requires the participation of both men and women, yet peatland restoration programs often overlook the gender dimension of peatland management.

Various international organizations, including the World Wildlife Fund (WWF) (2012), the United Nations *Economic and Social Commission for Asia and the Pacific* (UN ESCAP) (2017), and the World Bank (2018), and scholars such as Elmhirst (1998), Resurreccion (2008), and Watson (2006) recognize and promote the importance of gender and gender analysis in natural resource management. Elmhirst and Resurreccion (2008, p. 5) assert that "men and women hold gender-differentiated interests based on their distinctive roles, responsibilities, and knowledge" and therefore, "gender is a critical analytical concept for understanding the social and political dimensions of natural resource management and governance across a range of empirical settings" (Elmhirst and Resurreccion 2008, p. 3). WWF (2012) recognizes the need for gender sensitivity in natural resource management to ensure that projects and programs recognize the different roles and needs of men and women.

A considerable number of studies have also been conducted on women's roles in the agricultural and rural development of Indonesia. In discussing agricultural production in Java, Sajogyo (1983) focuses on women's time allocation in productive, reproductive, and decision-making work. Widiarti and Hiyama (2007)'s study of Citarik Village, Sukabumi, West Java demonstrates the considerable contribution of women in the *Pengelolaan Hutan Bersama Masyarakat* (PHBM), or Joint Forest Management Program, in land clearing, planting, and plant maintenance. Although some women decided to join the PHBM without their husband's permission, the study found that women's decision-making power within the family does not translate to the community level, because the decision-makers in village meetings are men, and women's access to knowledge is limited, as only men attend trainings (Widiarti and Hiyama 2007). Mugniesyah and Mizuno's study (2007) of women's access to land and their control over it in Sundanese communities with bilateral kinship systems reveals that: (1) These communities follow *sanak* values (customary law), a set of values concerning gender equity and the rights of sons and daughters to the household property and *sanak* values strongly influence peasant households in the allocation of their land through inheritance and grant systems; (2) *sanak* values lead to gender equality in access to and control over land among household members; and (3) gender equality in land ownership is also shown in the practice of the inheritance system, which is calculated through both the male and female lines. Using the Harvard Analytical Framework, Dewi et al. (2020) analyze the role of male and female farmers in the Special Purpose Forest Area of Parungpanjang, West Java and discover that: (1) Female farmers participate in all dimensions of productive, reproductive, and sociopolitical activities, while male farmers tend to limit their participation only to productive and sociopolitical activities; (2) the Special Purpose Forest Area of Parungpanjang does not grant official rights to female farmers to use the land (only a male head of household can register for such rights); and (3) gender-responsive policies and gender awareness programs among male farmers need to be strengthened. Studies on gender and natural resource management outside Java include Elmhirst et al. (2017), which examines oil palm plantations in Kalimantan through a feminist political ecology perspective. Other scholars, such as Villamor et al. (2015) and Villamor et al. (2014), analyze gender and land use change in Central Sumatra.

Despite the rich literature on gender and natural resource management, however, very little of it deals with peatland management specifically. Conversely, among the numerous studies devoted to understanding peatland restoration and management in Indonesia (such as Mizuno et al. 2016), few explore women's roles in these processes. Exceptions to this include Subono et al. (2020), which focuses on the involvement of women in peatland restoration in Central Kalimantan Province and finds that although women facilitators faced structural and cultural obstacles, an economic revitalization program they implemented strengthened the economic resilience of rural women's communities and changed gender relations. In a study of women's experiences in Central Kalimantan and Riau, Indrastuti (2020) reveals that although firefighting requires women's involvement, especially when it happens on their land or in their living spaces, women do not have access to the resources they need to prevent and fight forest and land fires.

Perhaps the most comprehensive study to date of gender roles in peat-based communities in Riau is Herawati et al. (2019). Modifying the Harvard Analytical Framework, it used a mix of qualitative and quantitative methods to study gender roles and livelihoods in seven villages and three districts of Riau Province from 2016 to 2018. It found that: (1) agricultural activities are significantly dominated by men, while women play a more significant role in domestic activities; (2) both men and women contribute equally to the social life of the community, in which women's participation and group membership is equal to men's; (3) low-income families tend to have higher gender equity in agricultural activities than high-income households; (4) the role of women in wealthier households is not in their physical contribution to the land, but is mostly in their role as decision-maker, indicating that women play a significant role in the livelihoods of both poor and rich families, but in different forms; and (5) community development interventions that involve women are recommended (Herawati et al. 2019).[1]

---

[1] Herawati et al. (2019, p. 854) does not analyze the possible influence of ethnicity on gender roles; instead, it provides only a general picture of the main ethnic groups of Riau Province, with the indigenous Malay making up 33% of the population, Javanese, 30%, Batak, 13%, and Minang, 12%.

Indonesia's Peatland Restoration Agency, the *Badan Restorasi Gambut* (BRG), carries out peatland restoration through the 3-R method, that is, by *rewetting* and *replanting* peatlands and by *revitalizing* the livelihoods of local communities (BRG 2018, p. 1). The BRG mobilizes community participation in peatland restoration through its *Kerangka Pengaman Sosial* (Social Safeguard Framework) and the Peat Care Village Program. Riau Province is one of the seven provinces targeted by the BRG for priority peatland restoration,[2] which stipulates that the area contains Peat Hydrological Units (*Kesatuan Hidrologis Gambut*, KHG) (BRG 2018, p. 1). As of 2019, there were 262 Peat Care Village Programs across the seven KHGs, with 49 located in Riau Province (BRG 2019a, p. 27). The BRG's Deputy Section for Education, Socialization, Participation, and Partnership has given special attention to women's roles and participation in peatland restoration. For example, 773 women's groups in Kalimantan have received assistance from the BRG to increase the added value of woven handicraft products made from grass or plants that mostly grow on peatlands (Sumartomjon 2019). This example illustrates how the BRG strives to provide gender-specific programs, an approach that aligns with global efforts that recognize the significance of the gender dimension in natural resource management. However, the specific contributions and potential of men and women remain under studied.

Given the lack of gender analysis in existing peatland literature and programming, this chapter examines the gender dimension more fully, providing new data and insights on community participation in peatland management in Riau, the center of peatland restoration in Indonesia. As a case study, it investigates the role of community members (both men and women) in peatland management in Rantau Baru, a designated Peat Care Village. The study sheds light on (1) the different levels of knowledge among men and women of *gambut*, or peatland, the sources of their knowledge, and involvement in training/socialization on peatland conservation, and (2) the roles of women and men (as farmers and fishers) in peatland management in Rantau Baru Village. By providing a more comprehensive understanding of male and female roles in peatland management in a Peat Care Village, it is hoped that the study can contribute to formulating more effective peatland programming that is suitable for all community members.

7.2 Method

7.2.1 Research Site: Location, Livelihoods, Population Make-Up

Rantau Baru Village has been introduced in earlier chapters of this book. It is located in Riau Province, which has the largest peatland area in Sumatra (Muslim and

---

[2]The other provinces targeted by the BRG for peatland restoration are: Jambi, Kalimantan Barat, Kalimantan Tengah, Kalimantan Selatan, Riau, Sumatera Selatan, and Papua.

Fig. 7.1 Flood in Rantau Baru Village in December 2018 (Photo by Dewi, December 2018)

Kurniawan 2008, p. 1), and in the Kiyap River KHG.[3] As it is on the edge of the Kampar River, the village is often flooded. This was observed by the author and research team in December 2018 as seen in Fig. 7.1. Such floods block access to the main road of the village and submerge the village entrance gate (pictured in green). Thus, during the flood season, often the only way to enter the village is by small boat through the *sekat kanal* (a canal bulkhead that also functions to prevent peatland fires), which runs parallel to the main road. It takes approximately 40 min to reach the village center from the entrance gate with this transport.

Rantau Baru Village is a fishing and farm-based community. As explained in Chaps. 4 and 5, the peat swamp forests are fishing grounds for communities living in the area. The presence of peasant fishermen among Malay (*Melayu*) Indonesians was recorded by Firth's study in 1946 (Firth 1946). Thus, the livelihood makeup of Rantau Baru Village is not surprising.

The total population of Rantau Baru Village in 2019 was 715 people, consisting of 379 men and 336 women living in 206 *kepala keluarga*, or households (BRG 2019b, p. 25). The majority of the population is Muslim; only five people adhere to Catholicism (BRG 2019b, p. 35). According to the village secretary, the community's origins are rooted in *suku Minangkabau*, or the Minangkabau lineage (interview, December 9, 2018), although the validity of this oral explanation still needs to be verified. In Chap. 3, Osawa and Banawan postulate that local people are descended from the intermingling of the Minangkabau, coastal Malays, and indigenous peoples who lived in the region prior to the arrival of these two groups.

The matrilineal society of the Malay Petalangan people in Rantau Baru Village, who make up 30% of the population, consists of transgenerational links through the maternal line, by which ancestral land and land rights pass from grandmother to mother, and then to granddaughter and her descendants in the female line (interview with village secretary, December 9, 2018). Moreover, each *Sialang* tree, where

---

[3]The BRG's 2019 Riau Province Yearly Action Plan is comprised of six KHGs, namely (1) Bengkalis Island, (2) Rangsang Island, (3) Indragiri River–River Cenaku, (4) Kiyap River–Kerumutan River, (5) Nilo River–Napuh River, (6) Rokan River–Kubu River (BRG 2018, p. 1).

honeybees make their nests, is highly regarded by the Malay Petalangan and is passed down from generation to generation through the female line (BRG 2019b, p. 37). Another characteristic of the Malay Petalangan's matrilineal society is the *ninik mamak*, or a male chief of the lineage, and *kepala suku*, or "prominent respected men" (BRG 2019b, p. 38; interview with the *ninik mamak* of Rantau Baru, December 2, 2020).

7.2.2   Research Methodology

This study relies primarily on data from a survey of all households in the village conducted during January–February 2020. The survey data is based on the responses of 77 women and 75 men. The survey was preceded by preliminary observations and interviews in the village in December 2018 and from August to September 2019. Following the survey, follow-up interviews were conducted online from November to December 2020 with three women from the village, namely the head of the Women's Farmers Group (*Kelompok Wanita Tani*, KWT), a member of the Empowerment of Family Welfare (*Pemberdayaan Kesejahteraan Keluarga*, PKK), and a fisherwoman, as well as prominent men in the village.

To explain the roles of women and men in peatland management in the village, I rely on the Harvard Analytical Framework (HAF) developed by the Harvard Institute for International Development in the USA in collaboration with the US Agency for International Development (USAID) Office of Women in Development in 1985. Also known as the Gender Roles Framework, the HAF provides a matrix for data collection at community and household levels and is comprised of three main tools, namely a "socio-economic activity profile" (identifying productive and reproductive tasks by asking "Who does what?"), an "access and control profile" of participants, and other "influencing factors" (March et al. 1999, p. 32). I used the HAF to better understand the basic features and differences of the gender roles and the specific needs of men and women in the village to help determine how to best ensure the optimal contribution of all community members to peatland management and restoration. In designing the survey for this research, I simplified the HAF and created three activity profiles for men and women, namely productive (socioeconomic), reproductive (household), and sociopolitical activities.

7.3   Findings and Discussion

7.3.1   Peatland Knowledge of Men and Women

Three questions in the survey assessed community member's knowledge of peatlands. First, men and women were asked if they could identify the differences between peatland soils (*tanah gambut*) and mineral soils (*tanah mineral*, or yellow

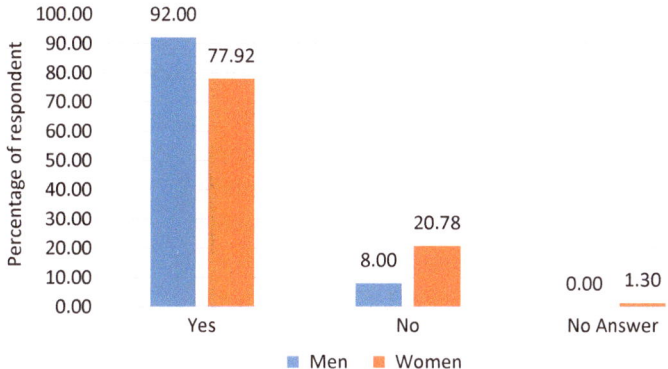

Fig. 7.2 Ability to differentiate between peat and mineral soils, by sex (Question 16)

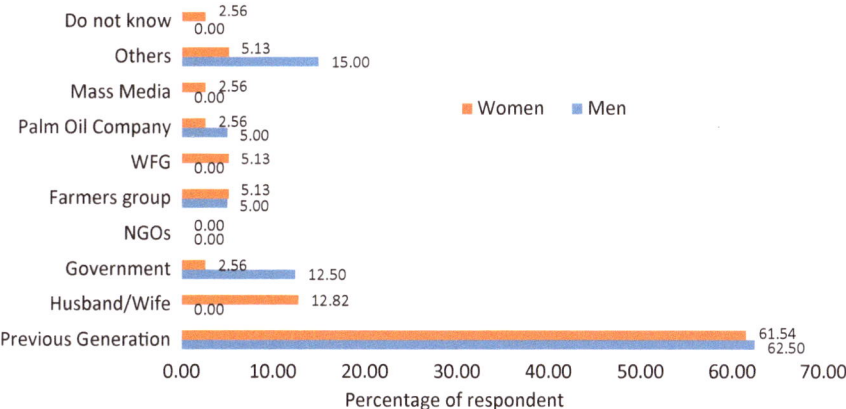

Fig. 7.3 Sources of knowledge about peatland cultivation, by sex (Question 22)

soil). Mineral soils in Rantau Baru Village are more suitable for agricultural activities than peat soils. Peat soils consist of vegetative material that has decomposed over the past thousand years and are always partially submerged in water (BRG 2019b, p. 13). Residential areas of Rantau Baru are usually close to mineral soils, where residents grow vegetables in small quantities for self-consumption, as mass cultivation is difficult in this flood-prone area. Villagers were asked, "Are you able to differentiate between peatland soil and mineral soil?" (Question 16). In Fig. 7.2, we see the response results, that 92% of men and 78% of women could differentiate the two types of soil.

Men and women were then asked about the source of their knowledge about peatland cultivation. The results are depicted in Fig. 7.3. Both men and women learned to cultivate peatland mainly from previous generations (63% of men and

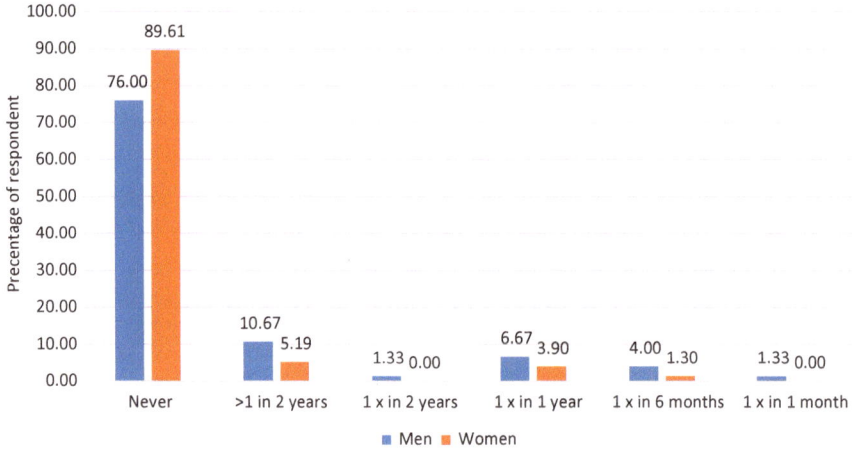

Fig. 7.4 Frequency of participation in training or socialization activities about peatland management, by sex (Question 23)

62% of women). The primary difference between men and women is in knowledge gained from the government and "other sources" (a 10% difference for each, with more men gaining knowledge from government and other sources than women). Also notable is that no men reported learning about peatland from their wives, but 13% of women reported gaining such knowledge from their husbands. Interestingly, no men or women reported gaining knowledge from nongovernmental organizations (NGOs). This data does not indicate a difference between men and women in the level of knowledge of peatland cultivation.[4]

Finally, men and women were asked about their participation in training and socialization activities about peatland management. The results, as shown in Fig. 7.4, show that while nearly 90% of women have never participated in such activities, 76% of men have never done so. Therefore, as of the time of the survey, more men in the Rantau Baru village had participated in peatland management training and socialization than women. The frequency of such participation is extremely low for both men and women, with only 11% of men having participated in training and socialization more than once in a span of 2 years; only 5% of women had attended with the same frequency. This gender gap remains more or less consistent across frequencies. This data points to an urgent need for more training and socialization about peatland management focusing on women in the village.

---

[4]Based on their qualitative interviews with the Rantau Baru villagers, Osawa and Binawan (Chaps. 3 and 6 in this book) assert that villagers did not begin using peatland for their livelihoods in earnest until the mid-1990s.

7.3.2   The Roles of Men and Women

By modifying the categories of the Harvard Analytical Framework, this section presents a profile of men and women's productive, reproductive, and sociopolitical activities in Rantau Baru Village.

7.3.2.1   Productive Activities: Peatland Cultivation and Fishing

The importance of gender issues in forestry and fisheries is inevitable, as these activities are not gender-neutral (FAO 2016). Gender-based segregation in forestry and fisheries is evident in the distribution of work between women and men, the invisibility of women's contributions, and women's limited access and control in decision-making. Although scholars note these tendencies and promote gender perspectives, many studies and policies in these sectors are gender blind (Colfer et al. 2016). Indeed, an World Bank et al. (2009) study notes that the role of women in the both the formal and informal forestry sectors has not been fully recognized or documented. In Indonesia, employment in the forestry sector is still dominated by men, and activities such as training, meetings, and campaigns are still directed at men and exclude women (Engelhardt and Rahmina 2011). Scholars agree that gender integration in forestry faces several obstacles and the sector lacks gender awareness (Mai et al. 2011, pp. 246–248). The following section explores role of men and women in peatland management, which is part of forestry and farming activities.

This section examines the different roles of men and women in the productive activities of Rantau Baru Village, an agricultural and fishery-based community. Villagers mainly use peatland for oil palm plantations, but they also cultivate chile, sweet potatoes, pineapple, banana, coconut, mango, guava, and rubber on peatland. Work related to the productive utilization of peatland in the village consists of: clearing the peatland for cultivation, fertilizing, harvesting, and selling the agricultural products. Table 7.1 depicts the participation of men and women in each of these activities according to survey respondents.

A majority of respondents (64%) reported that land preparation activities, namely clearing the peatland, are performed jointly by men and women. Half of the respondents also reported that fertilizing of the peatland was jointly done by men

Table 7.1 Productive peatland activities in Rantau Baru Village, by sex (Question 21)

Peatland Cultivation Activity	Women Only No. (%) of respondents	Men Only No. (%) of respondents	Men and Women No. (%) of respondents
Clearing peatland	1 (1)	24 (35)	44 (64)
Fertilizing		34 (50)	34 (50)
Harvesting	1 (1)	56 (81)	12 (17)
Selling	4 (6)	44 (64)	21 (30)

and women. Role differentiation was most striking in the activity of harvesting the fruit from the oil palms, with 81% of respondents reporting that this is done only by men, demonstrating the prominent role of men in this activity. A majority of respondents (64%) also noted that harvesting was done only by men. From these percentages, it can be said that although women participate in productive activities in peatlands by jointly clearing and fertilizing peatlands with men, men play a more predominant role, especially in harvesting and selling.

Programs to include or improve women's contribution to peatland management in the village are also lacking. According to Erna, the head of the Women's Farmers Group (*Kelompok Wanita Tani*, KWT) of Rantau Baru and a member of PKK Rantau Baru, monthly PKK meetings held before the pandemic discussed related topics, such as the environment, children's health, healthy food, and clean water. A women's training program to learn things like baking or making plates from *pandan* leaves or palm sticks was also conducted, but the majority of the women in that program quit and not many continued to use those skills (interview with head of KWT, November 23, 2020). Erna explained that in the previous 6 years, she had attended three *musrenbangdes*, or village development plan deliberation, where PKK is the only women's representative group, and not all PKK officers were invited. Interestingly, Erna has never discussed KWT's various problems in PKK or *musrenbangdes*, as PKK mainly discusses PKK's performance (interview with head of KWT, November 23, 2020). There are no meetings or trainings specific to agricultural activity, or a program for peatland restoration, for women in Rantau Baru.

A gender perspective can be promoted to ensure the participation of women in peatland restoration and management. For example, the Indonesia Climate Change Trust Fund (ICCTF) ICCTF implemented a program to reduce emissions in Indonesia through local activities, including in Dumai, Riau, which involved a special program for women to participate in peatland management through the creation of women's groups. The women's groups have been involved in peatland management by planting red ginger through agroforestry and fish farming through biofloc ponds, which have become productive activities not only to improve organizational skills, but also to increase economic value (Wagey 2018, p. 2). This example shows real action to improve women's contribution to and participation in peatland management in Riau.

In the fishery sector, the FAO (2016, p. 1) notes that 'women's engagement in fisheries can be viewed from social, political and technical perspectives, all of which show that the role of women is often underestimated." The FAO (2016, p. 3) further notes that "almost universally, women play key roles in the fishery industry and household livelihoods and nutrition. These women, estimated at approximately 90 million, are often invisible to policymakers who have traditionally assumed – mistakenly – that fisheries are largely a male domain." This inadequate recognition of women's contributions in fisheries hampers the development process.

As explained in Chap. 5, fishery production in Rantau Baru increases during the rainy season, which lasts 3–4 months. Work in the village's fishery sector consists of catching fish in rivers/lakes, processing fish by salting and smoking, and selling fresh

Fig. 7.5 Traditional method of salting and baking fish by the women of Rantau Baru Village (Photo Dewi, August 2019)

Table 7.2 Productive fisheries activities in Rantau Baru Village, by sex (Question 44)

Fisheries Activity	Women Only No. (%) of respondents	Men Only No. (%) of respondents	Men and Women No. (%) of respondents
Catching fish in river/lakes	2 (2)	44 (47)	48 (5)
Processing fish (salting/smoking)	45 (48)	4 (4)	44 (47)
Selling fish	35 (37)	17 (18)	42 (45)

fish and processed fish. Many of the fish caught in Rantau Baru are not sold fresh, for example, *baung*, *patin*, and *gabus*. These are processed in the village (see Fig. 7.5) and can fetch higher prices than fresh fish. This fish processing depends on the season, with only a few fisherfolk preparing salted fish in the dry season, which lasts from March to August, and many preparing smoked and salted fish during the rainy season, when the catch is abundant (BRG 2019b, p. 62).

Table 7.2 depicts the participation of men and women in the fisheries activities of Rantau Baru. While a slim majority of survey respondents (51%), report that catching fish in rivers and lakes is done by both men and women, nearly half of respondents (48%) note that processing the fish (salting and smoking) is done exclusively by women. Only four respondents reported that only men process fish, with the remainder of respondents (47%) reporting that both men and women process fish. Women also play a large role in selling fish, with 45% of respondents reporting that both men and women sell fish and 37% reporting that only women sell fish.

According to the results of this study, women in Rantau Baru participate in and contribute to the fisheries sector, particularly in the processing (salting and baking) of the fish. Overall, the sector has a more egalitarian gender role distribution pattern compared to peatland cultivation activities. Nearly half of respondents said that catching fish is done jointly done by men and women; women in the village play a major role in processing the fish; and a majority of respondents report that selling fish is done by both men and women.

The case of Ana, a 49 year-old Meliling fisherwoman from Sepunjung sub-village (*Dusun 1*), provides a sketch of daily life. After her husband became sick in 2004, every morning and evening Ana has used a *sampan* (traditional boat) and "net catches" (*jaring yang ditinggalkan*) to catch fish in the Kampar River. She then salts or smokes her catch (usually *baung* or *salai*) and brings the fish to the Kerinci Market every Sunday. In 2004 she could sell around 10 kg of salted fish per week at a price of Rp15,000 per kg, but in 2020 she was only able to catch and sell 4 kg of salted fish per week at a price of Rp20,000 per kg. She says that since 2017, the number of fish has been rapidly declining due to the increasing number of fisherwomen who use more advanced technology to catch fish. Ana is a member of a men's fisher group that fisherwomen are allowed to join. The group can write proposals to request funding support from the District of Pangkalan Kerinci, but there is no funding or support from the Rantau Baru Village Fund or from the District of Pangkalan Kerinci for fisher groups (interview with Ana, fisherwoman, December 10, 2020).

How can we understand the above data in the broader framework of gender and fisheries? Here, efforts by international agencies can be useful. Examples include the FAO's Sustainable Fisheries Livelihoods Programme (SFLP), as a holistic approach to gender analysis and the incorporation of local gender action planning through a bottom-up approach to advocating policy to support and empower women (FAO 2016, p. 8) and Weeratunge-Starkloff and Pant's (2011) collaboration with diverse women and men farmers and fishers, government agencies, and research institutions to create innovative research-based policies and interventions to close the gender gap in fisheries. The FAO (2007, p. 1) notes that in the small-scale fisheries sector, development policies have traditionally targeted women as fish processors, and fishery-related development activities have engaged men as exploiting, and sometimes managing, resources, whereas women have been excluded from planning "mainstream" fishery activities. The Rantau Baru fisheries sector aligns with Ogden (2017, p. 117), who notes that "although there are global patterns of women mainly gleaning and men mainly engaged in capture fisheries, researchers are revealing that women's involvement in fisheries dynamic and diverse over space and time."

Despite women's involvement in the productive activities of peatland cultivation and fisheries, gender analysis is often missing from development projects. In their evaluation of how gender was considered in 20 livelihood development projects implemented in coastal communities in Indonesia since 1998, Stacey et al. (2019) found that: (i) despite many projects reaching women, particularly with efforts to increase women's productive capacity through training and group-based livelihoods enterprises, 40% of the projects had no discernible gender approach, and (ii) only two of the 20 projects (10%) applied a gender transformative approach that sought to challenge local gender norms and gender relations and empower women beneficiaries, suggesting the need for greater understanding of the role of gender in reducing poverty and increasing well-being. Stacey et al. (2019) provides further evidence of the importance of gender equality and women's empowerment in small-scale fisheries and associated livelihood improvement programs.

The findings of this study offers new insights into the different roles of men and women in peatland cultivation and fisheries activity, which can hopefully contribute to a better understanding of the gender dimension of these sectors and a subsequent improvement in programs.

7.3.3   Reproductive Activities

Reproductive activities (in both agricultural and fishery-based families) mainly occur inside the household and commonly refer to domestic activities, as opposed to activities in the public sphere. The domestic activities (of which there are many) chosen for this study are childcare and managing household finances. A majority of respondents (61%) reported that childcare is performed by both men and women, while 39% reported that only women perform childcare. This finding indicates relative gender equality in the households of Rantau Baru Village when it comes to the activity of childcare, although the degree to which both men and women participate in this activity was not elaborated. When it comes to deciding how to use household funds, 59% of respondents reported that women alone manage household finances, and 34% report that finances are jointly managed by men and women. Only 7% reported that household finances are managed exclusively by men. This finding indicates that women in the village play a dominant role in family money management (Table 7.3).

To better understand the reproductive activities of fishery families in the village, I interviewed Santi, a Malay Datuk Tuo fisherwoman who is also the head of working group 3 (*kelompok kerja, Pokja*) PKK Rantau Baru 2020–2023. Santi said that she does all the same household activities as ordinary housewives (washing, cooking, and cleaning), but her fisherman husband never helps with these duties due to being busy. Santi said that both she and her husband both take care of child-rearing and manage the household money (interview with Santi, December 23, 2020). Similarly, the fisherwoman Ana said that her husband is willing to help wash dishes, but he does not wash clothes, except in the washing machine. Ana said that she and her husband always discuss various problems at home, including the allocation of money for their two daughters studying at university. Notably, Ana always tells her daughters not to be fisherwomen, but that it is acceptable to pursue a career in agriculture (interview with Ana, December 10, 2020). Rosa (19 years old), secretary

Table 7.3 Reproductive activities of men and women in Rantau Baru (Question 120)

Reproductive Activity	Women Only No. (%) of respondents	Men Only No. (%) of respondents	Men and Women No. (%) of respondents
Childcare	57 (39)	1 (1)	90 (61)
Managing household finances	88 (59)	10 (7)	51 (34)

Table 7.4 Management of household finances in Rantau Baru, by ethnicity and sex (Question 121 cross-referenced with ethnic identity of respondents)

Person(s) managing finances	Malay No. (%) of respondents	Non-Malay No. (%) of respondents
Husband	8 (7)	2 (6)
Wife	68 (58)	20 (65)
Both	42 (36)	9 (29)

of PKK Rantau Baru, also said that those in her generation no longer want to be fisherwomen or work in agriculture due to natural disasters and regular flooding (interview with Rosa, secretary of PKK Rantau Baru, December 10, 2020). In general, we can observe a trend of declining interest among younger generations in both fishing and farming. However, this does not necessarily mean that the role of women in productive activities will decrease, because women (especially young women) can still work outside these sectors and outside the village, for example as factory workers, and contribute to the productive activities of the family in this way.

The management of household finances by men and women in Rantau Baru was further examined according to ethnicity. By cross-referencing the ethnic identity of the respondents, we can differentiate the responses of Malay and non-Malay (namely Javanese, Batak, Minangkabau, and Nias) respondents. The differentiation between Malay (*Melayu*) and non-Malay (non-*Melayu*) is noted by Simulie (2002, p. 13) in which Malay including Deli Malay (*Melayu Deli*) and Riau Malay (*Melayu Riau*), in contrast to other ethnicities, such as Javanese, Sundanese, Bugis, Minangkabau. The results, presented in Table 7.4, reveal that a similar percentage of Malay (58%) and non-Malay (65%) respondents report that the wife in the family manages the household's finances. This indicates that generally, both Malay and non-Malay families have similar experiences: the wife is primarily in charge of money management in the family, while the husband has the supporting role. A slightly higher percentage of Malay families (36%) say that both husband and wife control money in the family, while only 29% of non-Malay families say the same. From these results we can conclude that (1) generally, women play a significant role in money management in Malay and non-Malay families, (2) the higher number of Malay families jointly managing the household's money may indicate more gender-equal relations in these families.

7.3.4 Sociopolitical Activities

Sociopolitical activities in Rantau Baru include attending village, neighborhood association, and community association meetings. The participation of men and women in these activities is depicted in Table 7.5. The majority of respondents (64%) report that only men attend these public meetings, while 32% report that both men and women attend, and only 4% report that only women attend such meetings.

Table 7.5 Sociopolitical activities of men and women in Rantau Baru (Question 120)

Sociopolitical Activity	Women Only No. (%) of respondents	Men Only No. (%) of respondents	Men and Women No. (%) of respondents
Attending village, neighborhood association, or community association meetings	6 (4)	91 (64)	46 (32)

This result indicates that women's participation in sociopolitical activity in the village lags behind that of the men. This is a significant finding because important decision-making related to women's needs takes place at such meetings.[5]

This finding could be explained by the matrilineage system of the Malay Petalangan community of Rantau Baru, in which although authority within a lineage or sub-lineage is in the hands of a *mamak* (a chief) or a *kepala suku*, (a prominent respected man), women gain high respect by possessing land or an inheritance, including a *Sialang* tree. In this system, women have less authority in the public sphere, where the mother's brother (*ninik mamak* or *mamak*) and prominent respected men in the lineage represent the family's voices and needs in public meetings.

Another indicator of sociopolitical activity is participation in women's groups. There are two women's groups in Rantai Baru Village, namely the PKK and the Women's Farmers Group (*Kelompok Wanita Tani,* KWT). Although Rantau Baru is a village dominated by fishing, there is no women's group related to fisheries. There are only men's fishery groups that fisherwomen are allowed to join (see the example of Ana who is a member of such a group as explained in Sect. 3.2.1 above).

The PKK has five main organizers (*pengurus*) and 23 members in Rantau Baru village; the village also has one *Posyandu* (integrated service for children) with three main organizers (BRG 2019b, p. 51). The religious group *Wirid Yasin* for women has approximately 25 members, while *Wirid Yasin* for men has 23 members (BRG 2019b, p. 52). As seen in Fig. 7.6, of the 76 female survey respondents, 47 have never attended a PKK meeting. The small numbers of women who attend PKK meetings more frequently indicates that only PKK committee members take an active role.

According to Rosa (secretary of PKK Rantau Baru), PKK management consists of one leader (the wife of the PKK village head), four vice-leaders, two secretaries, and two treasurers, which were elected by a *musyawarah* (collective decision) attended by approximately 40 women from three sub-villages in Rantau Baru. Some of these attendees are from the KWT. The PKK receives funds from the Village fund (*dana desa*) of Rantau Baru (see Chap. 8 in this book). During the last *musrenbangdes* meeting that Rosa she attended, there was no discussion about the

---

[5]This finding contradicts Herawati et al. (2019)'s study of gender roles in peat-based communities in Riau Province, which found that both men and women contribute equally to the social life of the community and that women's participation and membership in groups is equal to men's.

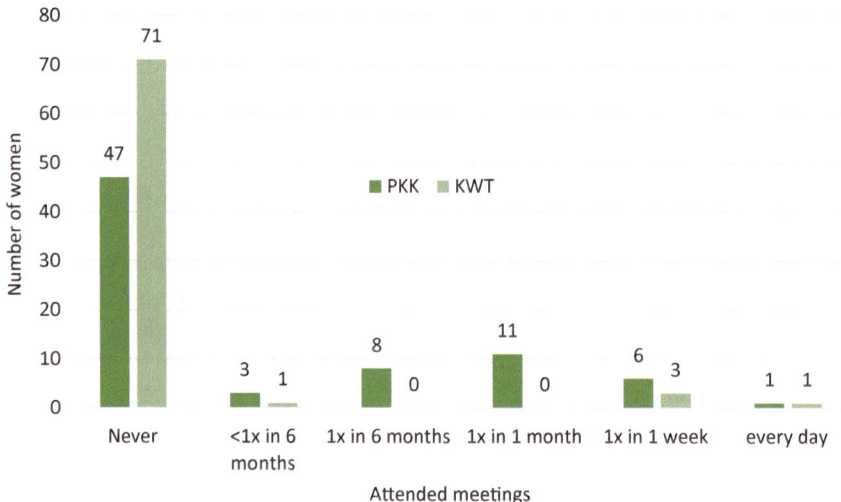

Fig. 7.6 Women's participation in women's group meetings in Rantau Baru Village

village's strategy for peatland restoration, although other institutions (private companies) have given it much attention and offered help in various ways, such as sharing pineapple seeds to be planted. There has so far not been any PKK meeting on peatland restoration (interview with Rosa, December 10, 2020).

My interview with Ibu Santi, the Malay Datuk Tuo head of *Pokja* 3 PKK Rantau Baru for the period 2020–2023, revealed that *Pokja* has approximately 20 women members, whose regular meeting is conducted once every 3 months. *Pokja* programs include greening (*penghijauan*) and planting herbs (such as *cabe, jahe, kunyit, kencur, daun kunyit*) by optimizing the land around the houses, and they encourage households in each neighborhood association (*Rukun Tetangga*, RT) to plant herbs around their houses for their own consumption. They plant in *tanah mineral*, or mineral soil, because they have no experience planting in peatland. *Pokja* 3 has never had a program to train their members to cultivate vegetables in peatland. According to Santi, no training is offered by *Pangkalan Kerinci District*; district officers only encourage villagers to cultivate herbs in mineral soil, not yet in peatland soil, and Pokja 3 and KWT have never worked together (interview with Santi, December 23, 2020).

Erna further explained that although she was also a member of PKK, the initial development of KWT did not involve PKK, because the PKK officers would not come to Dusun Seipebadaran, which is located rather far from the PKK officers' houses inside the village (interviews with Erna, September 1, 2019 and November 23, 2020). As the village head's wife generally leads PKK organizations, they generally have good access to resource support from the village government. The PKK organizations tend to be centralized and biased toward the interests of the village elite and the center of village administration.

Only five women respondents have ever attended a KWT meeting, with only 3 women attending once a week. In an interview with Erna, the head of KWT Rantau Baru, she explained that Agricultural Trainees (*penyuluh pertanian*) from the Agricultural Bureau of Pelalawan District created KWT in 2016 and she has been its head since then. While KWT was initially made up of 30 women, this number shrank to only five to ten women due to differing ideas among them, the challenges of cultivating in the peatlands, and exasperation with the floods, which often destroyed the KWT demonstration plots of chilies, long beans, and eggplant (interviews with Erna, September 1, 2019 and November 23, 2020).

Erna and her parents are originally from Rantau Baru (they called it *daerah bawah*), but they moved to Seipebadaran in 2013 and gradually learned to cultivate crops in peatlands, which often face failure. They moved to Seipebadaran because their former location often became flooded, which distressed their life, and her school-age son needed a more stable condition.

Some of KWT's initial program involved creating sustainable food houses (*pembentukan rumah pangan lestari*), such as chile nurseries, and creating *demplots* for chile and eggplant, for which they received assistance from the Agricultural Bureau of Pelalawan District, including seeds, equipment, and fertilizer in 2013; although, they no longer receive assistance. According to Erna, when the crops they plant are successful, they share them among the five women who are still active in KWT for family consumption and not for sale. Big floods are an ongoing challenge for KWT's planting. For example, of their efforts to cultivate chiles, long beans, eggplant, and kale (*kangkung*), only kale was successfully harvested, as the rest were flooded (interview with Erna November 23, 2020). The types of vegetables planted by Erna and her KWT friends are not very different from those grown on peatlands in Central Kalimantan (including mustard greens, kale, and cucumbers) and on peatlands in other areas, such as chilies, tomatoes, celery, leeks, long beans, corn, pineapple, and bananas (Harsono 2012, p. 31).

Erna explained that KWT members practice different planting techniques than those taught by Agricultural Trainee Officers, as the technique taught by the officers was not suitable for the surrounding peatlands. The women proceeded with their own method: They combined mineral soil (*tanah mineral*, bought from Pelalawan) and natural fertilizer (*pupuk kompos*), placed this mixture in the center of the plot, planted vegetable seeds (Chinese cabbage, chilies, peanuts, corn, eggplant, long beans) there, and surrounded this planted area with peatland soil. This technique was successful for growing mustard greens in the *demplot*, but in 2015, a big flood swept away the plants just before harvest. This exasperated the KWT members, who no longer wanted to do group gardening (interview with Erna, September 1, 2019).

During observation in 2019, Erna and four other members of KWT were planting peanuts (see Fig. 7.7). As peanut cultivation is simple, they do not need fertilizer, and they can be harvested within 3 months. The *demplot*, as well as the KWT members' individual initiatives to grow vegetables (chilies, peanuts, pumpkins) around their houses in polybags, are helpful in fulfilling their family's food needs. They often do not need to buy vegetables, especially chilies, when the prices are very expensive. As KWT only received assistance from Pelalawan District in 2013 and

Fig. 7.7 Rantau Baru KWT's *demplot* for peanut farming (Photo by Dewi, September 2019)

2016, Erna and the other four women in KWT run the farm—buy the seeds and manage the land—through *gotong-royong* (collective effort). She has encouraged women in Dusun Bawah to create a farming *demplot* like the one she made in Dusun Atas (Seipebadaran) (interviews with Erna, September 1, 2019 and November 23, 2020).

The farming technique used by Erna and her KWT friends on the peatlands is efficient and cheap. When the chili plants are exhausted, they can be replaced with other plants using the remaining mineral soil and disposing of the used peat soil. Additionally, seeds from previously harvested chilis can be planted for a following crop. Erna learned this this cultivation technique in a training on chile cultivation provided by Riau Province in 2017 and training on how to cultivate peatland by the BRG in Siak in 2018. Erna said that more training is needed for KWT members to learn more about the types of peatland in Rantau Baru, such as wet peatland (*gambut basah*) and dry peatland (*gambut kering*), and how to cultivate in the conditions of each type, as well as how to improve the economic conditions of their families by, for example, selling crafts (interviews with Erna, September 1 and November 23, 2020).

When studying gender and natural resources management, it is important to capture and present the voices, needs, and rights of women, which are often neglected in natural resource management policies and programs (Dewi et al. 2020). By listening to the voices of women farmers in Rantau Baru, this study discovered the potential of Women's Farmers Groups, such as KWT. Such groups apply local knowledge that is different from the techniques taught by agricultural officers and contribute to family food supplies. However, the study found that women in the village do not yet participate optimally in KWT. Therefore, more effort is needed to encourage women to participate in KWT and in socio political activities.

In addition, there is limited support for KWT from the village network, as no allocation from the village fund is used to support the activities of women farmers. The findings of this study indicate the need to provide more spaces for women farmers in the village to express, practice, and accommodate their locally based peatland management including via attending the village, neighbourhood association, or community association meetings. The local knowledge of women farmers in peatland cultivation is critical not only to ensuring full community participation in peatland management, but also represents a potential resilience that should be supported amidst the regular flooding of the Kampar River and the deforestation surrounding the village.

7.4 Conclusion

Based on observation, a household survey, and interviews in Rantau Baru Village, this study has four main findings. First, men were significantly more knowledgeable of peatlands than women. This was explained by the fact that the number of men who had participated in training or meetings for peatland conservation was higher than that of women, although there is initially no gender differentiation in family's socialization (for men and women) on peatland. This gender difference in peatland knowledge was not observed in Herawati et al. (2019) and therefore contributes a new insight.

Second, while men and women both contribute to "productive activities" in the village, either as farmers (by cultivating peatland) or fishers, peatland agricultural activities are dominated by men. According to the survey results, much of the peatland clearing, harvesting, and selling of peatland crops is done only by men and only a small percentage of women, demonstrating that men in the village play a more considerable role in peatland agricultural activity. This finding is consistent with Herawati et al. (2019), which also found that agricultural activity in peatland communities are dominated by men. However, this study finds that in the fishery sector, gender roles are more evenly distributed, with both women and men participating in the activities of fishing, processing, and selling and women dominating the activity of processing fish.

Third, women and men play complementary roles in "reproductive activities" of the household, namely childcare, while women have significant control over household finances. This latter finding may be explained by the matrilineal culture of many villagers, which places women in a position of high respect.

Fourth, in contrast to the above finding, women do not participate nearly as much as men in the public sphere of "sociopolitical activities," such as attending community, association, and village meetings. However, some women do belong to women's groups and some women farmers have developed their own techniques to produce food on peatlands based on local knowledge.

This chapter echoes Elmhirst and Resurreccion (2008, p. 5), who assert that "experiences of the environment are differentiated by gender," and thus "men and women hold gender-differentiated interests in natural resource management through their distinctive roles, responsibilities, and knowledge." Although men currently play the dominant role in peatland management in Rantau Baru Village, as part of the Peat Care Village Program, women have the potential to contribute to the restoration of peatland, especially through women's farmers groups, such as KWT. Any project or program on peatland restoration in Rantau Baru Village should recognize the basic features and differences of the gender roles and the specific needs of men and women in the village to ensure the optimal contribution of all community members to peatland management and restoration.

References

BRG (Badan Restorasi Gambut) (2018) Lembar pengesehan rencana tindakan tahunan restorasi gambut Provinsi Riau tahun 2019. BRG, Jakarta

BRG (2019a) Laporan tahunan restorasi gambut. BRG, Jakarta

BRG (2019b) Profil desa peduli gambut: Desa Rantau Baru Kecamatan Pangkalan Kerinci Kabupaten Pelalawan Provinsi Riau. BRG, Riau

Colfer CJP, Basnett BS, Elias M (eds) (2016) Gender and forests: climate change, tenure, value chains and emerging issues. Routledge, Oxon; New York

Dewi KH, Raharjo SNI, Desmiwati et al (2020) Roles and voices of farmers in the "special purpose" forest area in Indonesia: strengthening gender responsive policy. Asian J Women's Stud 26(4): 444–465. https://doi.org/10.1080/12259276.2020.1844972

Elmhirst R (1998) Reconciling feminist theory and gendered resource management in Indonesia. Area 30(3):225–235. https://doi.org/10.1111/j.1475-4762.1998.tb00067.x

Elmhirst R, Resurreccion BP (2008) Gender, environment and natural resource management: new dimensions, new debates. In: Resurreccion BP, Elmhirst R (eds) Gender and natural resource management: livelihoods, mobility and interventions. Earthscan, London; Sterling, pp 3–20

Elmhirst R, Siscawati M, Basnett BS et al (2017) Gender and generation in engagements with oil palm in East Kalimantan, Indonesia: insights from feminist political ecology. J Peasant Stud 44(6):1135–1157. https://doi.org/10.1080/03066150.2017.1337002

Engelhardt E, Rahmina (2011) Development of a gender concept for the forests and climate change programme (FORCLIME) in Indonesia. GIZ and FORCLIME, Jakarta

FAO (Food and Agriculture Organization of the United Nations) (2007) Gender policies for responsible fisheries: policies to support gender equity and livelihoods in small-scale fisheries. New directions in fisheries: a series of policy briefs on development issues, no 6. FAO, Rome. http://www.sflp.org/briefs/eng/policybriefs.html. Accessed 2 Apr 2021

FAO (2016) Promoting gender equality and women's empowerment in fisheries and aquaculture. FAO, Rome

Firth R (1946) Malay fishermen: their peasant economy. Kegan Paul, Trench, Trübner and Co., London

Harsono SS (2012) Mitigasi dan adaptasi kondisi lahan gambut di Indonesia dengan sistem pertanian berkelanjutan. Wacana Edn 27:11–37

Herawati T, Rohadi D, Rahmat M et al (2019) An exploration of gender equity in household: a case from a peatland-based community in Riau, Indonesia. Biodiversitas 20(3):853–861. https://doi.org/10.13057/biodiv/d200332

Indirastuti C (2020) Perempuan bertarung dengan api di lahan gambut: pengalaman perempuan desa di Provinsi Kalimantan Tengah dan Riau. J Perempuan 25(1):13–24

Mai YH, Mwangi E, Wan M (2011) Gender analysis in forestry research: looking back and thinking ahead. Int For Rev 13(2):245–258. https://doi.org/10.1505/146554811797406589

March C, Smyth I, Mukhopadhyay M (1999) A guide to gender-analysis frameworks. Oxfam GB, Oxford

Mizuno K, Fujita MS, Kawai S (eds) (2016) Catastrophe and regeneration in Indonesia's peatlands: ecology, economy and society, Kyoto CSEAS series on Asian Studies, vol 15. NUS Press; Kyoto University Press, Singapore; Kyoto

Mugniesyah SSM, Mizuno K (2007) Access to land in Sundanese community: a case study of upland peasant households in Kemang village, West Java, Indonesia. Southeast Asian Stud 44(4):519–544. https://doi.org/10.20495/tak.44.4_519

Muslim, Kurniawan S (2008) Fakta Hutan dan Kebakaran Riau 2002–2007: informasi atas perubahan hutan gambut/rawa gambut Riau, Sumatra – Indonesia. Jikalahari, Riau

Ogden LE (2017) Fisherwomen: the uncounted dimension in fisheries management. BioScience 67(2):111–117. https://doi.org/10.1093/biosci/biw165

Resurreccion BP (2008) Gender, legitimacy and patronage-driven participation: fisheries management in the Tonle Sap Great Lake, Cambodia. In: Resurreccion BP, Elmhirst R (eds) Gender and natural resource management: livelihoods, mobility and interventions. Earthscan, London; Sterling, pp 151–173

Safitri MA (2020) Membumikan ekofeminisme dalam restorasi gambut: kebijakan, aksi, dan tantangan. J Perempuan 25(1):1–12

Sajogyo P (1983) Peranan wanita dalam perkembangan masyarakat desa. Rajawali, Jakarta

Simulie HKRP (2002) Melayu dan Minangkabau bagaikan dua sisi mata uang. In: Bakry SY, Kasih MS (eds) Menelusuri jejak Melayu-Minangkabau. Yayasan Citra Budaya Indonesia, Padang

Stacey N, Gibson E, Loneragan NR et al (2019) Enhancing coastal livelihoods in Indonesia: an evaluation of recent initiatives on gender, women and sustainable livelihoods in small-scale fisheries. Marit Stud 18:359–371. https://doi.org/10.1007/s40152-019-00142-5

Subono NI, Pratiwi AM, Boangmanalu AG (2020) Aksi perempuan fasilitator desa dalam revitalisasi ekonomi kelompok perempuan di desa gambut: atudi kasus 3 desa di Kalimantan Tengah. J Perempuan 25(1):47–61

Sumartomjon M (ed) (2019) Badan Restorasi Gambut mulai melibatkan kaum perempuan untuk jaga lahan gambut. Kontan.co.id, 14 Feb. https://nasional.kontan.co.id/news/badan-restorasi-gambut-mulai-melibatkan-kaum-perempuan-untuk-jaga-lahan-gambut. Accessed 1 Apr 2021

UN ESCAP (United Nations, Economic and Social Commission for Asia and the Pacific) (2017) Gender, the environment and sustainable development in Asia and the Pacific. United Nations Publication, Bangkok

Villamor GB, Desrianti F, Akiefnawati R et al (2014) Gender influences decisions to change land use practices in the tropical forest margins of Jambi, Indonesia. Mitig Adapt Strateg Glob Chang 19:733–755. https://doi.org/10.1007/s11027-013-9478-7

Villamor GB, Akiefnawati R, van Noordwijk M et al (2015) Land use change and shifts in gender roles in Central Sumatra, Indonesia. Int For Rev 17(S4):61–75. https://doi.org/10.1505/146554815816086444

Wagey T (2018) Pengarusutamaan gender dalam tata kelola sumber daya alam: inisiasi kelompok perempuan dalam pengelolaan lahan gambut. In: Catatan Perjalanan Media Visit ke Lokasi Program ICCTF - UKCCU 2018. ICCTF, Jakarta

Watson E (2006) Gender and natural resources management: improving research practice. NRSP Brief, Hemel Hempstead

Weeratunge-Starkloff N, Pant J (2011) Gender and aquaculture: sharing the benefits equitably. Issues Brief 2011–32. WorldFish Center, Penang, p 12

Widiarti A, Hiyama C (2007) Prospek pelibatan perempuan dalam rehabilitasi hutan. In: Indriatmoko Y, Yuliani EL, Tarigan Y et al (eds) Dari desa ke desa: dinamika gender dan pengelolaan kekayaan alam. CIFOR, Jakarta, pp 83–91

World Bank (2018) Closing the gender gap in natural resource management programs in Mexico. World Bank, Washington, DC

World Bank, FAO, IFAD (International Fund for Agricultural Development) (2009) Gender in agriculture: sourcebook. World Bank, Washington, DC

WWF (World Wildlife Fund) (2012) Natural resource management and the importance of gender. WWF Briefing, Washington DC

Chapter 8
Village Initiatives for Fire Prevention and Peatland Restoration in Riau After the Enactment of the 2014 Village Law

Takuya Hasegawa

Abstract Since Indonesia enacted the Village Law of 2014, fiscal transfers from the central and district governments to villages have increased markedly. As a consequence, high expectations are placed on villages to initiate local-level approaches to peatland restoration. Inclusion of a wide range of community stakeholders in the processes of determining these initiatives is assumed to produce sustainable development outcomes. To analyze village initiatives and determine to what extent villages have earmarked parts of their budgets for environmental programs, this chapter examines the case of Pelalawan District in Riau Province. To examine how local communities have been involved in the process of planning such initiatives, it focuses on one village, Rantau Baru. Our study found that villages have started to plan and execute low-cost programs for environmental protection. However, budgeted programs for environmental protection accounted for only a tiny proportion of total village expenditure; therefore, these initiatives represent only small and gradual change. Moreover, power in decision-making processes tends to be limited to a few village officials. Our quantitative survey on participation in village development meetings also indicates that such meetings are dominated by local elites, to be more specific, by peatland owners and educated people. Given that Rantau Baru completely complies with the existing rules for community engagement, more innovative arrangements beyond existing regulations are needed to engage a wide range of actors in budget-making processes.

Keywords Village law · Peat Care Village Program (DPG) · Local-level solutions · Village development meeting (*Musrenbangdes*) · Community engagement · Village budgeting unit

T. Hasegawa (✉)
Asian Cultures Research Institute, Toyo University, Tokyo, Japan

© The Author(s) 2023

M. Okamoto et al. (eds.), *Local Governance of Peatland Restoration in Riau, Indonesia*, Global Environmental Studies,
https://doi.org/10.1007/978-981-99-0902-5_8

8.1 Introduction

Since Indonesia enacted the Village Law in 2014 (Law No.6/2014), fiscal transfers to villages have markedly increased. The central government launched the Village Fund (*Dana Desa*, DD) scheme to provide grants to all villages in the country. District governments, the second tier of local government,[1] were also obliged to significantly increase their existing grants to villages, which are known as Village Fund Allocations[2] (*Alokasi Dana Desa*, ADD). These newly increased fiscal resources provided villages with more discretion to plan and execute development projects. In utilizing their own budgets, villages gained the ability to pursue policies in various sectors, including environmental protection. On the restoration of peatland and fire prevention, as with many other issues, villages have been expected to take the initiative to reduce problems on their own.

Such expectations are presumably based on the following three assumptions. First, villagers live near plantations and forests and thus stand at the forefront of fighting forest fires associated with peatland degradation. Being among the most vulnerable to the adverse impacts of peatland degradation, villagers are expected to have a compelling need to respond to the problems (Wahyudi and Wicaksono 2020) and one of the villagers' demands is envisaged to be the sustainable use of peatlands (Hapsari et al. 2020, p. 3). Second, local communities are considered "more cognizant of the intricacies of local ecological processes and practices" than the state (Brosius et al. 1998, p. 158) and thus able to provide "local-level solutions derived from community initiatives" (Leach et al. 1999, p. 225). Third, when it comes to development planning, villages are considered more inclusive, or at least to have lower barriers to participation, than upper administration levels. One of the crucial arguments of participatory planning is that the inclusion of a wide range of community stakeholders in decision-making processes produces sustainable development outcomes (World Bank 2020, p. 12). All three assumptions point to local-level community engagement as the critical component to finding solutions to environmental issues and to achieving environmentally sustainable development.

Acknowledging the potential benefits of incorporating knowledge and opinions from villagers for peatland management, as well as the increasing importance of village budgets in terms of volume, the central government, international donors, and local governments have created various schemes to encourage villages to make

---

[1] Indonesia has a five-level administrative structure: central, provincial, district/municipality, sub-district, and village. There are two types of villages: *desa*, or "villages," and *kelurahan*, or "urban villages." While *desa* are given far-reaching autonomy, *kelurahan* remain under the firm authority of districts/municipalities. There are approximately 75,000 desa and 8400 *kelurahan* (as of 2019) in the country. This chapter does not discuss *kelurahan*, as they do not receive DD or ADD. Therefore, the villages referred to are *desa* in this chapter unless otherwise stated. This chapter also does not discuss municipalities where *desa* rarely exist.

[2] The central government established the ADD requirement with Government Regulation 72 of 2005. Despite this regulation, over 60 percent of districts only partly complied with the requirement or did not allocate any ADD at all until the enactment of the new Village Law (Antlöv et al. 2016).

development decisions for peatland restoration. The Peatland Restoration Agency (*Badan Restorasi Gambut*, BRG), the central government's unit in charge of peatland restoration between 2016 and 2020, was one of the most active institutions in offering such incentives. BGR has highly touted the achievements of its incentives, as will be discussed later.

Encouraged by various schemes, not a small number of villages have indeed taken initiatives to implement programs for peatland restoration. Village governments have become increasingly indispensable actors in peatland management. These initiatives have received scant attention in the study of peatland management, however. Many studies discuss the role of local communities (Mizuno et al. 2016), but rarely address village governments as key actors. While some studies discuss government policies, they focus on the role of central and district governments (Januar et al. 2021). These oversights may be attributable to the well-known infamy of village governance in Indonesia as ineffective, inefficient, and corrupt (Lewis 2015). Such biases may only emphasize the limitations of village-based peatland restoration initiatives. As an opening for further understanding of current peatland management, however, village-based initiatives merit further analysis.

In order to fill this void in the study of peatland management, this chapter discusses how village governments have utilized their own budgets for peatland restoration and fire prevention since the enactment of the new Village Law. It then examines to what extent local communities have been involved in the planning of such environmental protection projects, with the aim of developing discussion of the sustainability of these projects.

To examine the above points, this study adopts a case study approach and selects the case of Pelalawan in Riau Province. Pelalawan is in an area prone to forest fires and, as will be discussed later, is one of the pioneering districts attempting a new approach to encourage villages to implement environmental protection projects. To analyze community involvement, this chapter discusses the case of Rantau Baru village in Pelalawan, where our research team conducted a questionnaire survey of all households in the village on various issues, including their perceptions of peatlands and level of participation in village governance.

This chapter is organized as follows. Following an overview of the increased budgetary resources of villages and surrounding institutional arrangements[3] to encourage community involvement in development planning under the current village law system (Sect. 8.2), Sect. 8.3 discusses the schemes and regulations enacted at different administration levels to encourage villages to utilize their budgets for peatland restoration. Section 8.4 analyzes the expenditures of village governments on environmental projects in Pelalawan District in comparison to the highly promoted achievements of the BRG program's targeted villages. Section 8.5 examines whether Rantau Baru village communities are involved in the planning

---

[3] Following the classic work of new institutionalism, this chapter takes the view that "institutions are the rules of the game in a society or, more formally, are the humanly devised constraints that shape human interaction (North 1990, p. 1)."

process for environmental protection and identifies three interrelated challenges the village faces in incorporating the opinions of a wide range of community members. Finally, this chapter concludes by considering some implications of this analysis for future arrangements encouraging community-based peatland restoration programs.

8.2 Overview of Institutional Arrangements Under the New Village Law

8.2.1 Increased Village Budgets

The Village Law of 2014 has established two major revenue resources for village budgets: DD from the central government and ADD from district governments. The central government launched the DD program in 2015 and allocated IDR 21 trillion to approximately 74,000 villages. The annual amount of DD has continually increased from IDR 47 trillion in 2016 to IDR 60 trillion in 2017 and IDR 70 trillion in 2019.[4]

Parallel to DD, ADD to villages has also increased considerably. The new Village Law obliges district governments to provide ADD, which amounts to at least 10 percent of the national balance transfers they receive (after deducting special allocation grants[5]). Within a few years of this newly imposed obligation, almost all districts began allocating the required amount of ADD.[6]

Together DD and ADD account for approximately 80 percent of total village revenue (World Bank 2020). From these two resources alone, on average each village receives IDR 1.3 billion annually (2017–2019). Before the enactment of the new Village Law, because of limited budgets, many villages spent little on development programs, with most of their budgets spent on administrative costs, such as the salaries of village officials. In contrast, under the new arrangement, more than half of village budgets are spent on infrastructure (World Bank 2020). This suggests that, whether in the form of infrastructure, agricultural training, or others, villages have sufficient revenues to fund environmental protection measures, if they choose to do so.

Thus, as a result of newly introduced DD and increased ADD, the prerequisite to act on restoring peatland and fire prevention was upheld for villages. Village heads and officials, in consultation with village councils, decide how villages spend their budgets. The extent to which local communities are involved in such decisions is another important topic.

---

[4]The amount of DD increased again in 2021 to IDR 72 trillion.

[5]The national balance transfers to districts consist of three elements: general allocation grants, special allocation grants, and revenue sharing.

[6]Interview with the employee from the Ministry of Home Affairs (MoHA) in charge of guiding local governments on ADD payments, January 6, 2021.

8.2.2 Arrangements to Encourage Community Involvement

The new Village Law emphasizes participatory planning and institutionalized several new mechanisms to facilitate community engagement. The first of these is the village forum, or *Musdes* (*Musyawarah Desa*) in Indonesian, in which participants discuss a village budget plan. Villages are required to hold *musdes* in order to receive DD and ADD. They are organized by village councils each year around June (MoHA Regulation 114 of 2014).

Indonesia has had a similar village forum called *Musrenbangdes* (*Musyawarah Pembangunan Desa*, or village development planning meeting) in place since the 1980s. This pre-existing forum is organized by village officials each year around January (Law 25 of 2004 on National Development Plan). Previously, discussion in *Musrenbangdes* centered on which development programs village officials wanted the district government to accommodate in the district budget. After enforcement of the Village Law, participants of *Musrenbangdes* also began discussing village budgets.

The *Musrenbangdes* and *Musdes* thus overlap in many ways, even though they are organized by different parties at different times of the year. Nonetheless, at least twice a year, local communities can learn about village budgets and convey requests to their village government. Based on the outcome of discussions in *Musrenbangdes* and *Musdes*, each village government makes an annual village development plan called RKPDes, which eventually becomes a budget, called APBDes.[7]

The second new mechanism established by the Village Law is a requirement to organize a unit in charge of drafting a budget.[8] Village governments are required to include a few community representatives in this unit, which must comprise 7, 9, or 11 members.[9] The secretary of the unit should come from the community empowerment organization, known as LPM (*Lembaga Pemberdayaan Masyarakat*) (MoHA Regulation 114 of 2014). While the unit is headed by the village secretary and most members are village officials, one or two more community representatives other than LPM representatives are added to the unit in some villages.[10]

Despite these changes established by the Village Law, some studies question the "quality of participation" (e.g., Wijaya and Ishihara 2018). Examining *Musrenbangdes,* Damayanti and Syarifuddin (2020) indicate that community participation in village governance remains minimal, and only community leaders who

---

[7]The RKPDes is the general development plan that village governments must prepare every year. Based on this plan, village governments develop a draft budget. With approval of the village council, a village budget (APBDes) is finalized.

[8]The formal name is *tim penyusun* RKPDes (drafting RKPDes team).

[9]Villages can determine the number of the drafting budget unit members in consultation with the subdistrict office, unless it is stipulated by the district head regulation.

[10]Article 33 of the MoHA Regulation 114 of 2014 does not oblige village governments to include additional community representatives in the budget drafting unit other than the representative from the LPM.

Table 8.1 Regulations and schemes to incentivize village programs for peatland restoration

Relevant actor	Regulation and/or scheme	Intention
Ministry of Villages (MoV)	MoV regulation on the annual priorities of DD spending	Ensure that villages spend DD for peatland restoration.
BRG	Peat Care Village program	Dispatch a facilitator to the targeted villages to provide advice on village budget spending.
(Riau) province	*Bankeu* (funds transferred from provinces to villages)	Provide guidance on how to use *Bankeu* for peatland restoration.
(Pelalawan) district	Forestry ADD (a variation of ADD)	Regulate that forestry ADD is used for environmental programs.

tend to support the village government are engaged effectively. They charge that village fora are only formalities at which participants convey requests of programs in a one-way dialog and lack of transparency regarding the acceptance or rejection of their requests is noticeable.

Although one can easily find anecdotal accounts to support such arguments, it is necessary to examine community engagement yet again, given its significance to the sustainability of village development programs. The chapter will revisit this topic in Sect. 8.5.

8.3 Encouraging Villages to Initiate Environmental Protection

Recognizing the potential of dramatically increasing village budgets, central and local governments generally encourage villages to create different kinds of programs for environmental protection, peatland restoration, and fire prevention. The main promoters are the Ministry of Villages, Disadvantaged Regions, and Transmigration (MoV), BRG, provincial governments, and district governments (see Table 8.1).

8.3.1 MoV

MoV, which supervises the spending of DD, has gradually revised its regulations to encourage villages to spend DD funds on peatland restoration, in part due to its consultation with BRG. Two years after the launch of DD, MoV included peatland restoration for the first time in its list of annual priorities for DD funds[11] (MoV Regulation 22 of 2016). In seminars co-organized by BRG to explain this regulation,

---

[11] Past regulations referred to infrastructure and developing people's capacity to contribute to environmental protection, but not directly to peatland restoration.

MoV clearly sent the message that villages could spend DD to build infrastructure for peatland restoration.[12]

The Ministry's 2019 priorities referred specifically to peatland restoration (MoV Regulation 11 of 2019). This document stated that forest fire prevention justified new infrastructure for environmental protection, as well as the purchase of fire extinguishing tools and new programs to increase community capacities for fire prevention. In naming these priority items, the MoV directly facilitated villages to spend DD on peatland restoration.

8.3.2   BRG

On another front, BRG initiated the Peat Care Village Program (*Desa Peduli Gambut*, DPG) to encourage villages to launch programs for peatland restoration. In the 5 years leading to 2020, BRG used its own budget to build more than 7000 canal-blocking facilities and 15,000 deep wells across the country[13] (BRG 2021). Nevertheless, it had an eye on the ample funds in the hands of village governments available to complement its efforts and designed the DPG program for this purpose.

The DPG program began in 75 villages in 2017. The initial plan used the national government's budget to target 300 villages during 4 years. Gaining support from international donors and private companies at the halfway point, the program eventually covered 640 villages. The main feature of the DPG program involved dispatching a facilitator to live in each village for 6–10 months. By interacting daily with the community, the facilitators were expected to persuade villagers to become "peatland friendly." One of the key tasks was to advise and encourage government officials and other actors to include programs for peatland restoration in the village budget planning process[14] (BRG 2017).

8.3.3   Provincial Governments

Provinces provide financial grants called *Bankeu* (*bantuan keuangan*) to villages. In Indonesia's local governance system, provincial governments do not directly supervise villages, and are not obliged to provide villages with financial grants. The amount of *Bankeu* therefore varies among provinces and from year to year.

---

[12] One of the seminars was held in Jambi Province on November 5, 2016 (Media Indonesia 2016).

[13] In Riau Province, BRG built 1620 canal blocking facilities and 1419 deep wells in 5 years.

[14] In addition to providing advice on village budgeting, the facilitators handled a wide variety of tasks, including advising on participatory mapping and establishing environmentally friendly village-owned enterprises (BUMDes).

Provincial governments can encourage villages to restore peatland by providing guidance on how to use *Bankeu*. In the case of Riau Province, IDR 200 million was allocated to each village in its jurisdiction in 2019, with the stipulation that villages could use 10% of the allocation for disaster prevention, specifically to build guard stations for fire prevention organizations (known as MPAs[15]) and to purchase monitoring equipment for fire prevention, such as GPS devices, rubber rafts, transceivers, and binoculars.[16]

However, in Riau, this direction only lasted 1 year. In 2020, the amount of *Bankeu* was halved, and no stipulation was included regarding fire prevention. Although in many cases commitment from provinces was pliable, provinces occasionally play a role in guiding villages toward peatland restoration, as in the case of Riau.

8.3.4 District Governments

Districts can be key actors because of their supervisory role over villages in budget execution. Among more than 400 districts in Indonesia, Pelalawan District in Riau Province pioneered an innovative scheme to encourage villages to spend on environmental protection. In 2017, in addition to the conventional ADD, which is distributed almost evenly among the district's villages, Pelalawan introduced a specific ADD to be distributed proportionally according to the extent of a village's business development in oil and natural gas operations (hereafter referred to as oil) and forestry.[17] This arrangement was made in response to village dissatisfaction with taxes from the oil and forest industries going directly to the central and district government, with no increases to village revenues. These specific ADD were named the oil ADD and the forestry ADD.

Pelalawan District requires the forestry ADD to be used for environmental programs. The guidelines stipulate 11 valid uses of the forestry ADD (District Head Regulation 7 of 2017); two of these are directly related to peatland restoration and forest fire prevention. Others include community forest development, afforestation, protection of water resources, cultivation of terraced rice fields, and river clearing.[18]

---

[15]MPA is a village organization that engages in fire prevention activities and was first stipulated in a 2009 regulation by the Ministry of Forestry.

[16]Technical guidelines on *Bankeu* in 2019 issued by the Riau Provincial Office of Empowering Villagers and Villages.

[17]Pelalawan District categorized all the district's 104 villages into three groups and distributed specific ADD proportionally according to each group's classification (Hadi et al. 2017). In contrast, the conventional allocation of ADD has not made a significant difference among villages because only 10–40% of conventional ADD has been distributed proportionally based on population, land area, and the number of poor people.

[18]The 11 expense items are as follows: developing community forest (*hutan desa*); forest fire prevention; protecting forest and land; afforestation; cultivating terraced rice fields; forest

This allocation scheme is groundbreaking in institutionalizing arrangements to guarantee that a certain amount of money from each village budget is spent on environmental protection annually.

The introduction of this scheme owed much to the lobbying of Fitra Riau, a non-governmental organization (NGO) in Riau. Supported by international donors,[19] Fitra Riau was inspired by a scheme in Bojonegoro District in East Java Province, which introduced a specific ADD of oil for the first time in Indonesia. Fitra Riau delivered tenacious presentations on this subject to several districts in Riau beginning in 2014, eventually attracting Pelalawan to its proposal. Pelalawan district government and Fitra Riau even enhanced Bojonegoro's original scheme by introducing the forestry ADD for environmental protection. [20]

Inspired by this initiative of Pelalawan District, other districts have since attempted to develop their own allocation schemes with various approaches. In 2019, Jayapura District in Papua Province introduced a scheme called "the index of ecology" to examine the extent of villages' environmental protections.[21] The district allocated part of the ADD to villages based on this index to incentivize environmental improvements. Of the innovative district attempts discussed above, this chapter chooses Pelalawan as a case study.

8.4 Analysis of Village Budget Spending for Environmental Protection

Encouraged by various actors to spend funds on peatland restoration and fire prevention, how have villages changed their spending so far? Since there are few studies on this topic, BRG's report on its DPG program is valuable. Before analyzing Pelalawan's case, it is helpful to examine this report to make a comparison.

8.4.1 BRG's Analysis of Village Budget Allocations

Although BRG did not regularly collect detailed data on village budgets, in 2020 (its final year) it presented an analysis of the DPG program to the media (Tempo.co.

---

management; protecting water resources; river and stream clearing; peatland restoration; and building infrastructure for environment protection within villages' authority.

[19] The Ford Foundation supported Fitra Riau for 2 years from 2014, and the Asian Development Bank (with financial and technical aid from the United Kingdom) provided support for four additional years.

[20] Online interview with Triono Hadi, the executive director of Fitra Riau, April 30, 2021.

[21] Following Jayapura's lead, Siak (Riau), Nunukan (North Kalimantan), Kubu Raya (West Kalimantan), and Bener Meriah (Aceh) introduced the same scheme (Fitra Riau 2020).

2020). According to this analysis, 143 out of 525 villages covered by the DPG program (until that time) had allocated some of their budget for peatland restoration in fiscal year 2020. The total allocation of the 143 villages amounted to IDR 9 billion.[22] In detail, IDR 5 billion was allocated for building and managing infrastructure, such as canal-blocking facilities and deep wells; IDR 2.2 billion was allocated for improving capacity of community associations; IDR 1.4 billion was allocated for MPA activities; and IDR 0.5 billion was allocated for pilot farms of peatland-friendly crops. This published data may give hope for improved peatland management in the future.

Supporting this data, BRG presented several villages for their demonstration of good practices. Especially regarding canal blocking, BRG frequently promoted Tri Mandayan Village in West Kalimantan Province. The BRG facilitator of the DPG program in this village helped villagers build several canal-blocking facilities, with a portion of the cost covered by the village budget. In 2018, BRG sent this facilitator to the Katowice Climate Change Conference in Poland (COP24) to convey the achievements of the Indonesian government.[23]

There are two points to consider when analyzing BRG's data, however. First, BRG used data from village annual development plans (RKPDes) and not from final budgets (APBDes). Frequently, items budgeted in the RKPDes are dropped from the APBDes in the course of consultation with village councils. Second, BRG analyzed data of the 2020 fiscal year. To alleviate the economic damage from the COVID-19 pandemic, however, in that year, many initially budgeted expenditures were diverted to support urgent new needs, such as cash handouts for the poor.

Thus, there are good reasons noted above to be wary of accepting BRG's claims at face value. At the least, analysis of more detailed data is needed, but BRG has not yet released it. Programs for building and managing infrastructure should be examined with particular care. Given that there have been abundant funds from the central government and international donors to support villages to build such infrastructure as canal-blocking facilities and deep wells, it is reasonable to assume that villages usually do not bother to use their own budgets for this kind of expenditure. As will be described below, our analysis of Pelalawan supports this view. Unfortunately, the famous Tri Mandayan Village case trumpeted by BRG as a central achievement seems likely to be rather the exception than the rule.

---

[22] BRG claimed that IDR 16 billion was the total amount of the 143 villages' budgeted items and not instead of IDR 9 billion, but the former number includes items not directly related to peatland restoration, such as investment in BUMDes.

[23] Interview with the head of BRG, March 23, 2019.

8.4.2   Analysis of Village Budget Allocations in Pelalawan District

The following subsection discusses the allocation of village budgets on environmental protection in Pelalawan District. The district is a pioneer of the innovative ADD forestry scheme. Our analysis uses budget (APBDes) data from all 104 villages in the district between 2016 and 2019.[24] The author collaborated with Fitra Riau to analyze the data.[25]

Looking at the general trend of village budgets in Pelalawan, total revenue per village in 2016 averaged IDR 1.5 billion, rising steadily to an average of IDR 2.2 billion in 2019. Compared to the national average of IDR 1.4 billion per village in 2019, the villages in Pelalawan, a district rich in natural resources, were financially comfortable. As for revenue resources, DD accounted for 47% of total revenue and ADD for 44% (2016–2019). From 2012 to 2014, prior to enactment of the Village Law, the primary available revenue resource was ADD; it amounted to only IDR 0.4 billion per village, most of which was spent on administrative expenses. From 2015 onwards, villages spent an average of 30 percent of their budget on administrative expenses and 70% on development programs. In terms of fiscal structure, villages in Pelalawan have had considerable leeway to implement environmental programs in this period.

There are two points to be noted in this analysis. First, all programs related to agriculture, including purchase of farming equipment and training for farmers, are considered environmental programs.[26] The area of Pelalawan is primarily covered with peat soils, and as BRG strongly advocated, revegetation is essential for peatland restoration. Given the increasing pressure from the international community and the central government to improve peatland management, we here assume that most of the government agricultural programs take an environment-friendly approach, even though some specific programs may degrade the environment. Second, programs to build deep wells are not listed in this analysis because of the difficulty in the dataset in differentiating wells for drinking water and wells for rewetting peatland.[27] In any case, according to district government officials, villages rarely spend their budgets on building deep wells for rewetting peatland.[28]

---

[24] Data of the settlement of accounts is useful complementary data to verify whether villages implemented budgeted programs or not. However, because of limited availability, the settlement of accounts was not examined.

[25] Hadi Triono, the executive director of Fitra Riau, contributed greatly to this analysis.

[26] On the budget documents, it is impossible to differentiate the agriculture programs that aim to protect the environment from those that do not.

[27] As an example, in 2018, budgeted programs to build wells and install plumbing, which are not considered environmental programs in this analysis, amounted to IDR 2.7 billion.

[28] Online interview with the subdistrict head of Teluk Meranti and employees of the Pangkalan Kerinci Subdistrict office, May 25, 2021.

Table 8.2 Village environmental expenditures in Pelalawan District (2016–2019)

		2016	2017	2018	2019	
Expenditure for environment protection and fire prevention (million Rp.)		0	705	965	1839	
					773	1065
Financial source			DD and ADD	DD and ADD	DD and ADD	Provincial grants
Percentage of total village expenditures		0%	0.4%	0.5%	0.8%	
Total amount allocated (million Rp.) (% of total environment-related expenditures)	MPA activities	0 (0%)	422 (60%)	458 (48%)	557 (30%)	
	Planting trees, seedlings, flowers	0	135 (19%)	161 (17%)	156 (8%)	
	Agricultural training and education	0	53 (8%)	49 (5%)	24 (1%)	
	Guard station or devices for disaster countermeasures	0	0	18 (2%)	984 (54%)	

Source: Village expenditure data from 2016–2019 in Pelalawan District (modified by the author)

As Table 8.2 shows, villages did not allocate any of their budgets to environmental programs in 2016. They began to modestly budget for these programs in 2017 and continually increased this spending thereafter. The abrupt increase in funding environmental programs in 2017 was likely due to the introduction of the forestry ADD.

The forestry ADD is an extremely small part of total ADD[29] (conventional ADD plus the forestry ADD plus the oil ADD), however. During 2017–2019, it accounted only for an average of 1.5%. On average, each village received IDR 13 million from the forestry ADD per year. Furthermore, it should be noted that the total amount budgeted for environmental programs amounted to only approximately 60% of the allocated forestry ADD (2017–2018). This suggests that many villages did not comply with the district's guidance, partly because of the hesitance of district officials to strongly urge villages to follow the rules regarding the forestry ADD.[30] As a result, budgeted expenditures for environmental programs remained at an average of 0.6% of total expenditures.

Examining expenditures in detail, the figure that stands out is for MPA activities, which comprised 60% of total environmental expenditures in 2017. While this percentage gradually decreased over time, the total amount of expenditures for this item remained the same. The amount of funds provided for planting trees, seedlings, and flowers, which includes planting trees in the community forest and flowers for

---

[29] In contrast, the total amount of ADD of oil is IDR 18 billion annually, accounting for 19 percent of total ADD.

[30] Several village officials at the subdistrict office, which directly supervises village budgets, admitted this hesitancy in an online interview with the subdistrict head of Teluk Meranti and employees in the Pangkalan Kerinci Subdistrict office, May 25, 2021.

beautifying living areas, comprised an average of 15% of total environmental expenditures each year. Funds dedicated to agricultural training were small in comparison to the two items above, but again, training was also consistently funded each year. As for guard stations and devices for disaster countermeasures, spending increased considerably in 2019, in large part due to the guidelines provided by the province (*Bankeu*).

No villages have used their budgets to build canal-blocking facilities in Pelalawan. This is in line with the trend in Riau Province, where only a few villages did so.[31] The same applies to the building of deep wells for rewetting peatland, as discussed above.

Examining individual cases, the five villages in the district with the biggest spending on environmental protection budgeted an average of IDR 45 million per year. On the other hand, 14 villages in the district allocated zero funds for environmental protection during the 4 years of analysis. There is no indication that villages located in areas prone to forest fires tend to budget any more for fire prevention and peatland restoration than villages in other areas. For example, in Kuala Kampar Subdistrict, where most of the area is prone to forest fires, only three out of all eight villages budgeted for environmental programs in 2017 and 2018.

Village-based initiatives for environmental protection in Pelalawan therefore seem less impressive that those described in the self-promoting BRG report touting the DPG program. The change triggered by Pelalawan's new allocation scheme deserves attention, however. Mainly due to the introduction of the forestry ADD, villages gradually began to earmark certain funds for environmental programs. While these program budgets are modest in total amount, their significance should not be underestimated. Among the programs for MPA activities, one village has continually secured a sufficient budget not just for emergency operational costs in the case of forest fire, but also for daily fire monitoring activities.[32] Among programs for planting trees, seedlings, and flowers, one village secured a budget for daily activities of the organization to manage a community forest.[33] In this way, by virtue of the district's incentive, some villages became better able to address environmental issues by using their budgets.

Still, it is fair to underscore again that total village expenditures for environmental programs remained considerably small in Pelalawan. Even in the pioneering districts such as Pelalawan, the patterns of spending village budgets have changed only gradually toward environmental protection.

---

[31] Bagan Libur Village in Meranti District is known for building canal-blocking facilities with its budget, but this case is unique, because the village head is a former activist of an NGO working on environmental protection.

[32] This is the case of Pulau Muda Village. Interview with Triono Hadi, April 30, 2021.

[33] This is the case of Segamai Village. Interview with Triono Hadi, April 30, 2021.

Table 8.3 Environmental protection projects in the Rantau Baru budget (2017–2020)

Description of project	Budget allocation (million Rp.)	Fiscal year
Operating expenses of MPA	7	2017
Building guard station for community forest	18	2018
Seedlings for environment-friendly agriculture	42	2019
Devices for disaster countermeasures (i.e. GPS)	20	2019
Portable and lightweight fire pump (mini-striker)[a]	85	2020
Total	172	

Source: Expenditure data of Rantau Baru Village from 2017–2020 (modified by the author)
[a] Village office shifted this project to others in the supplementary budget

8.5 Case Study of Rantau Baru Village

8.5.1 Environmental Programs in Rantau Baru

Of the 104 villages in the district, Rantau Baru ranked eighth in the amount of spending for environmental programs from 2017 to 2019, and therefore could be considered environmentally conscious. Table 8.3, however, which shows budgeted items in detail, indicates there was no funding for environmental protection in the village for more than 1 year. For example, a program for MPA activities, the main budgeted item for environmental protection in Pelalawan, was only funded in 2017. Moreover, each of the projects is relatively standard and has not yet acquired a reputation from villagers for being effective in protecting the environment.

Meanwhile, Rantau Baru has received plenty of support for peatland restoration and fire prevention from external agencies, such as private companies and the central government. A global company, Asian Agri of the Royal Golden Eagle Group, which operates plantation businesses in the village, has consistently offered corporate social responsibility (CSR) programs.[34] The company implemented its *Desa Bebas Api* (the Free Fire Village Program) from 2016, providing an annual reward of IDR 100 million to villages in which no forest fires occurred for 1 year.[35] Using this grant, which it has received almost every year, Rantau Baru Village purchased a motorbike for patrols, a diesel fire pump, and, in 2020, three small fishing boats with outboard motors. The village planned to generate profits for MPA activities by renting out these fishing boats, establishing a village-owned enterprise (BUMDes)

---

[34] As well as providing CSR programs every year, the company often provides one-time help to Rantau Baru. For example, in 2020, it provided pineapple seedlings to be planted in the newly cultivated lands behind the village office.

[35] This program initially targeted nine villages, including Rantau Baru, in its first year. Asian Agri then expanded to 16 target villages in 2017 and 28 in 2021. It should be noted that another group company of Royal Golden Eagle, APRIL, is also implementing the same name CSR program in Riau Province, but the details of the support are different. While the name of Asian Agri's program is promoted in Indonesian, that of APRIL's program is promoted in English.

to this end. Asian Agri also strongly supports the activities of fire prevention patrols by hiring a villager for daily patrol. The chief of MPA assumed this position, thus becoming the only MPA member to receive a monthly salary for fire prevention activities, while all other MPA members serve as volunteers.[36]

BRG has also implemented several programs in Rantau Baru. The DPG program targeted the village in 2019, and a facilitator was dispatched to stay for 6 months. In the same year, BRG built 13 canal-blocking facilities and 50 deep wells, temporarily employing a group of villagers for each project. In the following year, BRG bore the operational costs for local management and maintenance of the facilities. This abundance of external support may partially explain why Rantau Baru Village has not yet launched consistent and innovative programs for environmental protection using the village budget.

8.5.2 Participatory Planning for Environmental Programs

It is important to examine to what extent communities were involved in decision-making process regarding spending for environmental protection measures and acceptance of programs to be implemented by external agencies, and to what extent implemented measures and programs reflect villagers' actual wishes. In terms of the institutional arrangements that encourage community engagement, Rantau Baru Village complies completely with regulations issued by the central government. It holds open fora on village development twice a year: *Musrenbangdes* around January and *Musdes* around September. The 11-person budget-drafting unit includes three community representatives. In addition to welcoming a representative from the LPM as the secretary of the unit, as stipulated by the MoHA regulations, the village government invited two additional community representatives to this unit: one from PKK, a nationwide women's group, and one from the youth community.

Beyond simply abiding by regulations regarding participatory planning, Rantau Baru shows a relatively high level of community participation, as evidenced in the responses to our questionnaire survey conducted in 2020.[37] It asked how frequently villagers interacted with key players in the village government. The same question was asked to villagers on Java Island in a survey conducted by Nishimura Kenichi's

---

[36]This created jealousy among MPA members, resulting in the MPA becoming almost dormant around the year of 2020. After the MPA proved itself incompetent in managing the fishing boats donated by Asian Agri, the boats were eventually handed over to the head of hamlets. Online interview with the chief of MPA and the crew leader of patrolling for fire prevention, January 24, 2020.

[37]We conducted a face-to-face questionnaire survey of all households in Rantau Baru Village in 2020. This survey data is analysed in various chapters in this book.

Table 8.4 Frequency of contact with key players in the village government in Rantau Baru and on Java Island (percent of total respondents)

		Not at all (%)	Less than once in a half year (%)	Once in a half year (%)	Once in a month (%)	Once in a week (%)	Every day (%)
Chief of hamlet	Rantau Baru	2	0	0	1	8	89
	Java Island	8	5	6	20	26	34
Village office staff	Rantau Baru	5	0	1	12	27	54
	Java Island	14	13	13	23	20	17
Village head	Rantau Baru	7	7	21	40	18	8
	Java Island	16	14	15	26	17	11
Village council members	Rantau Baru	9	1	1	9	19	61
	Java Island	45	10	10	14	8	6

Note: The total number of responses to the survey in Rantau Baru was 152, while the number of responses to the survey in Java Island was 3384
Source of data: Data of 2018 local governance survey (Java Island) and 2020 questionnaire survey (Rantau Baru)

team at Osaka University during May–June 2018.[38] Table 8.4 presents a comparison of survey responses in the two sites, illustrating the closeness of villagers with government officials in Rantau Baru. The Rantau Baru figures differ markedly from those from the survey in Java, particularly regarding contact with village council members.

Our survey also asked villagers how often they attended annual village development planning meetings. In exact terms, the survey asked how frequently villagers had attended *Musrengbangdes* during the past 6 years. To avoid confusion, our survey did not ask about *Musdes*. As with the previous question, the same question was asked to villagers on Java Island. As shown in Table 8.5, residents of Rantau Baru village attend *Musrenbades* more frequently than those of Java Island. In Rantau Baru village, approximately 30 percent of the villagers attend the forum at least once every 2 years.

---

[38] From May to June 2018, this academic team collaborated with the Indonesia Survey Institute (LSI) to conduct a face-to-face survey of 3420 randomly chosen people on Java Island. The survey selected 30 people from each district and city on the island. Some authors of this book, Takuya Hasegawa, Okamoto Masaaki, and Wahyu Prasetiawan, were also members of this project, which was named the "2018 Opinion Survey on Local Governance in Indonesia." A summary of the survey's results was published in Japanese (Kobayashi et al. 2019)

Table 8.5 Frequency of attending village development meetings among Rantau Baru and Java Island villagers (percent of total respondents)

	Not at all (%)	1–2 times in 6 years (%)	Once every 2 years (%)	4–5 times in 6 years (%)	Every year (%)
Rantau Baru (n = 152)	58	11	4	4	23
Villagers on Java Island (n = 3384)	79	3	2	2	13

Source of data: 2018 local governance survey (Java Island) and 2020 questionnaire survey (Rantau Baru)

Table 8.6 Correlation between peatland ownership and frequency of attending village development meetings in Rantau Baru

		Frequency of attending village development meetings					
		Not at all	1–2 times in 6 years	Once every 2 years	4–5 times in 6 years	Every year	Total
Own a piece of land on peat soil	Yes	26 (37.1%)	7 (10.0%)	5 (7.1%)	5 (7.1%)	27 (38.6%)	70 (100%)
	No	47 (75.8%)	7 (11.3%)	1 (1.6%)	1 (1.6%)	6 (9.7%)	62 (100%)
	Total	73 (55.3%)	14 (10.6%)	6 (4.5%)	6 (4.5%)	33 (25.0%)	132 (100%)

Cramer's $V = 0.429$, Non-respondent $= 20$

These differences presumably result partly from Rantau Baru's smaller population size and the compactness of the residential area. Rantau Baru has a population of only around 1100 people, smaller than in Riau Province, in which village populations average around 3900, and much smaller than on Java Island, where the average is around 5500. Community engagement is also likely facilitated by the relatively dense and long-established residential settlement alongside the Kampar River. A new residential area was developed around the village office building after 2005, but since it is located far from the river and most villagers' livelihoods traditionally depend on fishing, most local people, including village officials and community leaders, still prefer to live near the river. It is assumed that such living conditions facilitate a communicative environment among villagers and lower the barriers for villagers to engage in the village development meeting, *Musrenbangdes*.

In addition to population and settlement patterns, cross-tabulation analysis indicates that two other variables positively correlate with increased participation in *Musrenbangdes*. The first variable is the ownership of peatland. The cross tabulation in Table 8.6 indicates a statistically significant positive correlation between ownership of peatland and frequency of participation in *Musrenbangdes* (Cramer's $V = 0.429$).

It stands to reason that peatland owners, whose properties are vulnerable to threats of forest fire, have increased incentive to demand budgetary allocation for environmental programs. On the other hand, peatland owners can be categorized as village elites, and thus the correlation indicated above may merely reflect the fact that the meetings are dominated by elite. It should be noted that most households in this village have received 3 ha of peatland from the village office in the past 15 years, but many households sold their peatland for various economic reasons.[39] Therefore "peatland owners" in our survey are those who could afford to keep the peatlands until the day of our questionnaire survey.

The second variable affecting public participation is level of education. To simplify the cross tabulation of this analysis, categories regarding the frequency of attendance in *Musrenbangdes* were reduced to two groups: people who attended at least once every 2 years and those who attended less frequently than that. Table 8.7 presents this cross tabulation, indicating a statistically significant positive correlation between educational background and the frequency of attending *Musrenbangdes* (Cramer's $V = 0.329$). Although the correlation is less significant than in the first variable analysis, it does appear people with higher levels of education are more interested in attending development planning meetings.

In Chap. 9 of this book, Prasetyawan indicates that educated people tend to be more concerned with peatland degradation, and therefore more actively demand environmental protection programs. But as with ownership of peatlands, those with higher education levels can be categorized as village elites, and the correlation indicated in Table 8.7 may merely reflect elite dominance.

As will be discussed later, according to interviews with villagers, villagers rarely call for environment protection programs in development planning meetings. Instead, elite dominance in public meetings —to be more specific, dominance of peatland owners and educated people—seems like a more plausible interpretation regarding the two correlations discussed above. This dominance may hinder a wide range of villagers from expressing their requests in community fora, and therefore appears to explain why the relatively high participation rate of villagers in development meetings in Rantau Baru has not yet resulted in positive impacts on environmental protection.

8.5.3 Problems in Participatory Planning in Rantau Baru Village

After analyzing the results of the survey, our team conducted online interviews with 15 villagers between November 2020 and January 2021. In these interviews, the author focused on examining the quality of community engagement in the decision-making process and tried to determine why environment protection projects in Rantau Baru did not have significant local impacts and lacked continuity. Interviews identified three closely linked barriers to community engagement.

---

[39] The village office divided areas of peatland and provided 1 ha to each household in the village in 2004 and 2 ha in 2012. See Chap. 3 (Osawa and Binawan).

Table 8.7 Correlation between educational background and the frequency of attending village development meetings in Rantau Baru

		Educational background					
		Never went to school	Elementary school	Lower secondary school	Upper secondary school	Tertiary school	Total
Frequency of attending village development meetings	Once every 2 years or more	0 (0%)	21 (45.7%)	10 (21.7%)	14 (30.4%)	1 (2.2%)	46 (100%)
	Less than once every 2 years	11 (10.6%)	61 (58.7%)	20 (19.2%)	11 (10.6%)	1 (1.0%)	104 (100%)
	Total	11 (7.3%)	82 (54.7%)	30 (20.0%)	25 (16.7%)	2 (1.3%)	150 (100%)

Cramer's $V = 0.329$, Non respondent $= 2$

First, the budget drafting unit of Rantau Baru village, which is supposed to act as a mechanism for receiving requests from community members under the new village law, seemed not to function successfully. According to a section chief of PKK, her section never thought of conveying budgetary requests to the group's representative in the drafting unit when it planned to buy seedlings of peatland-friendly crops. Abandoning a plan to get funds from village budget, her section eventually decided to cover the costs on its own.[40] Thus, despite the arrangement to include community representatives in the budget drafting unit, these representatives rarely heard the opinions of community groups on whose behalf they were supposed to serve.

The second barrier is the difficulty of disseminating information regarding district policies to village residents. Surprisingly, none of the villagers interviewed knew about the specific ADD of forestry. These included two seasoned village officials in charge of drafting budgets and two village council members.[41] It should be noted that forestry ADD expenditures are stipulated annually in the district head's regulation on guidelines of spending ADD, one of the most basic regulations that all persons engaging in drafting village budget should ideally read through. Lack of knowledge about forestry ADD within the village government seems to suggest that dissemination of information from the district on budget regulations might be targeted to a select few in the village office. Even though the ADD scheme is innovative in encouraging villages to launch environment protection programs, its effectivity is limited since its very existence is not known by all village government officials.

Third, top-down decision-making is notable in village decisions regarding receipt of external support for peatland restoration. Generally, villages do not need to hold public fora to hear villager request for external support, as they must do in the process of budget planning. Ideally, the decisions regarding external support should be in line with the six-year village development plan (RPJMDes) and annual development plan (RKPDes), both of which are made through the development planning meetings (*Musrenbangdes*), in which communities are encouraged to participate.[42] However, in the case of Rantau Baru, it seems that the village office did not consider RPJMDes or RKPDes in these decisions. Usually, only the village head and village secretary have the final word on whether a certain external project is accepted or not.

This top-down decision-making process is clearly illustrated by the adoption of BRG programs.[43] In the interview survey introduced above, no senior village

---

[40] Online interview with the chief section of PKK, December 23, 2021. The plenary meeting of PKK elected its representative to the budgeting team, who is an ordinary member of PKK.

[41] Online interviews with the secretary of the village office, January 27, 2021; a senior official in charge of development and planning, December 18, 2020; a senior official and member of the budget drafting team, December 2, 2020; the secretary of the village council, December 4, 2020; and a council member, December 4, 2020.

[42] As in the process of planning RKPDes, villages must also hold *Musrenbangdes* for RPJMDes.

[43] As discussed before in this section, Rantau Baru accepted a BRG program to build 13 canal-blocking facilities and 50 deep wells in 2019. In addition to the process to accept BRG programs,

officials or council members could explain how the village decided to accept the BRG program to build infrastructure for rewetting peatland: construction of canal-blocking facilities and deep wells. Each of these individuals claimed that only the village head determined whether this program is significant to village development or not.[44] Since such infrastructure, especially canal blocking facilities, have been known to be scapegoated as a cause of floods and drag on village economies, it is important to achieve some consensus among villagers before they are implemented. The RPJMDes and RKPDes do provide good opportunity to discuss and reach consensus on all kinds of assistance programs. However, neither Rantau Baru's six-year development plan nor annual development plan mentioned construction of these peatland rewetting facilities at all. Although our questionnaire survey showed that villagers were not generally against those facilities,[45] the village office apparently skipped the community involvement process regarding the BRG program.

Thus, even in villages like Rantau Baru, where the village office complies with the regulations of central government to facilitate participatory planning, only a limited number of persons have been aware of important budgetary regulations and these few persons have tended to dominate the decision-making process, often making decisions without regard for what has been discussed in the development meetings. In this way, a wide range of community members have not yet been empowered to wield influence in the decision-making process regarding environmental policies and programs.

8.6 Conclusive Remarks

Enactment of Indonesia's 2014 Village Law placed much expectation on the villages to initiate solutions to peatland degradation and forest fire problems. In Pelalawan, the innovative forestry ADD scheme had an immediate effect on villages, as the spending for environmental programs in village budgets rose from zero to a certain amount. Although very few villages have spent their budgets on building infrastructure to rewet peatland, villages have begun to take initiative to launch various programs of modest cost, such as those for MPA activities, agriculture training, and management of community forests. Starting with such small programs may lead

---

the decision to recommend the chief of MPA to Asian Agri to be hired for patrols also follows the same pattern. According to the chief of MPA, this was due to the village head pushing for this decision. Online interview with the chief of MPA, January 24, 2020.

[44] Other than the village head, the village secretary might be another person to know about this decision-making process. But at the time of the interview, the former village secretary had passed away a year before.

[45] According to our questionnaire survey, 89% of villagers agree with building canal-blocking facilities and deep wells, and only 4% disagree. As to accessing the information about building them, at least 60% obtained information on these facilities once a year.

to local-level solutions for environmental issues, which attaches a positive value to the ample funding the villages have received.

However, given that budgeted programs for environmental protection never exceeded 1% of total village expenditure as in Table 8.4, it is fair to say that these village initiatives represent only gradual and small change. Since Pelalalwan is a pioneering district in encouraging its villages to spend part of their budgets on environmental protection, the progress in other districts is likely to be slower, although the situation may vary in other provinces. The disheartening progress in Riau Province casts a doubtful eye on the much-publicized data of the BRG on the achievements of the DPG program. It calls for further careful analysis to determine whether BRG used lenient criteria when identifying spending on peatland restoration.

Whether it is encouraged by this touted achievement of the DPG program or not, the central government has decided to continue this program, recently retitled as the Self-sufficient Peat Care Village (DMPG). The successor organization of BRG, the Peatland and Mangrove Restoration Agency (BRGM), will manage this program. The program was listed in the national five-year development plan from 2020 to 2025 and is to cover 675 villages for 5 years. This program ensures that village initiatives will receive a certain level of attention in years to come.

This study indicates that, to date, however, Indonesia still faces daunting challenges in the endeavor to engage and empower local communities in the restoration and sustainable management of peatland. This study highlights some of the difficulties in engaging a wide range of villagers in the processes of determining village initiatives for peatland restoration. Our quantitative survey on participation in village development meetings, complemented by follow-up interviews with villagers, indicates that development meetings are dominated by elites, specifically, peatland owners and people with higher levels of education. Examination of decision-making processes reveals that power and information regarding budget-making tends to be limited to a few village officials. Although Rantau Baru enjoys a higher rate of villager participation in development meetings than villages on Java Island, as a result of elite dominance, community engagement is not so different from other villages.

This picture raises concerns about the sustainability of village initiatives for peatland restoration, particularly since one of the key assumptions inspiring these initiatives is that the inclusion of a wide range of actors in the decision-making processes will lead to sustainable development outcomes. Our analysis also raises questions about the appropriateness of local-level solutions, as these initiatives may come from a less diverse point of view.

Given that Rantau Baru complies with the existing rules for community engagement and yet has not made substantive progress in environmental programs, more innovative arrangements beyond the existing regulations are needed to guarantee the engagement of a wide range of actors in budget-making processes. It is widely believed that people in general are reluctant to participate in development meetings because they reduce time available for livelihood activities. New policy schemes may therefore need to incorporate arrangements to compensate participants for participation. Such arrangements must also be in line with local practices and contexts. In any case, new arrangements to empower a wide range of people in development meetings are key to achieving community-based sustainable peat restoration.

References

Antlöv H, Wetterberg A, Dharmawan L (2016) Village governance, community life, and the 2014 village law in Indonesia. Bulle Indones Econ Stud 52(2):161–183. https://doi.org/10.1080/00074918.2015.1129047

BRG (Badan Restorasi Gambut) (2017) Pedoman pelaksanaan: program desa peduli gambut. BRG, Jakarta

BRG (2021) Gambut basah adalah anugerah: 5 tahun kerja BRG RI. BRG, Jakarta

Brosius JP, Tsing AL, Zerner C (1998) Representing communities: histories and politics of community-based natural resource management. Soc Nat Resour 11(2):157–168. https://doi.org/10.1080/08941929809381069

Damayanti RA, Syarifuddin S (2020) The inclusiveness of community participation in village development planning in Indonesia. Dev Pract 30(5):624–634. https://doi.org/10.1080/09614524.2020.1752151

Riau F (2020) Transfer anggaran kabupaten berbasis ekologi (TAKE): model insentif kinerja kampung mendukung kebijakan Siak Hijau. Fitra Riau, Pekanbaru

Hadi T, Usman TA et al (2017) Pengelolaan alokasi dana desa berbasis sumber daya alam: inisiatif Kabupaten Pelalawan-Riau mendorong redistribusi anggaran desa untuk perbaikan tata kelola hutan dan lahan. Fitra Riau, Pekanbaru

Hapsari N, Widiastomo T, Ardila J et al (2020) Towards sustainable and productive management of Indonesian peatlands: case studies of Indonesia's peat restoration agency (Badan Restorasi Gambut) interventions in Siak, Riau and Pulang Pisau. Central Kalimantan. Earth Innovation Institute, Berkeley

Januar R, Sari ENN, Putra S (2021) Dynamics of local governance: the case of peatland restoration in Central Kalimantan, Indonesia. Land Use Policy 102:art 105270. https://doi.org/10.1016/j.landusepol.2020.105270

Kobayashi J, Okamoto M, Hasegawa T et al (2019) 2018-nen Indoneshia no chiho jichi ishiki chosa (2018 opinion survey on local governance in Indonesia). Hogaku Zasshi 65(3–4):323–375. https://doi.org/10.24544/ocu.20210108-001

Leach M, Mearns R, Scoones I (1999) Environmental entitlements: dynamics and institutions in community-based natural resource management. World Dev 27(2):225–247. https://doi.org/10.1016/S0305-750X(98)00141-7

Lewis B (2015) Decentralising to villages in Indonesia: money (and other) mistakes. Public Adm Dev 35(5):347–359. https://doi.org/10.1002/pad.1741

Media Indonesia (2016) Dana desa dukung restorasi gambut. Media Indonesia, 7 Nov. https://mediaindonesia.com/nusantara/76003/dana-desa-dukung-restorasi-gambut. Accessed 16 Aug 2022

Mizuno K, Fujita MS, Kawai S (eds) (2016) Catastrophe and regeneration in Indonesia's peatlands: ecology, economy and society. Kyoto CSEAS series on Asian studies, vol 15. NUS Press, Singapore; Kyoto University Press, Kyoto

North DC (1990) Institutions, institutional change and economic performance. Cambridge University Press, Cambridge. https://doi.org/10.1017/CBO9780511808678

Tempo.co. (2020) Desa-desa alokasikan Rp.9 Miliar dalam APBDes untuk cegah Karhutra. Tempo.
    co., 29 June. https://nasional.tempo.co/read/1359322/desa-desa-alokasikan-rp-9-miliar-dalam-
    apbdes-untuk-cegah-karhutla. Accessed 16 Aug 2022
Wahyudi R, Wicaksono RL (2020) Policy forum: village fund for REDD+ in Indonesia: lessons
    learned from policy making process at subnational level. For Policy Econ 119:art 102274.
    https://doi.org/10.1016/j.forpol.2020.102274
Wijaya YA, Ishihara K (2018) The evolution of community-driven development policy and
    community preferences for rural development after the enactment of village law 6/2014: a
    case study of Indragiri Hulu regency, Riau Province, Indonesia. Seisaku Kagaku 25(2):45–69.
    https://doi.org/10.34382/00005203
World Bank (2020) Indonesian village governance under the new village law (2015–2018): sentinel
    villages report. World Bank, Jakarta

Chapter 9
Willingness to Pay for Environmental Conservation of Peat and Aquatic Ecosystems in a Cash-Poor Community: A Riau Case Study

Wahyu Prasetyawan

Abstract This study assesses the Willingness to Pay (WTP) for environmental conservation in a community with limited incomes that is experiencing peatland degradation. Rantau Baru village is a unique environment in which both aquatic and peatland ecosystems are equally dominant, but community members mainly depend on the freshwater ecosystem for livelihoods and income. Based on survey data from 152 households, the study uses a contingent valuation method (CVM) to measure how villagers value each ecosystem and an ordinary least square (OLS) method to measure the significance of factors influencing WTP for conservation of peatland and fishing areas. It finds that WTP for conserving fishing areas is closely associated with household expenditures, while WTP for conservation of peatlands is associated with education and weakly associated with household expenditures. Community members' WTP for the conservation of both peatlands and fishing areas is very much associated with their perception of these environments.

Keywords Willingness to pay (WTP) · Environmental conservation · Peatland · Poverty · Household economic condition

9.1 Introduction

Peatlands in Riau play an important role in the provision of ecosystem services, such as carbon storage, climate regulation, biodiversity, and water supply. However, forest fires that take place every year in the Riau peatlands have transformed them from carbon sequesters to carbon emitters, which has led to serious environmental and social problems. The aquatic ecosystem of Riau also provides vital services, yet oil palm plantations and human settlement are challenging the health of the province's rivers, streams, and oxbow lakes. The village of Rantau Baru, located on the

W. Prasetyawan (✉)
Syarif Hidayatullah Islamic State University, Tangerang Selatan, Banten, Indonesia

M. Okamoto et al. (eds.), *Local Governance of Peatland Restoration in Riau, Indonesia*, Global Environmental Studies,
https://doi.org/10.1007/978-981-99-0902-5_9

Kampar River, is unique in that these two ecosystems are equally dominant. While the village is rich in biodiversity and it remains an important carbon-stock area, ongoing environmental degradation poses a serious threat. How can environmental conservation be achieved, and who will pay for it?

This chapter assesses the factors influencing community members' willingness to pay (or WTP) for environmental conservation in Rantau Baru. These factors include: (1) household expenditures, (2) wealth, (3) education level, and (4) perceptions of the two distinct ecosystems. Willingness to pay (WTP) is mainly defined as the "sum of money the individual would be willing to pay rather than to do without an increase in some good such as an environmental amenity" (Freeman III et al. 2014, p. 9). Because such willingness is context-specific, numerous studies have analyzed WTP to determine significant influencing factors. Among these factors, income level is often considered critical (Vincent et al. 2014).

According to survey results from the 152 households in Rantau Baru, villagers primarily depend on the freshwater ecosystem for their livelihood, as most of their income-earning activities are related to fishing in the Kampar River, its tributaries, and local oxbow lakes (Nakagawa, Chap. 4; Nofrizal et al., Chap. 5). Indeed, nearly 70% of villagers are small-scale fishermen/fisherwomen who have a range of skills related to living in the freshwater environment,[1] and fish diversity is high (Elvyra et al. 2010). This environment is facing serious challenges, however, not only due to human activity, but also to rainy season floods. Approximately 50% of community members have a plot of land in the peatlands, but livelihood activity on this land is limited (Osawa, Chap. 6).[2]

In Riau, peatlands are complex socio-ecological landscapes characterized by various interests, conflicting types of resource usage, and overlapping land tenure claims (Mizuno et al. 2016). Exploitation of the peatlands began in the 1980s with the establishment of oil palm plantations in Riau. This precipitated a range of direct and indirect impacts on Rantau Baru, chief among which is the annual forest fires and resulting air pollution that remain an unresolved problem today. Both the peatland and freshwater ecosystems of Rantau Baru can be considered common resources (see Miller et al. 2020; Cosens 2018) and the environmental challenges to these common resources have directly affected the livelihood of the community members.

Rantau Baru is not only facing environmental degradation; it is also a community with limited cash income and low educational attainment. The majority of villagers can be categorized as living in poverty. Average expenditures per household per month are approximately Rp2,112,651 (US$150).[3] Considering an average household has four members,[4] per capita expenditure is Rp528,163 (US$37) per month. This is less than the per capita poverty line in Riau in 2020, which was about

---

[1] Survey result.

[2] Survey result.

[3] Survey result. In 2020 the average exchange rate of 1 US dollar to Rupiah was 14,105.

[4] According to the BPS (n.d.), the average number of family members in a household in Riau is 4.

Rp546,090 (US$39) per month. Therefore on average the members of this community are living below the poverty line. More than half of all heads of household (54%) have only a primary level education.[5]

Those living in Rantau Baru have interacted with the two distinct ecosystems of their homeland for generations, learning and adapting to environmental conditions to create specific opportunities and benefits from each. Perceptions of each ecosystem vary accordingly.

This chapter presents results of study based on a survey on the WTP for the conservation of peatlands and fishing areas in Rantau Baru. The study uses a contingent valuation method (CVM) to measure how villagers value each ecosystem and an ordinary least square (OLS) method to measure the significance of factors influencing WTP. The policy implications of the findings are then discussed in the context of conserving both peatland and aquatic ecosystems. This research deepens our understanding of context-specific factors that impact WTP and can inform the formulation of strategies for environmental conservation in cash-poor communities.

9.2 WTP and Poverty

The concept of WTP is derived from economics and employed in the economic valuation of environmental goods and services. The sum of money an individual is willing to pay for environmental conservation instead of retaining it to spend on other things is critical to WTP. Therefore, income either at an individual or household level can constrain WTP.

Scholarship is divided on the relationship between WTP and income level. While some argue that the relationship is not very clear, others assert that there is a strong connection between WTP and economic conditions, both in general and at a personal level. A group of studies using surveys to understand WTP for biodiversity conservation of domestic populations of specific countries have been largely unsuccessful in detecting a significant income effect (see Jacobsen and Hanley 2009; Lindhjem and Tuan 2012). In a similar vein, a few cross-country studies have considered the effect of national income on the creation of protected areas (Bimonte 2002; Dietz and Adger 2003), but findings on the significance of the effect have been mixed.

Holden and Shiferaw (2002, p. 91), however, find that poverty, as an indicator of low income, undermines conservation investment on land even when the community members are fully aware of the need for conservation. Other scholars have demonstrated that demand for nature conservation increases as income increases (Baumol and Oates 1979; Kahn and Matsusaka 1997; Diekmann and Franzen 1999; Franzen and Meyer 2010). Other studies suggest that public support for environmental protection tends to increase when the economy is doing well and tends to weaken during economic recessions (Scruggs and Benegal 2012; Bechtel and Scheve 2013;

---

[5] Survey result.

Kachi et al. 2015). A case study from Malaysia evaluates the relationship between rising household income to increased household WTP and forest protection actions by the government, thereby presenting explanations for the under-provision of forest protection relative to household income (Vincent et al. 2014). Several studies also find that despite a relatively strong connection between WTP and income, there may be diminishing marginal utility for environmental protection (McConnell 1997; Israel and Levinson 2004).

While this chapter confirms a connection between WTP and the income of community members, it argues that perceptions of ecosystems, which people rely on in their daily lives, also impact WTP. In this sense, WTP in a village that depends upon two different ecosystems is likely to be different compared to that of a village with only one environmental condition.

9.3   Methodology and Data

The WTP analyses for this study adopt the contingent valuation method (CVM), which is frequently used to estimate WTP (Cho et al. 2008). Conceived in the field of economics, CVM is used to value goods that are not traded in the market. The CVM was first employed by Davis (1963) to estimate the value that hunters and tourists place on wilderness area(s). The CVM's advantage is that it allows researchers to estimate values that are not directly linked to pragmatic use; for example, the wish of community members to conserve natural environments for future generations (Kopp 2005). CVM uses surveys "to obtain consumer responses to a hypothetical situation," with questions aiming to elicit preferences for public goods and paying attention to what respondents would be willing to pay for specified improvements to such goods (Mitchell and Carson 2005, p. 2).

Data for this study was collected by carrying out surveys of all 152 households in Rantau Baru village. The survey was conducted face-to-face with the heads of households. Two questions were asked about WTP:

1. *Jika ada program pengelolaan dan pelestarian lahan gambut di Rantau Baru yang dilaksanakan masyarakat agar hasilnya dinikmati Ibu/Bapak dan biayanya ditanggung bersama, maka seberapa besar Ibu/Bapak bersedia membayar iuran program tersebut per bulan?* (If there is a program for conservation of peatland in Rantau Baru which is carried out by the community for your benefit, and the cost will be shared, how much do you want to pay for the program monthly?)
2. *Jika ada program pelestarian sungai dan danau di Rantau Baru yang dilaksanakan masyarakat untuk keberlangsungan hasil tangkapan ikan agar hasilnya dinikmati Ibu/Bapak dan biayanya ditanggung bersama, maka seberapa besar Ibu/Bapak bersedia membayar iuran untuk program tersebut per bulan?* (If there is a program for river and small lakes conservation in Rantau Baru carried out by the community for your benefit, and the cost will be shared, how much do you want to pay for the program monthly?)

To determine WTP, an open-ended question was initially employed. Upon failing to collect clear information from this method, the study adopted ordinal categories to assign a specific value to payments. These ranged from less than Rp9,999 (less than US$1) to more than Rp50,000 (about US$5). A mean of the results was calculated and transformed into a natural log. Other than these ranges in terms of money, the choice for WTP was: want to pay/do not want to pay. The estimation of WTP then excluded community members who do not want to pay at all.

The WTP model used in this study follows Seller et al. (1985) and Ndebele and Forgie (2017) as:

$$WTP = f\ (Y, Z)$$

where Y is the expenditure variable and Z represents explanatory variables. Explanatory variables include socio-economic background, such as education level, boat ownership, and house ownership (see Zhongmin et al. 2003; Amirnejad et al. 2006; Lienhoop and Macmillan 2007; Kopnina 2012; Forlin and Chambers 2011), and perceptions of ecosystem conditions (Liu et al. 2021). The study adopted an ordinary least squares (OLS) estimation as follows:

$$WTP = \beta_0 + \beta_1 \text{ expenditure} + \beta_2 \text{ education} + \beta_3 \text{ boat ownership}$$
$$+ B_4 \text{ house ownership} + \beta_5 \text{perception} + \varepsilon$$

9.3.1 Dependent and Independent Variables

Table 9.1 depicts the variables of this study and the descriptive statistics are explained in Table 9.2. WTP is the dependent variable. The independent variables are: household expenditures, boat ownership, house ownership, education level, perception of peatland conditions, and perception of fishing area conditions. These independent variables were adopted to gain more information about the individual characteristics of the community members. Expenditure (Decancq and

Table 9.1 Variables

Variable	Description
Ln WTP peatland	Ln of average WTP for peatland
Ln WTP fishing areas	Ln of average WTP for fishing areas
Ln expenditures	Ln of household expenditures
Education	Level of education in years
Boat ownership	Yes = 1, no = 0
House ownership	Yes = 1, no = 0
Perception of peatland conditions	Good = 1, other = 0
Perception of fishing area conditions	Good = 1, other = 0

Table 9.2 Descriptive statistics

Variables	Mean	SD	Min	Max
Ln WTP peatland	9.2273	1.1596	0	11.2252
Ln WTP fishing areas	9.2673	0.8523	8.51699	11.2252
Ln expenditures	14.372	0.6663	12.2061	15.8949
Education (years)	7.2632	3.2485	1	16
Boat ownership	0.9342	0.2487	0	1
House ownership	0.7763	0.4180	0	1
Perception of peatland conditions	0.2763	0.4486	0	1
Perception of fishing area conditions	0.1513	0.35954	0	1

Lugo 2012), instead of income, was used to measure wellbeing, because community members can more easily remember their spending. Information on spending therefore provides a more accurate representation of reality as compared to income. Information on incomes is difficult to collect because community members may have more than one source of income and may be hesitant to disclose income. The study measured wealth with the use of two indicators: boat and house ownership. As boats are critical to the main livelihood of fishing, boat ownership, as opposed to rental, is a key indicator of wealth and well-being. House ownership is also a strong indicator of the economic situation of a household.

To better understand attitudes and their link to behavior, the study also measured (1) educational level and (2) perceptions of the current condition of each ecosystem. The latter relies on the hypothesis that if respondents consider that a particular ecosystem is damaged, the probability of them participating in its conservation will be higher.

This study also uses supplementary information about perceptions of peatland conditions. This information was collected by asking the following two questions:

1. *Apa penilaian Ibu/Bapak mengenai keadaan gambut dan lingkunganya di Rantau Baru. Apakah baik, mulai mengalami kerusakan atau rusak?* (What is your evaluation of the condition of peatlands and their surrounding environment in Rantau Baru, are they in good condition, beginning to be damaged, or heavily damaged?)
2. *Sebutkan alasan mengapa Ibu/Bapak memilih jawaban di atas* (State the reason why you have given the answer above).

The answers were then categorized and the results are presented in Tables 9.5 and 9.6.

9.4 Findings and Discussion

9.4.1 Valuing Ecosystems in Rantau Baru

Both the fishing areas and the peatlands of Rantau Baru hold economic value. To calculate this total value, we added the estimated direct use value and indirect use

Table 9.3 Estimated economic value of Rantau Baru's environment

Areas		Estimated value in US$/Year	Percentage of total value
Direct use	Fishing areas	470,487	21.1
	Recreational areas	38,897	1.7
Indirect use	Carbon sequestration areas	1,720,119	77.2
Total		2,229,503	

Source of data: survey, field observation, and Sugardiman and Rovani 2015

value of three main areas, those used for fishing, recreation, and carbon sequestration. The results are depicted in Table 9.3.[6] Information about the value of each ecosystem in Rantau Baru is additionally applied to estimate ecosystem services. This information highlights that Rantau Baru is an important area for carbon storage and is therefore necessary to protect. The continuing expansion of oil palm cultivation would put this area of carbon stock in jeopardy, especially as it encroaches on the secondary forests.

The main service provided by the environment of Rantau Baru is carbon sequestration, which accounts for more than 77% of the environment's total estimated economic value. Direct-use value from fishing and recreational areas accounted for just under 23% of the total value, although fishing activity certainly has relevant economic value. Rantau Baru has a rich aquatic ecosystem, with waterways covering 460 hectares. Environmental services are primarily provided by secondary forests, industrial forests, peatlands, and plantations. These peatland and aquatic ecosystems form Rantau Baru's environment.

9.4.2   Significance of the Factors Influencing WTP for Conservation of Peatland and Fishing Areas

Table 9.4 presents the correlation between five independent variables and villagers' WTP for conservation of peatland and fishing areas. We see from Column 2 of Table 9.4 that the WTP for peatland conservation is positively associated with education levels, with a significance level of 0.1%. A coefficient of 0.08 indicates that education level is associated with a 0.08-point increase in WTP for peatland conservation. This finding on the education variable and its association to WTP was expected. We can preliminarily conclude that formal education serves as a foundation for community members to understand and accept information provided by the government on peatland, mainly on peatland fires. While the survey finds that only 8% of community members attended specific socialization or training activities on peatland conservation, they gain knowledge and information from being exposed to

---

[6] See Appendix 9.1 for the detailed calculation of the estimation of the carbon sequestration area.

Table 9.4 WTP for conservation of peatland and fishing areas in Rantau Baru according to influencing factors[a]

Variable	Peatland Areas	Fishing Areas
Ln household expenditure	0.26577[+] (0.15925)	0.40373** (0.11848)
Education level	0.08719** (0.03282)	0.02827 (0.02405)
Boat ownership	0.39959 (0.41926)	0.24238 (0.31063)
House ownership	−0.06855 (0.24126)	0.04841 (0.17562)
Perception of peatland conditions	−0.38785[+] (0.22012)	
Perception of fishing areas conditions		0.01010 (0.19494)
Constant	4.53754[+] (2.32013)	2.97614[+] (1.72635)
Adjusted R-squared	0.08378	0.07721
Sample size	128	128

Significance codes: 0 '***' 0.001 '**' 0.01 '*' 0.05 '+' 0.1.
[a]See Appendix 9.2 for the assumptions of the OLS employed in these calculations

fire or haze. Formal education attainment can be viewed as supplementary to this experience. Members of the community with higher education levels may therefore have more enhanced understanding of the value of conservation. Following the rise in fires in Riau 2015, the central government, followed by the provincial and district-level governments, adopted dramatic measures to reduce the incidents of fire. Perhaps the program related most closely to the understanding of the peatland problem in Riau is that of the volunteer firefighters. The relevance of this volunteer activity is not only to fight peatland fires, but, importantly, to provide general knowledge about protecting this environment.

The WTP for peatland conservation also correlates to household expenditure and perceptions of peatland conditions, and these variables are both significant at 10%. Household expenditure correlates positively to WTP for peatland protection, with a coefficient of 0.26. The significance level for household expenditure is at 10%, and there is a weak association between WTP and household expenditure. A coefficient of 0.26 for the variable indicates that household expenditure is associated with a 0.26-point increase in WTP for protecting peatland. Perception of peatland conditions also correlates positively to WTP, although the association is not as strong as education level. It was expected that if peatland is perceived to be in a bad condition, WTP would increase. We found that WTP is weakly negatively associated with attitudes toward peatland protection. This result is not in line with the expectation. The effect of this variable is statistically significant at 10%. This finding indicates that WTP decreases with the perception that peatland conditions are bad. This result may be explained by the feeling among most community members that peatland conservation is beyond their capability, despite 70% of villagers assessing that the

Table 9.5 Assessment of Peatland Conditions by Rantau Baru Villagers

No.	Assessment of peatland conditions	Percentage of total population
1	Relatively good	28
2	Beginning to be damaged	36
3	Heavily damaged	34
4	Do not know	2

Source of data: Survey

Table 9.6 Reasons for Peatland Damage as Reported by Rantau Baru Villagers

No.	Reason for damage	Percentage of total population
1	Flood or fire	58
2	Farming	14
3	Less fertile	11
4	Others[a]	8
5	Do not know	10

[a] "others" includes logging and plantations
Source of data: Survey

peatlands are damaged (see Table 9.5). According to the survey results in Table 9.6, peatland damage is mostly caused by floods or fires. Community members have seen fires and floods on peatlands in their surrounding areas, which are owned by corporations, and report the cause of fires as stray cigarette butts of outsiders.

The WTP for conservation of fishing areas positively correlates with household expenditure. The effect of this variable is statistically significant at 0.1%. A coefficient of 0.40 indicates that household expenditure is associated with a 0.40-point increase in WTP for protecting fishing areas. Other variables, such as education, also have a positive correlation, but they are not significant in explaining WTP for fishing areas in the village.

As clearly seen in Table 9.4, WTP for conservation differs for peatland and fishing areas. Only one variable consistently explains WTP, and that is household expenditure. Despite this consistency, its significance level varies considerably, with a much higher significance level demonstrated in WTP for preservation of fishing areas compared to peatland. In other words, the association of WTP and household expenditure is weaker for peatland conservation compared to that of fishing area conservation. Two variables in particular can explain this difference; these are education level and perceptions of the two environments.

What do the findings mean? Why do the community members have different stances regarding WTP for peatland and fishing areas? What can we learn from these differences when considering natural resource conservation in the context of Indonesia as a whole? These questions are relevant to deepening understanding of the behavior of community members regarding WTP for conservation of natural resources.

As long as the community members strongly depend on the natural resources of the surrounding areas, WTP for conservation correlates to household expenditure. Survey data discussed in other chapters of this book demonstrate that Rantau Baru villagers depend on the surrounding environment for their livelihoods, and their basic survival as a community is strongly influenced by the capacity to benefit from the natural resources available to them. Fishermen and women adapt fishing technology by using material from nature as well as manufactured goods (such as nylon nets, hooks, and other fishing gear). One should also remember that Rantau Baru is relatively difficult to access, if not isolated. In the rainy season, the main road is flooded with water, and therefore cannot be used as a means of transportation. This geographical location limits the community members' access with outside world during the rainy season.

Yet despite this dependency on natural resources, WTP differs for peatland and fishing areas not only according to household expenditure, but also to education level. This difference can be explained by the environmental conditions and the daily life of small-scale fishermen/women in Rantau Baru. The household expenditure variable can be considered a proxy for the welfare of community members because, according to the finding of this study, WTP, to a varying degree, depends upon the level of welfare that people enjoy from the natural resources available to them—if their welfare increases, they are willing to give more money for environmental conservation. This applies to WTP for fishing areas, as the villagers to date have been able to gain more livelihood benefits from fishing areas than peatlands.

On the other hand, WTP for peatland conservation correlates highly to education level, while it does not explain WTP for fishing areas. In this study, the education variable represents the level of formal schooling attained by community members. The positive correlation of education to WTP for both peatland and fishing areas could indicate that the findings are in line with the expectation that education, to a varying degree, is related to WTP. However, the significance of education level varies for WTP for peatland and fishing areas. Education correlates to WTP for peatland at a relatively high degree of significance, while it is not significant enough to explain WTP for fishing areas.

The difference in the significance of education as an independent variable to explain WTP for peatland and fishing areas might also be related to the different use of and policies applied to these ecosystems. It is important here to mention that peatlands have received much more attention from the government in comparison to fishing areas, because the peatland fires cause air pollution. The dramatic forest fires of 2015 burned for weeks; it caused economic losses of more than US$16 billion (World Bank 2016) and led to the initial establishment of the Peatland Restoration Agency (*Badan Restorasi Gambut*, BRG) in 2015. In this regard, any information related to peatlands that is conveyed by the government likely facilitates community members' understanding of this environment, or in other words, higher education levels lead to increased knowledge of these environments. This may explain why the significance of the education variable differs for different environments. We may tentatively conclude that if the government gives more attention to the conservation of fishing areas, education as an independent variable to explain WTP of fishing areas may increase. However, this study did not collect information on education efforts or policies related to peatland and fishing areas.

The significance of household expenditure to WTP also differs according to ecosystem. Most villagers' main livelihood depends on fishing activities, while to some extent they can gain income from peatland by cultivating some agricultural commodities. Nofrizal et al. (Chap. 5) note that (1) the rivers and peat swamps of the village provide habitat for 44 fish species that have market value, (2) Rantau Baru village is a freshwater fish production site in Pangkalan Kerinci, and (3) fish products from the village are sold in markets in the capital city of Riau, Pekanbaru. According to data collected from the survey, approximately 70% of Rantau Baru villagers are fishermen/women, while only 3% work in the agricultural sector. We can therefore conclude that local livelihood very much depends on fishing areas, even though the village is also surrounded by peatland. The different connections that community members have to fishing areas and peatland furthermore forms the basis upon which they frame and understand their relationship to nature.

Community members catch fish in the Kampar River and its tributaries. The Kampar River also serves as a main transportation route for the fishermen/women to reach peat swamps, oxbow lakes, and smaller rivers. While villagers are very dependent on the rivers, peat swamps, and oxbow lakes in the surrounding areas, they have almost no control over the aquatic environments and during the rainy season, the village is prone to flooding.

We see from the findings in Column 3 of Table 9.4 that perceptions of fishing areas do not explain WTP for the fishing areas. The positive correlation is in line with the expected result, but it is not significant. This result therefore reveals that WTP for fishing areas cannot be solely explained by perceptions about the condition of fishing areas.

Therefore, it is indicated that the community members of the village are very much dependent upon the fishing areas, mainly the Kampar River and its tributaries. They can benefit from these environments with very limited intervention, if any. Villagers catch fish directly from the natural environment without farming. Nofrizal et al. (Chap. 5) explain that the availability of fish relies on existing natural conditions, with almost no conservation intervention. Villagers seem willing to accept the existing balance of threats and benefits that they receive from fishing areas. Various reasons explain this passive approach, including a lack of knowledge on how to control floods, and a lack of money, time, and energy. Government intervention to control flooding has resulted in the construction of a dam upstream of the village in Koto Panjang and when rain intensity is very high, the water is released, which in turn creates floods in Rantau Baru.

To further understand the community members' relationship to their aquatic environment, it is relevant here to compare this to their connection to peatland. WTP for peatland is associated with the community members' perception of peatland conditions, although the significance level is not high. Perception of peatland conditions is negative and indicates an inverse correlation. This most important finding is that WTP for peatland is closely related to the perception of peatland conditions.

Having mentioned that household expenditure could explain WTP for conservation of peatlands and fishing areas, this should be interpreted carefully in terms of possible environmental degradation. As this study finds, most community members work as fishermen/women on a small scale, which means that their livelihood is very much dependent upon nature. Improving expenditure by further exploiting the aquatic ecosystem would cause serious damage to the environment. This makes it nearly impossible to improve the expenditure of community members through increased exploitation of the aquatic ecosystem, as it has limited resources and can only support the livelihood of the community members at its current level. Other strategies for increasing income, such as recreational fishing (Nofrizal et al., Chap. 5), should be considered. In a similar line of argument, further exploiting the peatlands is also not likely. The plots of peatland owned by community members are relatively not very big, and the level of knowledge and capital they possess are relatively low. Ultimately, income that community members gain from nature is used for survival, which leaves little money to pay for conservation.

9.5   Conclusion and Policy Implications

The aim of this study is to understand the factors that determine Willingness to Pay (WTP) for conservation of both peatlands and fishing areas in an area experiencing peatland degradation. A household survey was conducted in a floodplain village that is surrounded by peatlands. A contingent valuation method, or CVM, was employed to measure how local villagers value the two ecosystems.

The findings reveal that community members' WTP for conservation of peatland areas differs from their WTP for conservation of fishing areas. WTP for peatland conservation correlates to education level, and, to a degree, household expenditure and perception of peatland conditions. WTP for conservation of fishing areas correlates almost exclusively to household expenditure. From these findings, it is safe to conclude that WTP is very much associated with household expenditures, with a differing degree of significance depending on the area to be conserved. These results are in line with other studies on the WTP of cash-poor people, which find that personal economic conditions correlate highly to WTP (Tilahun et al. 2011).

The contribution of this study, however, lies in its discussion of WTP for the conservation of two very different ecosystems in a community where all other factors are equal. It thus enables us to identify which specific factors influence WTP for conserving peatlands versus fishing areas. First, the study found that the correlation between household expenditure and WTP is higher for fishing areas than for peatland. Differences in WTP for the two distinct ecosystems are a result of community members' direct, long-term interaction with them. While most villagers depend on the aquatic environment for their livelihood, they have less control over it. In the rainy season, they may catch more fish, which directly improves their expenditures, but at the same time their village may be hit by a flood. The peatlands are less important to livelihood of the community members because their holdings are small and they have less knowledge, technology and capital to exploit peatlands.

Second, education level can explain WTP for peatland, while it does not explain WTP for fishing areas.

Third, this study finds that WTP is weakly negatively associated with perceptions of peatland conditions, or in other words, that WTP decreases with the perception that peatland conditions are bad. This result may be explained by a feeling among most community members that peatland conservation is beyond their capability, and some of the peatlands are too damaged to be restored. These findings are not in line with the expectation; they could indicate that policymakers have a steep barrier to overcome in convincing community members to participate in peatland conservation efforts and therefore have serious implications for conservation policies. Perceptions of fishing areas do not explain WTP for conservation of fishing areas. Conservation of peatlands and fishing areas seems to be beyond the reach of the community members, especially if they have to participate in terms of cash money.

While this study confirms that household economic conditions are relevant for the conservation of nature, this finding should be read cautiously when formulating policy, as most members of this community are cash poor. Relying on conservation directly from their income is difficult because they must allocate this income for daily spending. Improving income and expenditure levels is challenged by skill and natural resource limitations as well as climate fluctuations. In addition, the potential for agriculture on peatland holdings and participation in palm oil plantation activity is minimal, limiting the amount of income that can be gained from peatlands.

Given these conditions, government assistance for conservation is crucial. It is very clear that conservation cannot solely be the responsibility of the members of the community. While education level does correlate to WTP for peatland, WTP for fishing areas is not currently impacted by education level. These findings reveal that government efforts to address forest fires on peatland through socialization could have improved awareness of peatland conservation among community members. Therefore, similar education efforts for conserving fishing areas may increase awareness and WTP for fishing areas as well.

This study provides insights into WTP for conservation in a cash-poor community that has two dominant ecosystems. As a case study of one village, it contributes to our understanding of context-specific factors that impact WTP and can inform the formulation of strategies for environmental conservation in such communities. However, further research is needed across Riau Province to create a fuller picture of WTP for environmental conservation.

Appendix 1 Estimation of Economic Value

Estimation of TEV = DUV + IV
    Total economic value
    Direct use value
    Indirect use value
    DUV consists of recreational value

Table 9.7 Carbon Stocks in Rantau Baru Village

Covered areas 2019 (KLHK)	Area (ha)	Carbon stocks (ton C/ha)		Price (US$)
Water	460.386	n.a.	n.a.	n.a.
Swamp grove	3821.318	30	114,639.6	527,341.9
Secondary forest swamp	398.3551	75.7	30,155.48	138,715.2
Forest	130.7545	76.7	10028.87	46,132.79
Open land	177.9484	2.5	444.8711	2046.407
Residential	0.107697	4	0.430788	1.981625
Plantation	3470.949	63	218,669.8	1,005,881
Total values of carbon stock				1,720,119

The average price of carbon credits (US$ 4.6/tCO^2) is based on Hamrick (2016)
Source of data: Source: Sugardiman and Rovani 2015

1. Recreational value

RV = total visitors/year × total cost/visit
    Total visitors = average number of people/week × average visit/person

2. Fishing production value (FPV)

FPV = total small-scale fishermen × average annual net revenue from capturing fish
    Average annual net revenue = [total revenue/season (bad, fair, and good season) × trips undertaken each season – average cost/trip
    IUV (Table 9.7)
    Estimated value of peatland and carbon sequestration
    Carbon sequestration (CSV) = carbon sequestration rate × total area of covered lands with vegetation × price in carbon market

Appendix 2 WTP for Peatlands (Figs. 9.1, 9.2, 9.3, and 9.4)

Fig. 9.1 Residual versus fitted, normally distributed

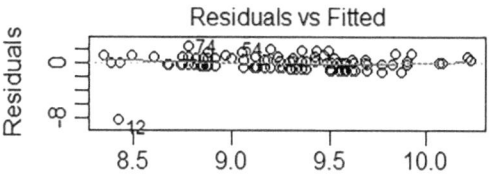

Fig. 9.2 Q-Q plot. (The data is normally distributed). WTP for fishing areas

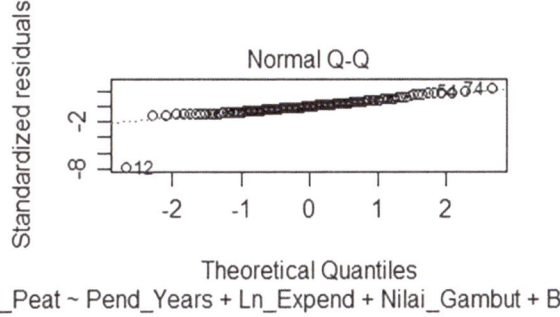

Fig. 9.3 Residual versus fitted is normally distributed

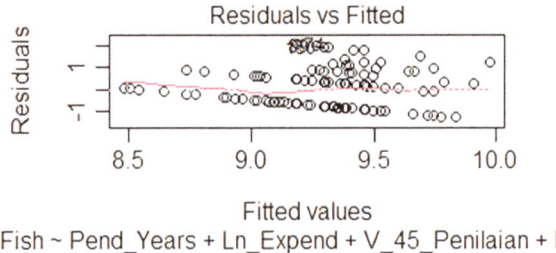

Fig. 9.4 Q-Q plot. (The data is normally distributed)

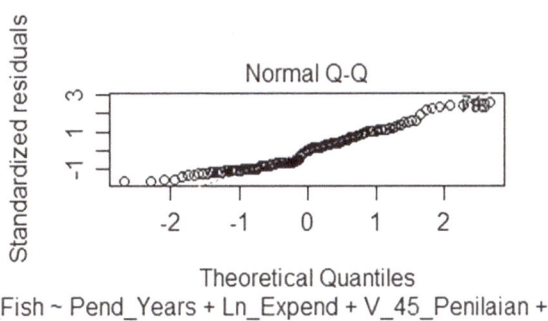

References

Amirnejad H, Khalilian S, Assareh MH et al (2006) Estimating the existence value of north forests of Iran by using a contingent valuation method. Ecol Econ 58(4):665–675. https://doi.org/10.1016/j.ecolecon.2005.08.015

Baumol WJ, Oates WE (1979) Economics, environmental policy, and the quality of life. Prentice Hall, NJ

Bechtel MM, Scheve KF (2013) Mass support for global climate agreements depends on institutional design. PNAS 110(34):13763–13768. https://doi.org/10.1073/pnas.1306374110

Bimonte S (2002) Information access, income distribution, and the environmental Kuznets curve. Ecol Econ 41(1):145–156. https://doi.org/10.1016/S0921-8009(02)00022-8

BPS (Badan Pusat Statistik) (n.d.) Rata-Rata Banyaknya Anggota Rumah Tangga 2010–2019. Accessed 14 October 2022

Cho SH, Yen ST, Bowker JM et al (2008) Modeling willingness to pay for land conservation easements: treatment of zero and protest bids and application and policy implications. J Agric and Appl Econ 40(1):267–285. https://doi.org/10.1017/S1074070800028108

Cosens B (2018) Governing the freshwater commons: lessons from application of the trilogy of governance tools in Australia and the western United States. In: Holley C, Sinclair D (eds) Reforming water law and governance. Springer, Singapore, pp 281–298. https://doi.org/10.1007/978-981-10-8977-0_13

Davis RK (1963) The value of outdoor recreation: an economic study of the Maine woods. Dissertation, Harvard University

Decancq K, Lugo MA (2012) Inequality of wellbeing: a multidimensional approach. Economica 79(316):721–746. https://doi.org/10.1111/j.1468-0335.2012.00929.X

Diekmann A, Franzen A (1999) The wealth of nations and environmental concern. Environ Behav 31(4):540–549. https://doi.org/10.1177/00139169921972227

Dietz S, Adger WN (2003) Economic growth, biodiversity loss and conservation effort. J Environ Manag 68(1):23–35. https://doi.org/10.1016/S0301-4797(02)00231-1

Elvyra R, Solihin DD, Affandi R et al (2010) Kajian aspek produksi ikan lais *Ompok hypophthalmus* di Sungai Kampar, kecamatan Langgam, kabupaten Pelalawan, provinsi Riau. J Natur Indones 12(2):117–123. https://doi.org/10.31258/jnat.12.2.117-123

Franzen A, Meyer R (2010) Environmental attitudes in cross-national perspective: a multilevel analysis of the ISSP 1993 and 2000. Eur Sociol Rev 26(2):219–234. https://doi.org/10.1093/esr/jcp018

Freeman AM III, Herriges JA, Kling CL (2014) The measurement of environmental and resource values: theory and methods, 3rd edn. RFF Press, Oxon

Forlin C, Chambers D (2011) Teacher preparation for inclusive education: increasing knowledge but raising concerns. Asia-Pac J Teach Educ 39(1):17–32. https://doi.org/10.1080/1359866X.2010.540850

Hamrick K (2016) Raising ambition: state of the voluntary carbon markets 2016. Forest Trends, Washington, DC

Holden ST, Shiferaw B (2002) Poverty and land degradation: peasants' willingness to pay to sustain land productivity. In: Barrett CB, Place F, Aboud AA (eds) Natural resources management in African agriculture: understanding and improving current practices. CABI Publishing, Wallingford, pp 91–102. https://doi.org/10.1079/9780851995847.0091

Israel D, Levinson A (2004) Willingness to pay for environmental quality: testable empirical implications of the growth and environment literature. Contrib Econ Anal Policy 3(1):1–29. https://doi.org/10.2202/1538-0645.1254

Jacobsen JB, Hanley N (2009) Are there income effects on global willingness to pay for biodiversity conservation? Environ Resour Econ 43(2):137–160. https://doi.org/10.1007/s10640-008-9226-8

Kachi A, Bernauer T, Gampfer R (2015) Climate policy in hard times: are the pessimists right? Ecol Econ 114:227–241. https://doi.org/10.1016/j.ecolecon.2015.03.002

Kahn ME, Matsusaka JG (1997) Demand for environmental goods: evidence from voting patterns on California initiatives. J Law Econ 40(1):137–174. https://doi.org/10.1086/467369

Sugardiman RA, Rovani R (eds) (2015) Buku kegiatan serapan dan emisi karbon. Kementerian Lingkungan Hidup dan Kehutanan, Jakarta

Kopnina H (2012) Education for sustainable development (ESD): the turn away from 'environment' in environmental education? Environ Educ Res 18(5):699–717. https://doi.org/10.1080/13504622.2012.658028

Kopp RJ (2005) Foreword. In: Mitchell RC, Carson RT (eds) Using surveys to value public goods: the contingent valuation method, 4th edn. Resources for the Future, New York and Oxon, pp xv–xvi

Lienhoop N, MacMillan D (2007) Valuing wilderness in Iceland: estimation of WTA and WTP using the market stall approach to contingent valuation. Land Use Policy 24(1):289–295. https://doi.org/10.1016/j.landusepol.2005.07.001

Lindhjem H, Tuan TH (2012) Valuation of species and nature conservation in Asia and Oceania: a meta-analysis. Environ Econ Policy Stud 14(1):1–22. https://doi.org/10.1007/s10018-011-0019-x

Liu J, Zao Y, Jan SC (2021) Environmental perceptions and willingness to pay for preservations: evidence from beach destinations in China. Int J Tourism Res 23(5):1–13. https://doi.org/10.1002/jtr.2442

McConnell K (1997) Income and the demand for environmental quality. Environ Dev Econ 2(4): 383–399. https://doi.org/10.1017/S1355770X9700020X

Mitchell RC, Carson RT (2005) Using surveys to value public goods: the contingent valuation method, 4th edn. Resources for the Future, New York and Oxon

Miller MA, Middleton C, Rigg J et al (2020) Hybrid governance of transboundary commons: insights from Southeast Asia. Ann Am Assoc Geogr 110(1):297–313. https://doi.org/10.1080/24694452.2019.1624148

Mizuno K, Fujita MS, Kawai S (2016) Catastrophe and regeneration in Indonesia's peatlands: ecology, economy and society. In: Kyoto CSEAS series on Asian studies, vol 15. NUS Press/Kyoto University Press, Singapore/Kyoto

Ndebele T, Forgie V (2017) Estimating the economic benefits of a wetland restoration programme in New Zealand: a contingent valuation approach. Econ Policy Anal 55:75–89. https://doi.org/10.1016/j.eap.2017.05.002

Scruggs L, Benegal S (2012) Declining public concern about climate change: can we blame the great recession? Glob Environ Change 22(2):505–515. https://doi.org/10.1016/j.gloenvcha.2012.01.002

Seller C, Stoll JR, Chavas JP (1985) Validation of empirical measures of welfare change: a comparison of nonmarket techniques. Land Econ 61(2):156–175. https://doi.org/10.2307/3145808

Tilahun M, Mathijs E, Muys B et al (2011) Contingent valuation analysis of rural household's willingness to pay for frankincense forest conservation. In: Paper presented at the EAAE 2011 congress change and uncertainty: challenges for agriculture, food and natural resources, ETH Zurich, Zurich, 30 August–2 September 2011

Vincent JR, Carson RT, DeShazo JR et al (2014) Tropical countries may be willing to pay more to protect their forests. PNAS 111(28):10113–10118. https://doi.org/10.1073/pnas.1312246111

World Bank (2016) The cost of fire: an economic analysis of Indonesia's 2015 fire crisis. In: Indonesia sustainable landscapes knowledge Note no 1. World Bank Group, Washington, D.C.. https://documents.worldbank.org/en/publication/documents-reports/documentdetail/7761014 67990969768/the-cost-of-fire-an-economic-analysis-of-indonesia-s-2015-fire-crisis. Accessed 29 March 2021

Zhongmin X, Guodong C, Zhiqiang Z et al (2003) Applying contingent valuation in China to measure the total economic value of restoring ecosystem services in Ejina region. Ecol Econ 44(2–3):345–358. https://doi.org/10.1016/S0921-8009(02)00280-X

Chapter 10
The Value of Participatory Mapping, the Role of the *Adat* Community (*Masyarakat Adat*), and the Future of the Peatlands

Akhwan Binawan and Takamasa Osawa

Abstract Indonesian central and local governments have not made serious effort to recognize and protect the rights of the *adat* community (*masyarakat adat*; indigenous/traditional community) or *adat* law community (*masytarakat hukum adat*; customary law community), despite the Indonesian Constitution of 1945 and many national laws enacted to recognize such rights. In practice, government institutions facilitate conversation of traditional forest lands to other uses, often plantations. Rantau Baru is one *adat* community where customary territorial management has been eroded by outsider interests, including concession companies, leading to social conflict and environmental damages, including peatland fires. By presenting maps produced from the perspective of the Rantau Baru villagers, this chapter explores the difficulties that *adat* communities face regarding government mapping policies and suggests the significance of participatory mapping projects to re-establish sustainable *adat* community management of customary lands.

Keywords *Adat* community · Participatory mapping · Mapping policies · Land grabbing · Deforestation · Rantau Baru

10.1 Introduction: Politics of Mapping in Indonesia

A map does not simply present an accurate description of geospatial information. The information on the map is selected, arranged, omitted, and categorized by the map maker for a particular purpose. Maps may be created to enable future spatial

A. Binawan
Perkumpulan Ara Sati Hakiki, Jl. Kayu Putih, Perumahan Athaya IV, Pekanbaru, Riau, Indonesia

T. Osawa (✉)
Institute of Liberal Arts and Science, Kanazawa University, Kanazawa, Ishikawa, Japan
e-mail: tosawa@staff.kanazawa-u.ac.jp

© The Author(s) 2023
M. Okamoto et al. (eds.), *Local Governance of Peatland Restoration in Riau, Indonesia*, Global Environmental Studies,
https://doi.org/10.1007/978-981-99-0902-5_10

211

plans. As the producer's intention penetrates the map, in this sense, a map becomes the optimal state of geography and space according to the producer's ideal geographic image. Others may have different optimal images, however, and when these different images compete, it creates friction, disputes, and negotiation (Okamoto et al., Chap. 2). Which image eventually dominates is usually decided by political power. Those with power can plan the use of space, delineate the relevant boundaries, designate the areas, and implement the policies to change the space. Conversely, those with less power are bounded, restricted, and dominated by the image imposed by the powerful.

In Indonesia, the central government's geographical image is significantly reflected in geospatial planning, the creation of maps, and actual land use. In particular, the Basic Forestry Law of 1967[1] designated most areas of the "Outer Islands" (such as Sumatra and Kalimantan) as state forest zones, and in the 1980s the government created maps accordingly (Okamoto et al., Chap. 2; Peluso 1995). Both the new forestry law and new maps conferred legal power to the central government to actualize its image of the forest zone to benefit its own interests. After the collapse of Suharto's regime in 1998, some authority over spatial planning was transferred to the provincial and district governments. The designation of the forest zone did not dramatically change, however, and the local governments have since promoted industrial use of the forest zone to develop the economy of their respective regions. This has led to various ongoing environmental problems (Mizuno and Kusumaningtyas 2016).

During the mapping process following from the Forestry Law of 1967, the central and local governments ignored the significance of the forest areas as living space for the people who have lived there for generations and managed the natural resources and environments. People who have used the resources and environment for several generations are known as the *adat* community (*masyarakat adat*; indigenous/traditional community) or *adat* law community (*masyarakat hukum adat*; customary law community).[2] Their legal position was first recognized by the Dutch colonial government, and the Indonesian Constitution of 1945 and many national laws recognize and protect the right of the *adat* community to customary territories. This right has been ignored in numerous places, however. The government and government-sponsored companies have continuously exploited the resources and environments in customary territories. If the government claims the land use of a specific area in the state forest zone, the *adat* community's protest against it is rejected and they cannot receive adequate compensation. The government may not intervene even to rectify illegal land grabbing. When companies cause environmental problems, *adat* communities are unable to affect degrading land use practices and

---

[1] *UU No. 5 Tahun 1967 tentang Ketentuan-ketentuan Pokok Kehutanan*

[2] *"Adat* law community" is a legal term that views a community as a legal governance unit. The term is used in national laws, including the 1945 Constitution. *"Adat* community" is a more general expression that indicates indigenous, traditional, or customary communities in the region (see Moniaga 2007 and following sections).

do not receive proper support or compensation, even though their livelihoods are directly affected by the damage (Duncan 2004; Salim 2017). Although the government has primary responsibility for creating maps that delineate customary territories, central and local governments have not complied with this responsibility. Government entities consistently act as if there are no *adat* communities in the forest zone. *Adat* communities are therefore in a predicament that is caused by government mapping and spatial planning that is biased in favor of powerful interests (e.g. Duncan 2004; Henley and Davidson 2007; Li 2007).

Participatory mapping, or mapping from the residents' perspective, is one of a few measures that enables *adat* communities to challenge the unilateral spatial planning and mapping. Previously, *adat* communities could not objectively show their historical land use or customary territory because their knowledge and practices were handed down through oral communication and practices. Recent developments in geospatial information technology enable local communities to create graphic visualizations of their traditional land use and territories in the form of modern maps. Alleging a lack of adequate technology and competence, the central government has not recognized local community land use and resource rights claims based on independently created maps (Okamoto et al., Chap. 2). In this context, mapping can be highly effective for and valuable to *adat* communities, and it can reinforce their land entitlement. Mapping from the local residents' perspectives can prevent exploitation of the forest areas, promote local care for the forests, and mitigate environmental damage.

This chapter explores the historical and political position of the *adat* community and the present situation of Rantau Baru village based on maps created in collaboration between the authors and the villagers. First, we explore the background of this mapping project by describing the historical and political position of the *adat* community at national and local levels. Second, we describe the mapping process and the resulting map, analyzing the gap between the community-produced map and those issued by the government. Third, we visualize current land use in Rantau Baru using the participatory map, exposing recent deforestation and peatland degradation caused by the expansion of oil palm plantations. Finally, we suggest some ways that the *adat* community could sustainably use land and natural resources.

10.2 Background of the Mapping: Adat Community and Their Customary Land

10.2.1 The Conceptualization of the Adat Community Before the Independence of Indonesia

Adat is translated literally as "custom" or "tradition." During the Dutch colonial period, customary practices had legal standing in the governance of the Indonesian archipelago. Before the nineteenth century, while the Dutch East India Company applied Dutch civil law to Europeans, Christians, and urban residents living in the archipelago, people in the rural areas were subject to customary rules and jurisprudence. During the first half of the nineteenth century, the Dutch colonial government gradually reinforced the control over the archipelago while adopting this system of "legal pluralism" to govern the colony. For example, an 1824 statute proclaimed that all natives of the Indonesian archipelago, including the urban residents of Java, were subject to the customary justice system (Li 2007, pp. 44–45). In 1848 and 1849, the Dutch legal code was formally implemented, and Europeans living in the archipelago were subject to its code. The Government Regulation for the Netherlands East Indies of 1854 categorized all inhabitants as either "Europeans," "natives," or "foreign Orientals" (who were mainly ethnic Chinese) (Fitzpatrick 2007, p. 133). The Dutch law was not applied to the natives, who comprised 95 percent of the total population, as they continued to adhere to the judicial system based on customary courts (*adat* courts) in heavily Dutch-influenced areas and customary practices in other rural areas (Fasseur 2007, pp. 50–53; Henley and Davidson 2007, p. 19).

In several places, customary governance practices had specific names according to local contexts, such as *dresta* among the Balinese, *aluk* among the Toraja of central Sulawesi, and *adat* among the Minangkabau of western Sumatra. These practices originally included not only legal rules governing human relationships, but also customary rules concerning spiritual relationships with the natural environment (see Osawa, Chap. 6). At the beginning of the twentieth century, Cornelis van Vollenhoven and his colleagues at the Leiden School abstracted, conceptualized, and elaborated on the significance of these customary practices as *adat* or *adat* law (*hukum adat* in Indonesian; *adatrecht* in Dutch). Although the term *adat* covers customary practices in terms of law, art, rituals, and ways of life, the term *adat* law focuses on the juristic aspect of *adat*. The Leiden scholars saw *adat* as a total worldview that enabled people to harmoniously govern their community, nature, and spiritual world that was fundamentally different from the European legal system, and they insisted that the communities maintaining the customary practices or the "*adat* law community" (*adatrechtsgemeenschap*) should be governed under their own *adat* law (Burns 1989, p. 8, 56). In particular, the scholars linked *adat* law with land rights—the "right of allocation" or "right of avail"—to territory, which was translated into the native language of Indonesian archipelago as *hak ulayat* (Henley and Davidson 2007, p. 20).[3] Meanwhile, the colonial government regarded uncultivated rural land as "wilderness" in the Agrarian Decree of 1870 and related regulations, and leased it out to European and Chinese enterprises (Fitzpatrick 2007, p. 133; Henley and Davidson 2007, p. 20; Mizuno and Kusumaningtyas 2016, pp. 41–42). The Leiden scholars claimed that *adat* communities held the right of allocation over these lands and tried to protect the communities and their land from any destruction caused by competition with capitalist interests (Henley and

---

[3] The term *hak ulayat* was adopted from the Minangkabau. The land that is associated with the right is called *ulayat* land (*tanah ulayat*). In the Pelalawan kingdom, the territory of each *pebatinan* was called *tanah wilayat* (Henley and Davidson 2007, p. 20; see also Osawa and Binawan, Chap. 3).

Davidson 2007, p. 20; Li 2007, p. 49). Although it is debatable whether the Leiden scholars' attempts were successful, their approach was accepted by the government in that *adat* laws in each area were not subsumed under the Dutch national law and became the de facto basis for settling local disputes (Henley and Davidson 2007, pp. 19–21; Li 2007, pp. 50–51).

The concept of "*adat* law community" and respect for *adat* were passed down through the eras of Japanese occupation and post-independence Indonesia. During the Japanese occupation (1942–1945), the Japanese Propaganda Department (*Sendenbu*) praised local customs as the embodiment of Asian traditional and ancestral values and unity, which they suggested were the opposite of the European values (Bourchier 2007, p. 116). Supomo, who is known as the "father of Indonesia's constitution," and who also studied at Leiden University, viewed *adat* as the legal and social basis of Indonesia and included the concept of the "*adat* law community" in the Constitution of 1945 (Henley and Davidson 2007, pp. 20–21, Bourchier 2007, pp. 116–117). Article 18B of the Constitution notes, "The State recognises and respects traditional communities [*masyarakat hukum adat*] along with their traditional customary rights as long as these remain in existence and are in accordance with the societal development and the principles of the Unitary State of the Republic of Indonesia, and shall be regulated by law."[4]

10.2.2   The Exploitation of Adat Land and Adat Revivalism

During Sukarno's regime (1957–1966), the concept of "*adat* law community" was partly respected but largely ignored during the process of state formation. The Basic Agrarian Law of 1960[5] is alleged to be based on *adat* and recognize the customary land right, or *hak ulayat*, in places where it still exists. However, simultaneously, the law provides that customary rights should be amended when they collide with the national law and proclaims that all land in Indonesia is under the state's right of control (*hak menguasai negara*) (Fitzpatrick 2007, p. 137). The right of control allows "the grant of rights to uncultivated and/or non-residential untitled lands without obtaining the consent of the relevant local communities and without triggering the legal obligation to pay 'adequate' compensation to holders of expropriated titles" (Fitzpatrick 2007, p. 137). Under the state's right of control, *hak ulayat* is typically ignored.

Suharto's "New Order" regime (1967–1998) was characterized by its highly authoritarian polity and ambition to develop the economy. The government aimed to exploit natural resources in rural areas, especially in forestlands, and implemented policies that constrained the land rights of rural populations. The Basic Forest Law of 1967 designated 143 million hectares, or three-quarters of the nation's total land

---

[4] The English translation was quoted from the Constituteproject.org (2021, p. 8).

[5] *UU No. 5 Tahun 1960 tentang Peraturan Dasar Pokok-Pokok Agraria.*

area, as a state forest zone (Okamoto et al., Chap. 2). This law recognized the existence of *adat* law community and a "function" (*fungsi*) of the forest that was managed by them, i.e. *hutan adat* (customary forest). However, this *hutan adat* function was only granted to a few select areas, and was regarded as part of the state forest. In implementing the Consensus-Based Forest Land Use Planning (*Tata Guna Hutan Kesepakatan*) during the 1980s, the government created a 1:500,000 scale map in which all forest land was categorized as conservation forest (*hutan konservasi*), protection forest (*hutan lindung*), production forest (*hutan produksi*), or convertible production forest (*hutan produksi yang dapat dikonversi*) (Chakib 2014, p. 13; McCarthy 2006, p. 5; Peluso 1995, p. 389; see also Okamoto et al., Chap. 2).

During these processes, the government bounded the space, categorized the land, and tried to actualize its optimal state of land in the forest area in order to facilitate its exploitation. The Basic Forestry Law designated forests that had been used by local communities as state forest zone, and the residents were alienated from the communal forest that they had inherited from their ancestors (Peluso 1995; Siscawati 2014). For instance, when the government established national parks and nature reserves, local people who had lived in those areas were forced to move out (Duncan 2004, pp. 102–103). Additionally, the government granted permits for land utilization or exploitation of forest areas to timber, pulp, oil palm, and mining companies, allowing them to legally encroach on the customary lands of local residents. Lands at the edges of settlements, including uncultivated fallow swidden fields or forests used by locals for logging, gathering, and hunting were especially vulnerable to expropriation.[6] Although land concessions provided companies and the government with considerable economic profit, it scarcely contributed to the economy of local communities.[7] Furthermore, industrial use of the forests often caused environmental problems, and the local communities suffered from the damage. Local protests against land grabbing and lack of compensation were often suppressed with violence and human rights abuses (Duncan 2004, pp. 101–103).

Suharto's regime collapsed in 1998, and Indonesia entered an era of decentralization and democracy (1998-present). Echoing the rise of the transnational movement of the 1990s to protect the rights of indigenous peoples, the value and validity of *adat* and the *adat* community was revaluated. The Archipelagic Alliance of Adat Communities (*Aliansi Masyarakat Adat Nusantara*, AMAN), an umbrella organization of local non-governmental organizations (NGOs) that aimed to gain recognition and protection of indigenous rights, was established in 1999, and it adopted the

---

[6]In the process of Regional Physical Planning Programme for Transmigration implemented in the late 1980s, while government land-use planners recognized local people's *adat* claim to certain forest trees and plants producing products (such as rattan, fruit, rubber trees and honey), they did not recognize the forest land as the territory of the local community (Peluso 1995, p. 392).

[7]During the 1980s, the government promoted timber logging in the forest zone. Although some residents were employed as loggers and profited from the work (see Okamoto et al., Chapter 2; Peluso 1995), it was temporary employment. In some places, such as Rantau Baru, residents did not receive any profit from such logging (Osawa and Binawan, Chap. 3).

term "*masyarakat adat*" as the translation of the transnational concept of "indigenous peoples" to refer to the *adat* community. Local authorities, activists, and locals began to defend the rights of *adat* communities which were suppressed during Suharto's regime.[8] Henley and Davidson (2007) refer to this movement as "*adat* revivalism."

In the movement to esteem democracy, decentralization, and *adat*, many laws enacted after 1998 codified the definition of and respect for the *adat* community. For example, the 2009 law to protect and manage the environment[9] defines "*masyarakat hukum adat*," or the *adat* law community, in their relationship with a specific local environment for generations and obliges the government to respect and protect their *adat* in the implementation of environmental policies. The Village Law of 2014,[10] which reinforces the authority of administrative villages, also defines the *adat* law community as possessing traditional territory and fulfilling one or more of the following four conditions: (1) a shared identity; (2) a customary governance system; (3) a customary property or object; or (4) traditional norms.[11]

In particular, the Constitutional Court's decisions in 2011 and 2012 (No. 45/PUU-IX/2011 and No. 35/PUU-X/2012, respectively) on the legality of the state forest according to the Basic Forestry Law of 1967 and its successor, the Forestry Law of 1999,[12] were epochal. Decision No. 45 of the 2011 decision stipulates that state forest can only be legally established following realization of proper procedures regarding designation, boundary delineation, mapping, and determination (*penetapan*) by the government. In 2009, the Department of Forestry had complied with these procedures in designating only eleven percent of the so-called state forest zone; by 2014, the designated state forest area had rapidly extended to 58 percent out of all forest (Safitri and Nagara 2015, p. 2). Decision No. 35 of 2012 reinforce the legal standing of *hutan adat* and claims that the *hutan adat* is no longer subsumed under state forest zone. It provides the head of a district or city (*bupati* or *walikota*) the authority to issue a decree (*surat keputusan*) or a regional regulation to designate an *adat* area (*wilayah adat*) within their jurisdiction (Vinolia 2021; Warman 2014). With this decision, the government claim over customary forests as part of the state

---

[8] The terms "indigenous peoples" and "*masyarakat adat*" were first mutually associated through the meetings of the Indigenous Peoples' Rights Advocacy Network (*Jaringan Pembelaan Hak-hak Masyarakat Adat*) in South Sulawesi in 1993 (Moniaga 2007, pp. 281–282).

[9] *UU No. 32 Tahun 2009 tentang Perlindungan dan Pengelolaan Lingkungan Hidup.*

[10] *UU No. 6 Tahun 2014 tentang Desa.*

[11] In addition to these two laws, numerous other laws mention the *adat* community or *adat* law community, including the Forestry Law of 1999 (*UU No. 41 Tahun 1999 tentang Kehutanan*), the Law on Human Rights (*UU No. 39 Tahun 1999 tentang Hak Asasi Manusia*), the Law Concerning the Areas on Coasts and in Small Islands of 2007 (*UU No. 27 Tahun 2007 tentang Wilayah Pesisir dan Pulau-Pulau Kecil*), and the Law on Local Governance (*UU No. 23 Tahun 2014 tentang Pemerintahan Daerah*).

[12] *UU No. 41 Tahun 1999 tentang Kehutanan.*

forest has collapsed. Following Decision No. 35, the Ministry of Home Affairs (*Kementrian Dalam Negri*, KDN) issued Regulation No. 52 in 2014,[13] which includes a guideline for recognizing and protecting the rights of the *adat* law community and provides district/city heads with the authority to issue decrees and/or regulations to designate the existence of the *adat* community and their *wilayah adat* within jurisdiction. This guideline notes that district/city heads should identify and determine the *adat* community by establishing a consultative team to address issues regarding the *adat* community and their rights.

Despite such laws and decisions, recognition of the land entitlement of the *adat* community still has a long way to go. While the various laws repeatedly affirmed the existence of the *adat* community and rights, new laws did not provide concrete policies to support *adat* communities or amend the existing forest policies. Although a bill that comprehensively ensures the rights of *adat* communities at the national level has been discussed in the National Parliament since 2013, it has not been passed yet (Arizona and Cahyadi 2013; Nugraha 2019).[14] District and city governments do not always act to designate *adat community* and *wilayah adat* even though they have the power to do so. Furthermore, even if the local government recognizes *wilayah adat*, it is necessary to obtain recognition from the central government agencies if the *wilayah adat* overlaps with an area that any central ministry has designated as having specific functions. For example, when a customary area overlaps with the state forest zone, it is necessary to obtain a decree from the Ministry of Environment and Forestry (*Kementrian Lingkungan Hidup dan Kehutanan*, KLHK), after which the area is recognized as *hutan adat*.[15] These procedures become much more difficult and complex to realize when the customary territory overlaps with other stakeholders' lands, such as a company's concession area or other customary territories. The community is required to settle any dispute before applying for recognition. Ultimately, it is very cost- and time-consuming for local communities to actualize *wilayah adat* and *hutan adat*.

10.2.3 Recognition of Adat Communities and Lands in Riau

Before the Independence of Indonesia, many communities in Riau managed their customary land and natural resources in and through their traditional institutions and historically accumulated knowledge. Such territories have various names, including *kebatinan* among the Talang Mamak living along the Indragiri River, *batin* among

---

[13] *Peraturan Mentri Dalam Negri No 52 Tahun 2014 tentang Pedoman Pengakuan dan Perlindungan Masyarakat Hukum Adat.*

[14] *Rancangan Undang-undang Pengakuan dan Perlindungan Masyarakat Hukum Adat.*

[15] In addition, when the *wilayah adat* overlaps with coasts and waters, the Ministry of Marine Affairs and Fisheries has the authority to issue the decree. When the community applies for the recognition of agricultural land as *wilayah adat* and their *hak ulayat*, the Ministry of Agrarian Affairs and Spatial Planning/National Land Agency has the authority.

the Sakai along the Mandau River (a tributary of the Siak River), *luhak* in the upstream areas of the Rokan River, *kenegrian* in Kampar and Kuantan Singingi Districts, and *pebatinan* in the midstream areas of the Kampar River in Pelalawan District. Most of these customary areas have been categorized as within the state forest zone. During Suharto's regime, the central government leased these territories out to mining, acacia, and oil palm companies. After 1998, although some communities attempted to re-claim their customary territories and protested against land concessions and related decisions, most situations did not change. Even after the Constitutional Court's decisions No. 45 of 2011 and No. 35 of 2012 and KDN Regulation No. 52 in 2014, many difficulties remain.

At the provincial level, the Riau government has not actively addressed the rights of customary communities. In Riau, designation of the state forest zone began in 1986 with the Decision of the Minister of Environment and Forestry No. 173 and was completed with the Decision of Minister of Environment and Forestry No. 903 in 2016.[16] During this process, although the government should have fulfilled the procedures required by the Constitutional Court's decision No. 45 of 2011 (designation, boundary delineation, mapping, and determination), to date, not all procedures have been completed. In this way, the government has implemented forest policies in an ambiguous and seemingly arbitrary manner (see also Okamoto et al., Chap. 2).

Additionally, KDN Regulation No. 52 in 2014 is ignored in Riau. Following the regulation, in 2015, the Riau government enacted Regional Regulation No. 10 of 2015 concerning customary land rights and use, [17] which was expected to recognize and protect customary land rights. However, this regulation did not restrict new heavy mining operations in customary territories and was regarded as friendly toward the companies and their interests. After representatives of several customary communities in Riau litigated a judicial review of the regulation, it was rejected by the Supreme Court in 2018 (WALHI 2018). In 2018, the Riau government enacted Regional Regulation No. 14 concerning the recognition of the existence of *adat* law communities in environmental protection and management.[18] This regulation is based on KDN Regulation No. 52 of 2014 and notes that the Riau government has established an advisory committee to address issues related to the *adat* community. The committee has not yet been established, however. The Riau provincial government has continuously prioritized the utilization or exploitation of province resources in its spatial planning (see Okamoto et al., Chap. 2), and in the process has failed to fully protect the rights of its local communities.

---

[16] *Keputusan Menteri Kehutanan No. 173/ Kpts-II/ 1986 tanggal 6 Juni 1986 tentang Penunjukan Areal Hutan di Wilayah Provinsi Dati I Riau; Keputusan Menteri Lingkungan Hidup dan Kehutanan Nomor 903/ MENLHK/ SETJEN/ PLA.2/ 12/ 2016, 07 Desember 2016 tentang Kawasan Hutan di Provinsi Riau.*

[17] *Peraturan Daerah Provinsi Riau No. 10 Tahun 2015 tentang Tanah Ulayat dan Pemanfaatannya*

[18] *Peraturan Daerah Provinsi Riau No. 10 Tahun 2015 tentang Pedoman Pengakuan Keberadaan Masyarakat Hukum Adat dalam Perlindungan dan Pengelolaan Lingkungan Hidup.*

At the district level, some governments in Riau have recognized the rights of customary communities and issued related decrees and regulations. With the support of national NGOs, residents of Kampar District have tried to obtain formal recognition of *wilayah adat* and *hutan adat* in their customary territory, or *kenegrian*, since 2012. They specified areas that have been historically used by the local communities, obtained boundary agreements from various stakeholders, and created documents and maps to demonstrate historical community land use. Due to their long-term efforts, the head of Kampar District eventually issued a decree to recognize their customary territory as *wilayah adat* in 2019. However, as the area mostly overlapped with the state forest zone, it was necessary to also obtain a decree from KLHK. In 2019, the residents, activists and Kampar government submitted an application to the KHLK requesting that the status of state forest zone in seven places, for a total 10,318 hectares, should be designated as *hutan adat*. In 2020, the KLHK issued a decree recognizing only two places as *hutan adat,* a total of only 408 hectares (Vinolia 2021). This is still today the only case of *hutan adat* in Riau recognized by KLHK. In addition to Kampar District, the Siak government has recognized eight customary villages (*desa adat*), which gives the *adat* communities more discretion over village governance, and passed a related regulation in 2015 (Vinolia 2021).

Other districts, such as Pelalawan, have not positively engaged in the recognition and protection of *adat* community rights or their *wilayah adat.* Even though the KDN Regulation No. 52 obliges the local government to organize a committee to address issues related to the *adat* community, this committee has not yet been established in Pelalawan, and no *adat* community or *wilayah adat* has been recognized in the district. Neither has the district government made any formal maps to clarify customary lands. The lack of formal maps may result in land and boundary disputes between communities.

10.2.4 Rantau Baru and Its Status as an Adat Community

According to the map issued by KLHK in 2016, a large part of Rantau Baru village overlaps with the state forest zone.[19]

Figure 10.1 depicts the customary territory of Rantau Baru and the state forest zone. The area bounded by the orange and black dotted line is the customary territory that Rantau Baru villagers inherited from their ancestors as identified through our participatory mapping, which will be discussed later. The state forest zone with the designated function of "convertible production forest" (*hutan produksi yang dapat dikonversi*, HPK: drawn in pink) overlaps most of the customary territory. The convertible production forest is designated as state forest. This space is reserved

---

[19]The base map of the state forest zone was based on decision KLHK No. 903 in 2016 (Menteri Lingkungan Hidup dan Kehutanan 2016).

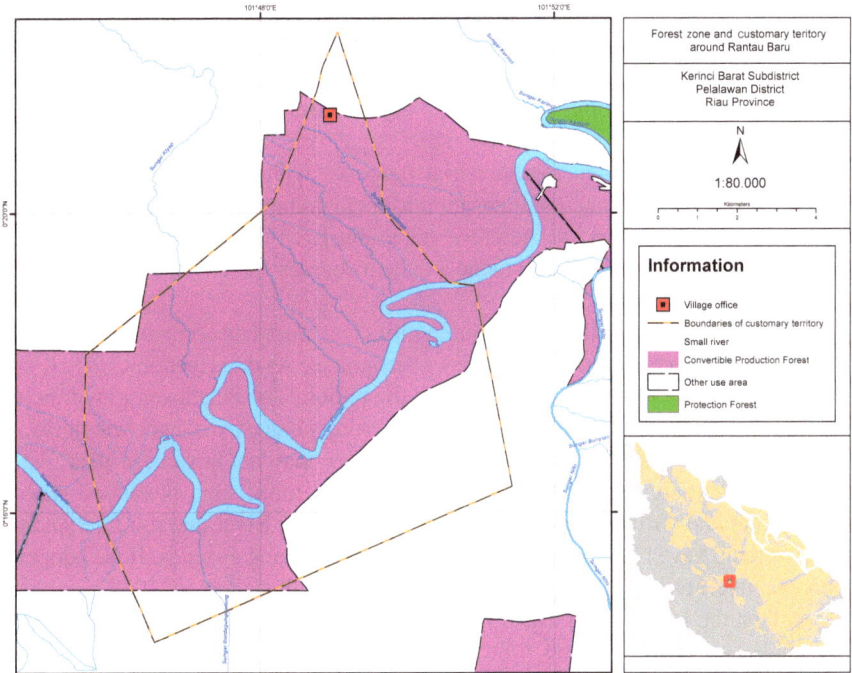

Fig. 10.1 Rantau Baru and the state forest area

Table 10.1 The area of Rantau Baru in the forest policies (Source: Menteri Lingkungan Hidup dan Kehutanan 2016)

No	Status of the zone	Function of the zone	Area (ha)
1	Forest zone	Convertible production forest	6085.1
2	Non-forest zone	Other use areas	2041.9
Water surface area			466.6
Total area			**8593.6**

for the development of transmigration, settlements, agriculture, and plantations, although people cannot use it as dwelling space or farmland without first obtaining permission from the KLHK. This convertible production area contains settlements of Rantau Baru villagers, their *sialang* forests, wasteland burnt by frequent fires, and oil palm plantation owned by the villagers, urban residents, and companies (see Osawa and Binawan, Chap. 3). The customary territory also includes "other use areas" or "non-forest area" (*areal pengunaan lain*, APL: drawn in white) that can be used for purposes other than forestry and may legally be owned by individuals and companies. Most of the APL areas around Rantau Baru have been designated as the state land, and this APL area has been leased out to oil palm and acacia companies who use it for plantations (Table 10.1).

A strict reading of legal regulations regarding convertible production forest indicates that villagers' use of these lands for traditional livelihood activities can be regarded as a violation of the law. Local customary use is therefore criminalized, as on these lands customary uses must be specifically approved by the central government.[20] Even in slightly less strict legal perspective, the government is authorized to use the area for industrial purposes without respect for customary uses or providing adequate compensation to local communities. The land rights of the Rantau Baru villagers are therefore vulnerable even though they have inherited the land from their ancestors and managed the resources and environment according to customary practices.

Rantau Baru is one of the "*adat* communities" or "*adat* law communities" that the 1945 Constitution and other national laws recognizes. As described in Chap. 3, they at the position of the indigenous people in the region. The Pelalawan kingdom and Dutch colonial government recognized their territory as *wilayat* land, which is land managed by *hak ulayat*. Although historically they may not have used peat hinterland for agriculture (Osawa, Chap. 6), they logged timber and possessed *sialang* trees in peatland territory. The tributaries of the Kampar River that run freely across the hinterland have been used for transportation and fishing grounds. If we adopt the more recent criteria of "*adat* law community" in the Village Law of 2014 mentioned above, then the following is true: (1) they have a shared identity as a member of the Rantau Baru community and followers of the *Adat Melayu Patalangan*; (2) the customary governance system of their *adat* heads, or *ninik mamak*, and their assistants is based on their *adat*; (3) their traditional territory is inherited from their ancestors; and (4) their norms are based on a matrilineal system, through which they have managed riverine resources and protected *sialang* areas (see Osawa and Binawan, Chap. 3). Nevertheless, they have not been recognized by the government as an *adat* law community nor are their rights to their customary territory protected by government forest policies, which have sought, first and foremost, to utilize or exploit forest areas to benefit vested interests.

Some scholars point out that the concepts of "*adat* law community" and "*adat* community" have been idealized by Leiden scholars and became an ideology in Indonesia (Burns 1989; Bourchier 2007; Henley and Davidson 2007; see also Osawa, Chap. 6). Whatever the origin may be, the protection of the "*adat* law community" is codified in the 1945 Constitution and many laws of post-independence Indonesia. Nevertheless, the government has not implemented concrete policies to follow these laws. On the contrary, it is the government's *modus*

---

[20]This occurred in a Sakai community in Bengkalis District. In 2020, a panel of judges sentenced a Sakai man, who had opened 0.5 hectare of land for planting tubers in an acacia forest, to one year in prison and a fine of Rp. 200 million. The land was a boundary area with concession land owned by an acacia company. The judge assessed that the man was proven to have cut down forests without having permission from the authorities, which violated Law No. 18 of 2013 concerning the Prevention and Eradication of Forest Destruction (*UU P3H No.18 Tahun 2013 tentang Pencegahan dan Pemberantasan Perusakan Hutan*). He was imprisoned for approximately seven months (Kompas 2020).

operandi to avoid implementing policies to recognize and protect *adat* law communities.

10.3   Mapping in Rantau Baru

10.3.1   Counter- and Participatory-Mapping

Counter- and participatory-mapping are techniques that enable *adat* communities to challenge the government's denial of customary land rights. The maps created counter colonial and postcolonial dispossessions and are intended to integrate multi-sector interests, protect public interests, and promote legal assurance and justice for local communities (Radjawali et al. 2017, pp. 818–819). In Indonesia, counter-mapping was first conducted in 1992 by World Wildlife Fund (WWF) in order to protect the forest environment in Kayan Mentarang, East Kalimantan. Local residents there participated in the mapping process (Dewi 2016, p. 97; Peluso 1995). In 1996, the Indonesian government implemented Regulation No. 69/1996 concerning public participation in spatial planning,[21] and following this regulation, Indonesian activists established the Network for Participatory Mapping (*Jaringan Kerija Pemetaan Partisipasi*, JKPP) to promote local residents' participation in mapping and spatial planning. After establishment of the AMAN, creating counter- and participatory-maps became an important strategy to strengthen *adat* claims across the archipelago (Dewi 2016, p. 97; Radjawali et al. 2017, pp. 821–823).

There are several criticisms of the strategy to create counter- and participatory-maps. One of the main criticisms is that the mapping leads to "'freezing' dynamic social processes which are referred to as 'customary law'" (Peluso 1995, p. 400). The maps depict an area's boundaries, territories and resources in a way that rejects the ambiguity, flexibility and dynamism that are essential to customary land management. Indeed, Rosita Dewi (2016) point out that creating participatory maps brought negative impacts to *adat* communities in Merauke District, Papua, where the rigid boundaries on the maps caused fragmentation and conflict among *adat* communities, and the identification of land users resulted in accelerating land sales to outsiders. Second, criticism is related to the politics of mapping. Mapping is embedded in spatial planning procedures and in claiming the legal validity of land occupation in Indonesia, and the central government does not incorporate the maps created by local residents into the spatial planning (Okamoto et al., Chap. 2). The counter- and participatory-mapping is therefore seen as an inefficient method to change the government policies. A third criticism is related to the power generated through mapping. The mapping process may not involve all people in a community, and the resulting map may reflect the interest of some people over others (Fox et al.

---

[21] *UU No. 69 Tahun 1996 tentang Pelaksanaan Hak Dan Kewajiban, Serta Bentuk Dan Tata Cara Peran Serta Masyarakat Dalam Penataan Ruang*

2008), possibly alienating local minorities such as woman and the poor (Radjawali et al. 2017, p. 820).

Attending to the significance of these criticisms or limitations of counter and participatory mapping, we believe that the negative impacts of our participatory mapping project in Rantau Baru is not so grave. As for the first criticism, the customary territory of Rantau Baru was historically recognized by the Pelalawan kingdom (Osawa and Binawan, Chap. 3). The rough location of the boundaries has been shared not only with Ranatu Baru villagers but also the members of the neighboring villages, and the ambiguity, flexibility and dynamism of land use have already been limited in this area. As for the government recognition of participatory maps, although the central government has not recognized the maps created by local residents, the maps can be judicially admitted as evidence and administratively adopted as spatial planning materials at local government levels (Radjawali et al. 2017, pp. 828–829). Finally, as for the power generated through mapping process, while creating a map to show land use in the village (see Fig. 10.4), we avoided identifying individual land holdings because this may cause unintended negative consequences in the community (cf. Okamoto et al., Chap. 2). Instead, we focused on identifying companies that legally or illegally used the land in Rantau Baru traditional territory.

We suppose three concrete usages of the resulting maps. First, the maps can be used as tools to help the community negotiate with the government, companies, and other villages during land disputes. During the previous three decades, Rantau Baru customary territory has been legally and illegally encroached upon by oil palm companies and neighboring villages, which remains a threat in the future. The maps will contribute to defense of customary territory in informal negotiations or in court proceedings. Second, in Pelalawan, the district government has not designated *wilayah adat* and *adat* communities. The maps can be an agent to drive the local government to implement related policies and can act as a reference material to designate the *wilayah adat* and *adat* community in the future. Finally, visualizing the reality of land use enables Rantau Baru villagers to gain new perspective on the detailed situation of the territory, through which it may also be possible to improve understanding of other related issues, such as peatland degradation and ways of raising local standards of living (cf. Chakib 2014, pp. 48–49). The maps and mapping process could positively affect the way residents use their lands and improve future decision making regarding the protection and sustainable use of the natural surroundings.

In any case, the mapping process is the first step to change the difficult position of the *adat* community. The maps created should not be regarded as ultimate solutions to their struggle, but as tools to propel their struggle against inequality and injustice in land rights (Dewi 2016, p. 102; Radjawali et al. 2017, p. 820). It is noteworthy here that the positive effects of mapping will be realized in continuous negotiation and collaboration among various stakeholders such as Rantau Baru villagers, local government, companies, neighboring villagers, and activists/scholars like us (see Conclusion of this and Conclusion chapter of this book).

10.3.2   Participatory Mapping and Customary Territory

To create the map of Rantau Baru, we adopted a participatory mapping method in which residents play an important role in the mapping process. This process abstracts their perspective on space, and the resulting map reflects their worldview or concept of cultural space (Pramono 2014). While various kinds of maps can be produced depending on a main concern, we focused on traditional territory and land use. These themes were chosen by the villagers who participated in the mapping, who comprised the incumbents of the matrilineal *adat* institutions (for example the *ninik mamak* and *anak jantan/ betina,* see Osawa and Binawan, Chap. 3), officials of the administrative village office, and ordinary villagers. They recognized that their customary territory has been eroded by oil palm companies, as well as land disputes with a neighboring village, and rapid changes in the local environment. They pointed out the need to visualize these issues and create the maps. The first map (Fig. 10.2) can be used as a reference to negotiate the village boundaries with neighboring villages, companies, and the government, and to apply for *wilayah adat* and *hutan adat* in the future. The second map (Fig. 10.3) can be used to understand the present land use and environment in their territory. We created the maps between November 2020 and January 2021.

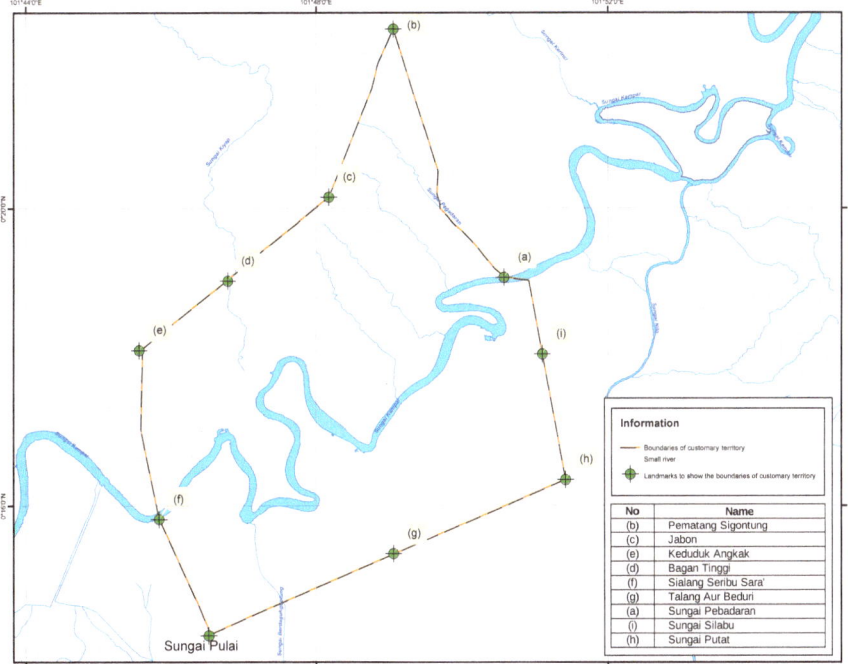

Fig. 10.2 Customary Territory of Rantau Baru

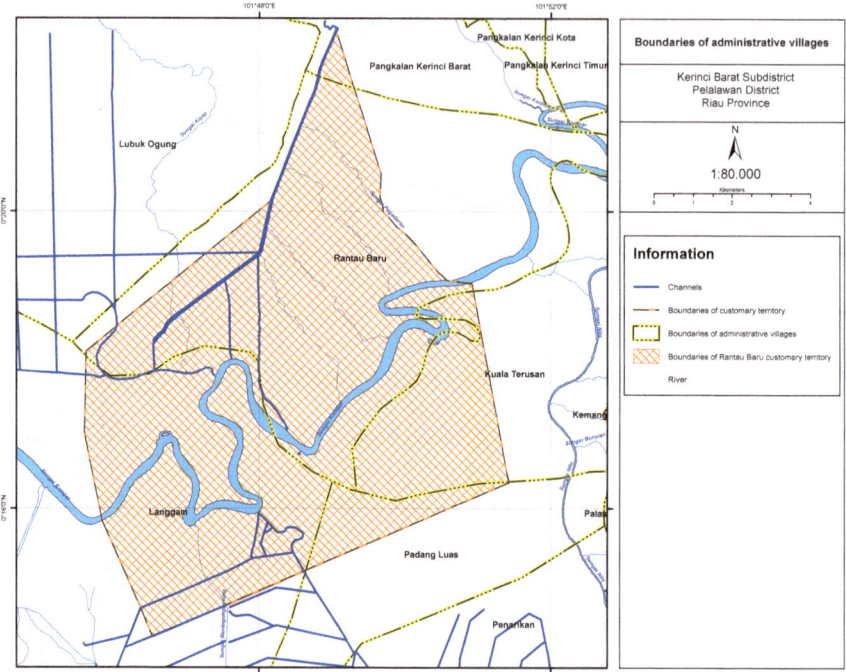

Fig. 10.3 Gap between the customary territory and the administrative village in Rantau Baru

Usually, people do not recognize the geospatial information of their living space and boundaries in the form of modern maps. Rantau Baru villagers are no different. They perceived space and boundaries in relation to natural and artificial landmarks, such as rivers, trees, roads, and buildings. It was necessary to place this information into a modern base map. We used the map of Rupa Bumi Indonesia (1:50,000 scale), which is a formal map issued by the Indonesian Geospatial Information Agency (*Badan Informasi Geospsial*, BIG), as the base map. This choice follows a Indonesian Geospatial Information Agency guideline for drawing maps of *wilayah adat*.[22]

First, we visited the main landmarks in the village territory with the villagers and checked the names of the points using a global positioning system (GPS) device. Then, we inputted the points into a base map after a discussion with the villagers. If necessary, we returned to the points to confirm their positions. After repeating these procedures, we drew the boundaries of the customary territory on the working map and identified the land use in the space. During this process, it was important to refer to high-resolution satellite imagery (Citra Satellite SPOT 7, 2017) as a supporting tool for the working map. The imagery includes information about natural and manmade landmarks such as rivers, canals, roads, vegetation, and buildings. At

---

[22] *Peraturan Badan Informasi Geospasial No. 12 Tahun 2017 tentang pedoman Pemetaan Wilayah Adat.*

the beginning, the informants struggled to align their understanding of the space with the imagery. However, they learned how to indicate the correct points on the images.

In creating the map of their customary territory, the identification of landmarks depended on the memories of elder and *adat* leader in the village. Originally, the boundaries of *pebatinan* were recorded and remembered in the form of folk song (*tombo*) in each *pebatinan*. Although folk song is changeable and the number of singers has decreased, the boundaries of *pebatinan* are clearly remembered by the elders and leaders. There is a general consensus about these customary boundaries between Rantau Baru villagers and those in neighboring villages because the *adat* leaders of *pebatinan* have traditionally been in communication with each other, and have negotiated territorial access. Nevertheless, some boundaries are ambiguous, which can cause conflicts with neighboring villages, as mentioned later.

In Rantau Baru, like other regions in Riau, the boundaries of customary areas are formed by lines that connect one point to another point. The "points" are named based on natural objects in the area, which typically include hills, rivers (particularly river mouths), and vegetation. For example, on the northern bank of the Kampar, the traditional area of Rantau Baru borders that of Pebatinan Kerinci (see Fig. 10.2).[23] The boundary begins with the mouth of the Pebadaran River (a) (Muara Pebadaran), then travels upstream to Pematang Sigontung (b), which is at the top of a hill. Then, Rantau Baru traditional area borders that of Pebatinan Sekijang, and the boundary turns toward Jabon (c), where there is a stand of *jabon*, or burflower trees (*Neolamarckia cadamba*). The boundary then passes Bagan Tinggi (d), which is a fishing ground in the Bokol Bokol River, and reaches Keduduk Angkak (e), a large thicket of the evergreen shrub *Melastoma candidum*. Keduduk Angkak is the boundary point with the customary area of Kepenghuluan Langgam.[24] Individual *sialang* trees are also important landmarks and boundary points. Sialang Seribu Sarak (f) marks the spot where a large *sialang* tree once grew. Although it has since died, the villagers remembered this point clearly.

During the mapping process, we could not identify all the landmarks at the correct points. This is because some vegetation and natural landmarks were lost due to expansion of oil palm plantations in the last twenty years. For example, there was formerly a patch of thatch screwpine (*pandanus odorifer*) at Talang Aur Berduri (g), but it has been completely replaced by oil palms. The Kampar tributaries of Sungai Putat (h) and Sungai Silabu (i) have been lost because canals were constructed across the tributaries, and the oil palm companies prohibit entry to the area. In these situations, we identified points depending on the memory of villagers and examination of high-resolution satellite imagery.

Figure 10.2 depicts the customary territory of Rantau Baru and was created using the participatory mapping method. The geographical landscape of Rantau Baru is

---

[23] Pebatinan Kerinci was divided into several administrative villages (*desa*), such as Lubuk Ogung and Kerinci Barat.

[24] Kepenghuluhan Langgam is a customary district that was recognized by the Pelalawan kingdom, like other *pebatinan* in the region.

between 101° 45′31.385 East Longitude–101° 51′23.276 East Longitude and 0° 14′ 13.748 North Latitude–0° 22′26.386 North Latitude. The topography is lowland with an altitude of 12 meters above sea level. The area of the customary territory is 8450 hectares and is crossed by the Kampar River, which occupies 458 hectares in the territory. The customary territory of Rantau Baru borders those of Kepenghuluan Langgam, Pebatinan Sekijang, Pebatinan Terusan, and Pebatinan Penarikan.

10.3.3   Ill-Defined Boundaries and Land Categories

In Riau, there is usually a significant gap between the boundaries of customary territories and an administrative village. This gap is caused by the different mapping processes. As mentioned above, although a customary territory can be identified through a detailed social investigation in a community, such procedures are not included in the mapping process of an administrative region. Administrative boundaries are delineated based on data from the district government and fixed by a decree and regulation stipulated by the district head under the control of the KDN. These boundaries would have been drawn based on a rough location survey and questionable assumptions.

For example, in Rantau Baru, a clear gap appears when we overlay the customary territory (*wilayah adat*) and administrative village boundaries on the Rupa Bumi Indonesia map created by the BIG. In Fig. 10.3, the area with a meshed pattern indicates the customary territory, and areas within the black and yellow dotted lines are administrative village boundaries. As seen in the map, the customary territory of Rantau Baru extends significantly south of its administrative boundaries and overlaps with areas within the administrative boundaries of Langgam, Kuala Terusan, Padang Luas, Penarikan and Pangkalan Kerinci Barat.

At present, the central government is promoting the creation of more detailed maps in accordance with the One Map Policy (Okamoto, Chap. 2; Dheny, Chap. 12). The KDN regulation No. 45 of 2016 regarding the determination of administrative village boundaries indicates that the boundaries should be drawn on the base map to a scale of 1:10,000. However, this map is not yet available. Even if it is completed, the boundaries of customary territories based on local agreements would not be reflected on the map, because BIG's mapping procedure does not involve social surveys that elucidate the boundaries of the customary territories. The lack of clear boundaries and adequate area designation on the base maps has resulted in land grabbing in the region.

10.3.4 Land Disputes Caused by Inaccurate and Unintegrated Maps

Lack of clear boundaries has caused land conflicts between Rantau Baru and a neighboring village. In approximately 2010, the village office of Pangkalan Kerinci Barat issued land and compensation letters (SKTs and SKGRs, respectively)[25] and sold approximately 200 hectares of land at the northern edge of Rantau Baru's customary territory to an oil palm company, Guna Dodos (see Fig. 10.4 and Table 10.2). This area includes both the forest and non-forest zones (see Figs. 10.1 and 10.4). In the state forest zone, it is prohibited to establish oil palm plantations. In the non-forest zone, in order to establish an oil palm plantation it is necessary to obtain a cultivation permit (*Hak Guna Usaha*, HGU) issued by the National Land Agency (Badan Pertanahan Nasional, BPN). Although Guna Dodos did not obtain the permit, it succeeded in establishing and running the oil palm plantation (see Sani 2015). Although the Rantau Baru villagers raised concerns about the ambiguous boundaries at the offices of Pangkalan Kerinci Barat and Guna Dodos, the company continues to operate.

Lack of adequate land titles has also accelerated land grabbing. In 2010, Kelompok Tani Bakti Bersama, a company based in Pangkalan Kerinci, began clearing approximately 200 hectares of land at the center of the Rantau Baru customary territory (see Fig. 10.4 and Table 10.2). This angered Rantau Baru villagers, and they accused Bakti Bersama of land grabbing. Although the sub-district office tried to mediate the conflict, it could not be resolved. A hearing was held at the Pelalawan District Assembly, after which the assembly dispatched a special team to the location (Fitri 2019; Terkini 2019). The team was also unable to resolve the dispute, because neither the KLHK nor the Pelalawan district government had granted land rights to the Bakti Bersama or Rantau Baru villagers. The legal affairs bureau of KLHK in Riau stated that the disputed land was categorized as a convertible production forest within the state forest zone—an area in which people cannot claim land rights. Additionally, the district government did not recognize the villagers' land rights because in 2005, the district head decided that the area around the disputed land was a "residential relocation area"[26] from which residents should be removed, because the area often suffered from seasonal floods (see also Osawa and Binawan, Chap. 3). Based on these designations, the government did not intervene in the land conflict, and the land remains as oil palm plantation under the control of Bakti Bersama, though with no established legal basis.

Lack of clear boundaries and adequate land designation may also lead to creation of new boundaries for specific political ends. This occurred in Riau in 2011 when the local communities of Padang Island within the Meranti Islands District protested a

---

[25]Land letter (*Surat Keterangan Tanah*, SKT) and compensation letter (*Surat Keterangan Ganti Rugi*, SKGR). See Osawa and Binawan, Chap. 3.

[26]*Surat Keputusan Bupati Pelalawan No KPTS/413.2/DKS/XII/2005/852 tanggal 28 Desember 2005 tentang Penetapan Relokasi Penduduk Kawasan Rawan Banjir.*

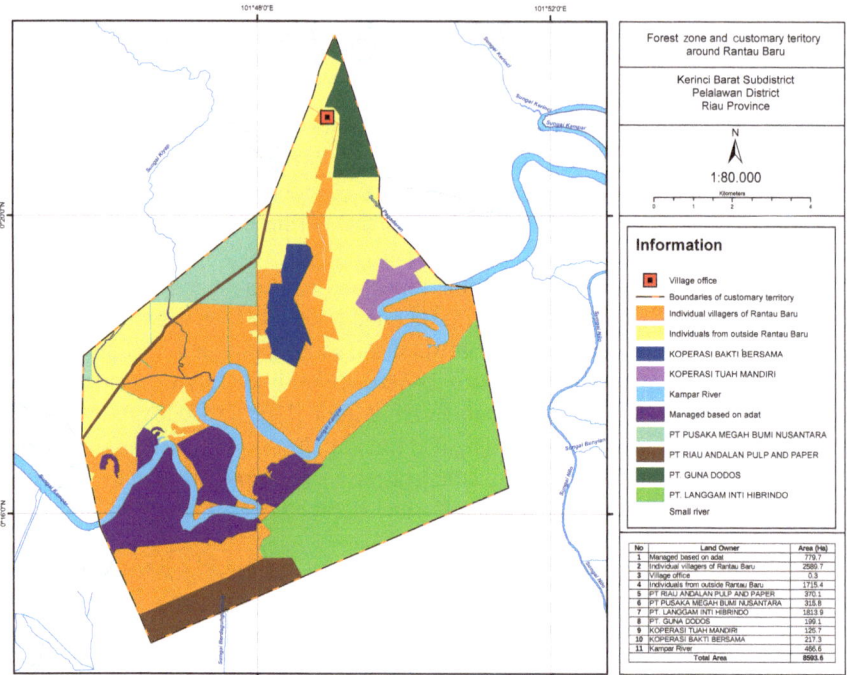

Fig. 10.4 Land users in Rantau Baru

Table 10.2 Land area according to land user

Land user		Area (Ha)
1	Managed based on *adat* (purple)	779.6
2	Individual villagers of Rantau Baru (orange)	2589.7
3	Village office	0.3
	Subtotal	**3369.6**
4	Individuals from outside Rantau Baru (yellow)	1715.4
5	Riau Andalan pulp and paper (RAPP) (brown)	370.1
6	Pusaka Megah Bumi Nusantara (emerald green)	315.8
7	Langgam inti Hibrindo (light green)	1, 813.9
8	Guna dodos (dark green)	199.1
9	Koperasi Tuah Mandiri (pink)	125.7
10	Koperasi Bakti Bersama (navy)	217.3
Total area managed by outsiders		**4757.3**
Total land area		**8126.9**
11	The Kampar River (purple)	466.6
Total area		**8593.6**

permit to establish an Industrial Plantation Forest (*Hutan Tanaman Industri*, HTI) granted to the acacia company Riau Andalan Pulp and Paper (RAPP). During the process of the negotiations between the villagers of Lukun Village and the Ministry of Forestry, ministry officials provided a map in which Lukun was divided into two administrative villages, Lukun and Tanjung Bunga. Nobody in the region knew that the administrative village Tanjung Bunga existed, however. The Ministry of Forestry invented a fictitious village, reduced the village area of Lukun, and tried to use the area subtracted from it for an acacia plantation (Salim 2017). This case shows that the government can unilaterally produce village maps according to their interests.

Since the Basic Forestry Law of 1967 government designation of forest production zones has typically ignored the territory of the *adat* communities traditionally inhabiting these zones. As the designation prioritizes the economic interests of the central and local governments, governments have not clearly established *adat* community boundaries and territory on official maps. This has allowed them to avoid conflicts challenging government interests and left open the possibility of manipulating boundaries and territories according to situational needs and interests. This inconsistent and situational approach has undermined the legitimacy of governance of rural forest areas in Indonesia. The lack of clear boundaries and territories has allowed outsiders to usurp customary *adat* community lands. Counter-mapping using a participatory method can expose the results of the government's situational and inconsistent management and the lawless situation in rural areas. Moreover, mapping can be seen as an attempt to reconstruct the customary territories, demonstrate the legitimacy of land use by the *adat* community, and recover inclusive land governance in rural areas.

10.4   Land Use and Environmental Management

10.4.1   Mapping Land Use in Rantau Baru

As shown in Fig. 10.1, part of Rantau Baru's customary territory is categorized as "other use" area, and oil palm and acacia companies have obtained permission to use some of these areas. Because other parts of the territory are categorized as convertible production forest, territorial land use rights have been the object of deal-making and possession through SKTs and SKGRs (see Osawa and Binawan, Chap. 3).

Figure 10.4 depicts the various land users in Rantau Baru. The orange-colored area is managed by Rantau Baru individuals. Although some hold SKTs, others have used the land customarily without official title. This orange area includes villager homesteads, rubber and oil palm gardens, broad swathes of uncultivated swampy grassland, and small forests. The purple area is the *sialang* area (*kepung sialang*) and the Kampar River managed by three matrilineal groups (see Osawa and Binawan, Chap. 3). While not colored on the map, the narrow area (0.3 hectares; see Table 10.2) along the boundaries of the territory is regarded as land of the administrative village. Other areas are controlled by village outsiders. The yellow area is land used by individuals from outside the village who bought it by obtaining SKTs and SKGRs from the village office or villagers (Osawa and Binawan, Chap. 3). This area includes grasslands and oil palm gardens. The remaining areas are used by oil palm and acacia companies. The areas in navy at the center of the territory and deep green at the north edge are used by Bakti Bersama and Guna Dodos, respectively. The brown area along the south edge of the territory is used by RAPP for acacia plantations.

Based on Fig. 10.4, Table 10.2 calculates the area managed by each kind of actor. Villagers manage 41 percent (or 3369.6 ha) of the total land area of Rantau Baru (8126.9 ha), while 55 percent (4757.3 ha) of village land is controlled by outsiders. Until approximately 1990, almost all the customary territory was under village management and control. However, since 1990, oil palm and acacia companies have encroached on the territory, which has resulted in deforestation of the customary territory.

10.4.2 Deforestation and Peatland Problems

The expansion of oil palm and acacia plantations in and around the customary territory began in the 1990s. At the end of the 1980s, the industrial road was constructed, and the area became the target of oil palm plantation extensions. In 1992, the Minister of Forestry issued a decree that changed the function of the convertible production forest (an area of 7087 ha) to an "other use" area for Pusaka Megah Bumi Nusantara, an oil palm company of the Asian Agri Group (Eyes on the Forest 2015a).[27] This company operates a 316 ha oil palm plantation at the north-western edge of the customary territory (see Fig. 10.3 and Table 10.2). In 1995, the Minister of Forestry issued a decree to change 1296 ha of the forest zone to other use areas for Langgam Inti Hibrindo, an oil palm company of the Provident Agro Group (Eyes on the Forest 2015b). This company obtained a HGU in 1999 from BPN (No. 110/HGU/BPN/99). Their oil palm plantation occupies the southeastern area of the customary territory. These concessions were completed without any negotiations with the Rantau Baru community, and the villagers did not receive any compensation for the land. These companies deforested the areas and constructed numerous canals to drain peatland (see Figs. 10.2, 10.3, and 10.4). Several researchers have demonstrated that the construction of canals and water blockages in peatlands influences the surface water and groundwater levels and can cause peatland fires (Jaenicke et al. 2011; Susilo et al. 2013). In Rantau Baru, the customary territory has suffered from frequent forest and peatland fires since the latter half of the 1990s. The burnt space became swampy grassland/bush. Since the 2000s, part of the opened

---

[27] The report by Eyes on the Forest (2015a, b, c), an alliance of local NGOs in Riau, states that they have not confirmed the HGU of this company.

Table 10.3 Land surface types in the customary territory (1990–2019) (Kementerian Lingkungan Hidup dan Kehutanan 2019)

No	Land surface type	Area (ha)		
		1990	2000	2019
1	Water	460.41	441.73	
2	Swampy grassland/bush	1742.39	4367.14	3823.77
3	Secondary swampy forest	6217.63	493.79	398.36
4	Opened land	61.63	472.17	177.95
5	Plantation	0	2434.04	3472.46

land has been sold to urban residents and companies and used as oil palm plantations. Additionally, some Rantau Baru villagers planted oil palm seedlings in these areas. However, oil palm cultivation in peatland in this region is difficult, as the land often suffers from peatland fires and seasonal floods (Osawa and Binawan, Chap. 3).

Forested area in the customary territory declined dramatically between 1990 and 2000. Table 10.3 depicts the land surface types in Rantau Baru's customary territory (which indicate land use). The Department of Forestry (and KLHK in 2019) release the data of the land surface types after analyzing satellite images. We superimposed the map of the customary territory on the KLHK maps and calculated the areas according to each use.

Although Rantau Baru had 6217 ha of forests in 1990, this had been reduced to less than 400 hectares by 2019. Conversely, the plantation area has dramatically increased from zero ha in 1990, to more than 3400 ha in 2019. It is noteworthy that the 1742 ha of swampy grassland/bush in 1990 had expanded to 4367 ha by 2000. This is related to forest fires, which repeatedly occurred during the 1990s; the swampy forest was burnt, converting it into swampy grassland/bush. Part of this area was in turn converted to plantations in 2019.

Before the 1990s, Rantau Baru was covered by thick forests, and fires were rare. Rantau Baru villagers protected their *sialang* areas and did not over-exploit the natural resources. However, after the construction of the RAPP road, the expansion of oil palm plantations, and repeated peatland and forest fires, forested areas have dramatically decreased. The fires create a haze that harms villagers' health. The production of honey in the *sialang* area is also decreasing due to the deforestation and haze. These environmental problems have been caused by government and company exploitations of peat swamp forests, which are made possible by the inconsistent and situational mapping of customary territories.

10.5   Conclusion: Suggestions for a New Governance System

When Riau suffered from large peat forest fires in 2015, I, Akhwan Binawan, co-author of the present paper, was in Pekanbaru and could not imagine how wide the peatland fires had spread. I remember how hard it was to breathe for almost 2 months as the air was polluted by haze. Visibility was very low, and my vision could only penetrate the smog for 50 meters. However, the situation in Rantau Baru was worse. According to the villagers, visibility was less than 20 meters. Almost all the villagers developed bronchitis. While the *sialang* areas did not suffer from the fires, the haze would have impacted the bees in the *sialang* trees. At night, villagers could see the light of the fires in the hinterlands, and the fires may have come within several dozens of meters of the settlement. In Rantau Baru, the fires were closer to the settlement area than in Pekanbaru. Eyes on the Forest (2015a, b, c), a NGO network, conducted field investigations just after the fires and confirmed the burned peatlands in several oil palm and acacia plantations around Rantau Baru, including Langgam Inti Hibrindo, and Pusaka Megah Bumi Nusantara.

The damage to the Rantau Baru *adat* community can be measured not only by the encroachment on their customary territory, but also by the fires and haze that occur close to their settlement. Although they lost their customary territory and their living environment is getting worse, the villagers have not received compensation equivalent to their losses. It is necessary to reinforce their land titles and make it possible for them to once again manage the land and resources. When the concession permissions of the companies have expired, the land should be returned to management by the Rantau Baru *adat* community.

One way to use the returned land is to manage and control the land based on traditional knowledge or local wisdom. As Osawa (in Chap. 6) points out, local traditional wisdom has been idealized and might be unable to resolve all environment problems. However, before the 1990s, Rantau Baru villagers protected and managed *sialang* forests and the peat environment without significant problems. If land rights are established, villagers can invest money and labor into the land from a long-term perspective, grow the *sialang* trees, and regenerate the forests. Support from the government, companies, and NGOs such as Hakiki which supports maintenance and expansion of honey collection and horticulture in Rantau Baru, would be essential in this project.

It would be impossible for the villagers to manage the vast returned area as *sialang* forest. Furthermore, as mentioned in Chap. 3, some villagers have sold land use rights to outsiders in order to improve their short-term livelihood prospects. It is essential to generate the villagers' livelihoods and raise their living standards so that they can reconsider this short-term perspective. In the meantime, however, some part of *adat* lands may be leased out to companies or urban residents to operate oil palm or acacia plantations. The main focus for now should be to create a governance system in which the villagers themselves can choose the land use, and the village office or villagers can receive rent or share the profit from plantations or any other use. By establishing such a system, we can expect that the villagers would not sell land rights to outsiders so easily and would be interesting to positively engage in the environmental management of the customary area.

Such suggestions require creation of integrated and detailed maps with accurate and useful land designations. Although participatory map-making takes time, it should be completed in order to improve governance of rural lands, natural resources, and environment in Indonesia. We hope that the participatory maps created in Rantau Baru will contribute not only to protecting land rights, but also to creating an integrated and detailed map that recognizes the customary territory of the Rantau Baru community in the future.

References

Arizona Y, Cahyadi E (2013) The revival of indigenous peoples: contestations over a special legislation on *masyarakat adat*. In: Hauser-Schäublin B (ed) Adat and indigeneity in Indonesia: culture and entitlements between heteronomy and self-ascription. Göttingen University Press, Göttingen, pp 43–62

Bourchier D (2007) The romance of adat in the Indonesian political imagination and the current revival. In: Davidson JS, Henley D (eds) The revival of tradition in Indonesian politics: the deployment of adat from colonialism to indigenism. Routledge, Oxon, pp 113–129

Burns P (1989) The myth of adat. J Leg Plur 28:1–127. https://doi.org/10.1080/07329113.1989.10756409

Chakib A (2014) Civil society organizations' roles in land-use planning and community land-rights issues in Kapuas Hulu regency, West Kalimantan, Indonesia. Working paper 147. CIFOR, Bogor. https://doi.org/10.17528/cifor/005426

Constituteproject.org (2021) Indonesia's Constitution of 1945, reinstated in 1959, with amendments through 2002. https://www.constituteproject.org/constitution/Indonesia_2002.pdf?lang=en. Accessed 29 Dec 2021

Dewi R (2016) Gaining recognition through participatory mapping? The role of adat land in the implementation of the Merauke integrated food and energy estate in Papua, Indonesia. ASEAS 9(1):87–105. https://doi.org/10.14764/10.ASEAS-2016.1-6

Duncan CR (2004) From development to empowerment: changing Indonesian government policies toward indigenous minorities. In: Duncan CR (ed) Civilizing the margins: southeast Asian government policies for the development of minorities. Cornell University Press, Ithaca, pp 86–115

Eyes on the Forest (2015a) Pemantauan pembakaran hutan dan lahan di perkebunan PT Pusaka Megah Bumi Nusantara. https://www.eyesontheforest.or.id/uploads/default/report/Pusaka_Bumi_Megah_Nusantara_edit.pdf. Accessed 29 Nov 2021

Eyes on the Forest (2015b) Pemantauan pembakaran hutan dan lahan di perkebunan PT Langgam Inti Hibrindo. https://www.eyesontheforest.or.id/uploads/default/report/Eyes-on-the-Forest-Laporan-Cek-Lapangan-PT-Langgam-Inti-Hibrindo-Karhutla-Desember-2015.pdf. Accessed 29 Nov 2021

Eyes on the Forest (2015c) Pengecekan lapangan di 37 lokasi terdeteksi titik api 2015: investigasi Oktober–November 2015. Laporan Investigatif Eyes on the Forest Desember 2015. https://www.eyesontheforest.or.id/uploads/default/report/Eyes-on-the-Forest-Laporan-Investigatif-Pembakaran-hutan-lahan-di-37-lokasi-Riau-Desember-2015.pdf. Accessed Nov 29, 2021

Fasseur C (2007) Colonial dilemma: Van Vollenhoven and the struggle between adat law and Western law in Indonesia. In: Davidson JS, Henley D (eds) The revival of tradition in

Indonesian politics: the deployment of adat from colonialism to indigenism. Routledge, Oxon, pp 50–67

Fitri (2019) Penyidik polda Riau cek lahan 300 ha bersama Koptan Bhakti Bersama dan warga. Marwah Kepri, 12 Sep. https://marwahkepri.com/2019/09/12/penyidik-polda-riau-cek-lahan-300-ha-bersama-koptan-bhakti-bersama-dan-warga/. Accessed 22 Dec 2021

Fitzpatrick D (2007) Land, custom, and the state in post-Suharto Indonesia: a foreign lawyer's perspective. In: Davidson JS, Henley D (eds) The revival of tradition in Indonesian politics: the deployment of adat from colonialism to indigenism. Routledge, Oxon, pp 130–148

Fox J, Suryanata K, Hershock P et al (2008) Mapping boundaries, shifting power: the socio-ethical dimensions of participatory mapping. In: Goodman MK, Boykoff MT, Evered KT (eds) Contentious geographies: environmental knowledge, meaning, scale. Ashgate Publishing, Oxon, pp 203–217

Henley D, Davidson JS (2007) Introduction: radical conservatism – the protean politics of adat. In: Davidson JS, Henley D (eds) The revival of tradition in Indonesian politics: the deployment of adat from colonialism to indigenism. Routledge, Oxon, pp 1–49

Jaenicke J, Englhart S, Siegert F (2011) Monitoring the effect of restoration measures in Indonesian peatlands by radar satellite imagery. J Environ Manag 92(3):630–638. https://doi.org/10.1016/j.jenvman.2010.09.029

Kementerian Lingkungan Hidup dan Kehutanan (2019) Data penutupan lahan Kementerian Lingkungan Hidup dan Kehutanan. http://webgis.menlhk.go.id/. Accessed 20 Nov 2021

Kompas (2020) Kisah Pak Bongku warga Suku Sakai, Dipenjara gara-gara tanam ubi, bebas karena asimilasi. Kompas, 13 Jun. https://regional.kompas.com/read/2020/06/13/09134611/kisah-pak-bongku-warga-suku-sakai-dipenjara-gara-gara-tanam-ubi-bebas-karena?page=all. Accessed 29 Nov 2020

Li TM (2007) The will to improve: governmentality, development, and the practice of politics. Duke University Press, Durham

McCarthy JF (2006) The fourth circle: a political ecology of Sumatra's rainforest frontier. Stanford University Press, Redwood City

Menteri Lingkungan Hidup dan Kehutanan (2016) Keputusan Menteri Lingkungan Hidup dan Kehutanan Nomor 903/ MENLHK/ SETJEN/ PLA.2/ 12/ 2016. In: tentang Kawasan Hutan di Provinsi Riau. Kementerian Lingkungan Hidup dan Kehutanan Republik Indonesia, Jakarta

Mizuno K, Kusumaningtyas R (2016) Land and forest policy in Southeast Asia. In: Mizuno K, Fujita MS, Kawai S (eds) Catastrophe and regeneration in Indonesia's peatlands: ecology, economy and society. Kyoto CSEAS series on Asian Studies, vol 15. NUS press, Singapore; Kyoto University Press, Kyoto, pp 19–68

Moniaga (2007) From bumiputera to masyarakat adat: a long and confusing journey. In: Davidson JS, Henley D (eds) The revival of tradition in Indonesian politics: the deployment of adat from colonialism to indigenism. Routledge, Oxon, pp 275–294

Nugraha I (2019) RUU masyarakat adat masuk prolegnas 2020, berikut masukan para pihak. Mongabay, 13 Dec. https://www.mongabay.co.id/2019/12/13/ruu-masyarakat-adat-masuk-prolegnas-2020-berikut-masukan-para-pihak/. Accessed 24 Nov 2021

Peluso NL (1995) Whose woods are these? Counter mapping forest territories in Kalimantan, Indonesia. Antipode 27(4):383–406. https://doi.org/10.1111/j.1467-8330.1995.tb00286.x

Pramono AH (2014) Perlawanan atau pendisiplinan? Sebuah reflecsi kritis atas pemetaan wilayah adat. Wacana 33:199–233

Radjawali I, Pye O, Filtner M (2017) Recognition through reconnaissance? Using drones for counter-mapping in Indonesia. J Peasant Stud 44(4):817–833. https://doi.org/10.1080/03066150.2016.1264937

Safitri MA, Nagara G (2015) Mendesaknya kaji ulang peraturan: pokok-pokok pikiran untuk perbaikan regulasi pengukuhan kawasan hutan di Indonesia. Policy Paper, vol 1. Epistema Institute, Jakarta

Salim MN (2017) Mereka yang dikalahkan: perampasan tanah dan resistensi masyarakat Pulau Padang. STPN Press, Yogyakarta

Sani A (2015) Polisi paksa PT Guna Dodos ikut padamkan lahan yang terbakar. Merdeka.com, 24 Sep. https://www.merdeka.com/peristiwa/polisi-paksa-pt-guna-dodos-ikut-padamkan-lahan-yang-terbakar.html. Accessed 23 Dec 2021

Siscawati M (2014) Masyarakat adat dan perebutan penguasaan hutan. Wacana 33:3–24

Susilo GE, Yamamoto K, Imai T et al (2013) Effect of canal damming on the surface water level stability in the tropical peatland area. J Water Environ Technol 11(4):263–274. https://doi.org/10.2965/jwet.2013.263

Terkini (2019) Masyarakat Rantau Baru minta Pemda Pelalawan serius. Detikperistiwa.com, 16 Apr. https://www.detikperistiwa.com/news-110993/masyarakat-rantau-baru-minta-pemda-pelalawan-serius.html. Accessed 21 Dec 2021

Vinolia I (2021) Jalan terjal pengakuan hutan adat di Riau. Ekuatrial, 19 Jan. https://www.ekuatorial.com/2021/01/jalan-terjal-pengakuan-hutan-adat-di-riau/. Accessed 24 Nov 2021

WALHI (2018) Ancaman tambang dan perampasan tanah adat dalam Perda Riau Nomor 10/2015 dibatalkan Mahkamah Agung. Walhi, 22 Aug. https://www.walhi.or.id/index.php/ancaman-tambang-dan-perampasan-tanah-adat-dalam-perda-riau-nomor-10-2015-dibatalkan-mahkamah-agung. Accessed 28 Nov 2021

Warman K (2014) Peta perundang-undangan tentang pengakuan hak masyarakat hukum adat. https://procurement-notices.undp.org/view_file.cfm?doc_id=39284. Accessed 26 Oct 2021

Chapter 11
The Inequity Implications of Peatland Conservation Policies

Maho Kasori

Abstract Implementation of conservation policies can result in negative impacts and exacerbate existing disparities, yet studies of these risks in peatland communities are minimal. This chapter identifies the equity implications of peatland conservation policies on local communities and suggests appropriate policy directions. The study is based on field work and a survey of 63 randomly selected households (22% of total households) in a multi-ethnic village of Riau Province, Indonesia that has both peatland and non-peatland. Analysis reveals that the make-up and efforts of the local firefighting group, the *Masyarakat Peduli Api* (MPA), do not include all villagers, suggesting that only specific villagers make an effort to participate in peatland conservation activity. Participation in Livelihood Improvement Programs is also limited, with high-earning households that own non-peatland benefiting the most, thus threatening to accelerate existing economic disparities. There is a strong need for policies and programs that mitigate and correct these disparities while taking into account the diverse nature of peat communities and fires.

Keywords Peatland conservation · Community firefighting · Livelihood improvement · Conservation equity

11.1 Introduction

Peatland fires in Indonesia first became a serious problem during the El Niño events of 1997–1998. Human activity, such as burning to clear land and cigarette littering, can trigger large-scale peatland fires in dry conditions. An analysis of fire hotspots in 2009 conducted in 2013 found that 78% of fires began outside forest areas and the main reason for the spread of these fires was the development of land for agriculture (Okamoto 2013).

M. Kasori (✉)
Graduate School of Asian and African Area Studies, Kyoto University, Kyoto, Japan

© The Author(s) 2023
M. Okamoto et al. (eds.), *Local Governance of Peatland Restoration in Riau, Indonesia*, Global Environmental Studies,
https://doi.org/10.1007/978-981-99-0902-5_11

As worsening peatland fires continued to raise international attention in the 2000s, the Indonesian government enacted laws and regulations to manage and conserve peatlands. These include Environmental Protection and Management Law No. 32/ 2009, which prohibits all clearing of land by fire (either for opening up new land or managing existing fields), also known as the "zero burning policy," and the 2011 Presidential Directive on the Promotion of Forest and Land Fire Management,[1] which stipulates the roles of local and national government entities[2] in forest and land fire prevention (Okamoto 2013).

While laws to conserve peatland were being developed, Indonesia's largest ever peatland fire broke out in 2015, triggering multiple fires and causing haze-induced illnesses across the country and in neighboring countries. According to World Bank (2016), Indonesia's agriculture and forestry sectors "sustained estimated losses and damages of US$8.8 billion (Rp120 trillion) in 2015." The 2015 fires were expected to cause "additional losses of about US$800 million per year for the next three in the case of estate crops (e.g., palm oil, rubber, and coconut) and five years for forests." The report noted that, "Damages to estate crops affected companies and small-holder farmers. Costs to food crops (US$1.7 billion) translate into lower incomes for farmers and possible impacts on food security."

Realizing the need for cross-ministerial coordination and action to better manage peatland, President Joko Widodo (Jokowi) inaugurated the Peatland Restoration Agency (*Badan Restorasi Gambut*, BRG) in 2016. The BRG targeted more or less two million hectares[3] of degraded and drained peatland for restoration using the "3Rs" approach, or Rewetting dry peatlands, Revegetating tree cover, and Revitalizing local livelihoods and economies. The BRG was under the direct control of the president from 2016 to 2020, and as of December 2020, an estimated 1.4 million hectares of peatlands had been restored.[4]

Since the BRG's establishment, studies have focused on evaluating the results of the 3Rs approach. Research on peatland governance has demonstrated both positive and negative impacts on local livelihoods depending on socioeconomic environments, traditional livelihoods, and the ethnicity of individuals. For example, the zero-burning policy of 2009 extended to traditional farmers practicing small-scale fire-based agriculture, forcing them to abandon their cheap and easy land management method and look for alternatives (Nurlia et al. 2021). Smallholders who did not

---

[1] Improved Forest and Land Fire Control (Executive Order No. 16/2011).

[2] Such as the Ministry of Forestry, the Ministry of Agriculture, the Ministry of the Interior, the Ministry of the Environment, and the National Disaster Management Agency.

[3] The restoration target area was revised several times. According to the latest decision of the Peat Restoration Agency about the restoration target are (No. 16/2018), the area is 2.67 million ha.

[4] In 2020, the Peat and Mangrove Restoration Agency (*Badan Restorasi Gambut dan Mangrove*, BRGM) was launched with the aim of restoring the remaining 1.2 million hectares (thereby meeting the BRG's original target), as well as an additional 0.6 million hectares of mangroves. Like the BRG, the BRGM hopes to not only restore peatlands, but also to improve the livelihoods of local people living in them. Another aim of the BRGM is to cooperate with stakeholders, including ministries, local governments, researchers, national and international NGOs, and local people.

have large machinery needed to pay for labor to cut and stack all biomass on designated strip lines (Murniati and Suharti 2018; Watts et al. 2019). Although the BRG introduced a method to manage land without burning (*Pengelolaan Lahan Tanpa Bakar*, PLTB) in some villages, Daeli et al. (2021) found that overall, the policy negatively affected swidden farmers' livelihoods. However, based on a field survey of swidden farmer communities in West Kalimantan, the researchers (Daeli et al. 2021) found that the impact of the zero-burning policy varied according to the ecological landscape and alternative livelihood opportunities of each location. Although previous research was conducted in less diverse swidden farmer communities, contemporary peatland communities are much more diverse due to an influx of migrant populations. In Riau Province, the percentage of migrants from Java increased significantly between 1980 and 1990 as a result of Indonesia's transmigration policies (Koizumi and Nagata 2018). An analysis of the 2000 and 2010 population censuses reveals a rich ethnic and livelihood diversity in Riau Province (Koizumi and Nagata 2018). Thus, there is a need to further investigate the impacts of the zero-burning policy from the viewpoint of diversities as well.

The acceptance or rejection of peatland management based on the 3Rs approach also differs among ethnic groups and individuals due to different livelihoods and the distribution of costs and benefits. Ward et al. (2021) find that indigenous households are more likely to support rewetting projects than transmigrant households, while Knieling (2020) points to a general lack of interest in peat conservation by the community that was studied. As part of its Revitalization strategy, the BRG assumes that "the cultivation of peatland-friendly crops will improve the livelihoods of local people." However, local livelihoods in peatland areas are not only on-farm (crop cultivation and wage labor on oil palm plantations) but also off-farm (including gold mining and fishing); it is therefore necessary to pay policy attention to such off-farm livelihoods (Silvianingsih et al. 2020; Nurlia et al. 2021; Januar et al. 2021). As an example, Thornton et al. (2020) demonstrates that conservation of the fishing environment supports an important livelihood culture in peatlands. Indeed, the 3R approach should pay ample attention to fishing activities, which are a traditional livelihood of the Malay living in peat swamp forest areas.[5]

A review of existing literature reveals both negative and positive impacts of peatland governance on local people. However, these studies are often not based on empirical research and tend to examine a single policy or a single ethnic community. This study aims to contribute to the literature by examining a multi-ethnic peatland community where the population engages in multiple livelihoods. It investigates the success of fire prevention programs and the accessibility and benefits of livelihood programs to empirically uncover any disparities in the impacts of peatland conservation governance in Indonesia.

---

[5] The local Malays did not open, or clear, peatlands until after migrants settled in peat swamp forests and the environmental and social conditions dramatically changed (Furukawa 1992).

11.2   Research Site and Methodology

11.2.1   Location and Geographical Features of the Study Site

The study site is R Village, located in Siak District, Riau Province, which faces the
Straits of Malacca and is 154 km from the capital city of the province (Fig. 11.1). The
R Village area was part of S Village until 2010, when it became an independent
village (Fig. 11.2). The total area of R Village is 16,803 ha (BPS Kabupaten Siak

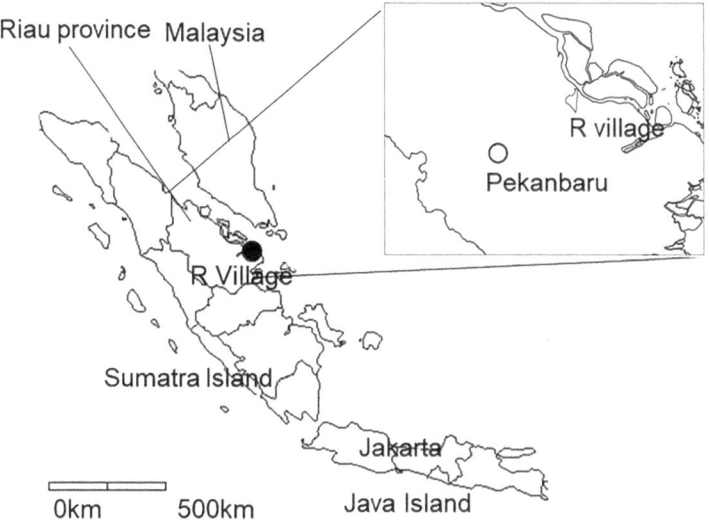

Fig. 11.1 The location of the study site

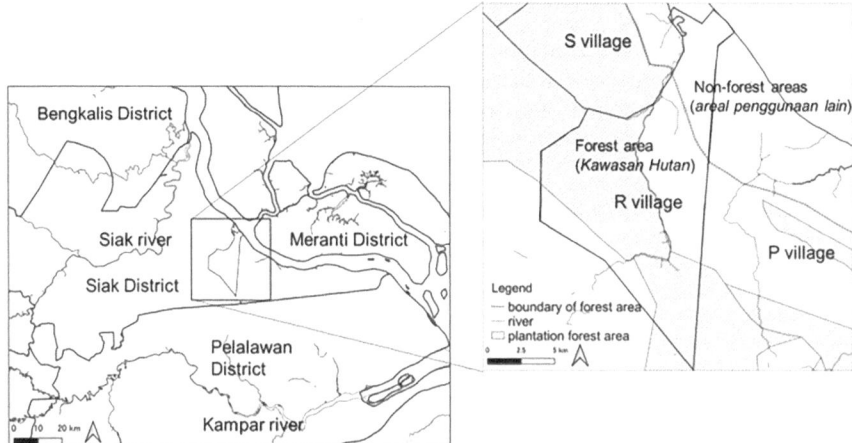

Fig. 11.2 An overview of the area around the study site

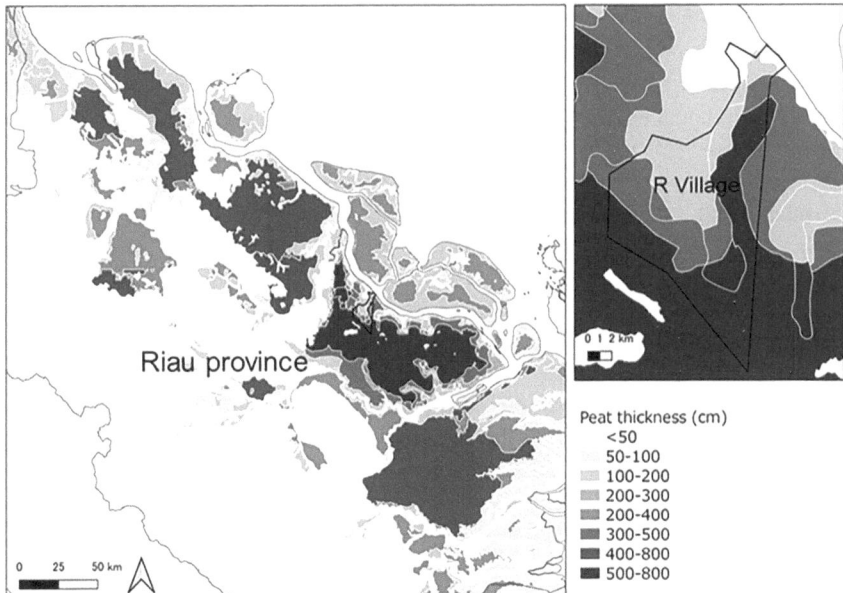

Fig. 11.3 Distribution of peatland in the study site

2017), and the total population amounts to 869 people, who live in 224 households (BPS Kabupaten Siak 2019). The village is selected for its ethnic diversity and distribution of both peatlands and non-peatlands. In addition, the village's community-based fire control group (*Masyarakat Peduli Api*, MPA) has been attracting attention from outside the village in recent years due to its active peatland conservation activities through cooperation with external actors, which has become more active and appreciated since the BRG was established.

There are both mineral and peat soils in R Village. Mineral soil is located in the coastal area and river basins, and peat soils become thicker as one moves inland from the coastal area (Fig. 11.3). All lands in Indonesia are classified into one of two types. The first is forest areas (*kawasan hutan*, KH). This type represents an area of about 124 million ha, or two-thirds of Indonesia's landmass, and it falls under the administration of the Ministry of Environment and Forestry. The second group is non-forest areas (*areal penggunaan lain*, APL), which covers an area of about 64 million ha and is under the administration of the Indonesian National Land Office (*Badan Pertanahan Nasional Republik Indonesia*, BPN) (Siscawati et al. 2017). The settlements and farmlands of the local people in R Village are located along the coast and are designated as APL lands. About 500 ha of private oil palm plantations are also located in the APL zone. These oil palm plantations are owned by Chinese people living outside of R Village, and local villagers have worked as agricultural day laborers on the plantations since around 2000. The remaining inland forest and acacia plantations are designated as KH.

11.2.2 Ethnic Diversity of the Study Site

Malay people in the study site distinguish themselves based on the time of their migration. Indigenous Malays, or those whose families have lived in the village for several generations, are simply described as *"Orang Melayu"* (Malay people) or *"Orang Melayu yang asli di kampung sini* (indigenous Malay of this village)." On the other hand, "newcomer" Malays, who migrated from the middle and upper reaches of the Kampar River during the past few decades, are called *"Orang Melayu Kampar* (Kampar Malay people)" and are distinguished from the indigenous Malays in the study site.

In this chapter, I refer to the indigenous Malay as "local Malay" and the migrant Malays as "Kampar Malay" according to the local categorization. Official Indonesian statistics do not recognize this distinction, but locals differentiate people from the two groups based on their birthplace, personal history, and language intonation. This chapter follows this local distinction. In recent years, marriage between the local Malay and Kampar Malay have become common, and ethnic boundaries are blurring. However, the distinction between the two based on differences in accent still remains. In addition, Malays who migrated from outside village upon marriage in the past few decades also live in the village. They called themselves *"Orang Melayu,"* not *"Orang Melayu Kampar,"* but they are not indigenous to the study site. Thus, I refer to these more recent migrant Malays as "non-local Malay" to distinguish them from the "local Malay" and the "Kampar Malay."

The major ethnic groups of the village are the local Malay and the Javanese, who migrated to the village voluntarily. The Javanese and Malays are both Muslims and there are many cases of intermarriage among these groups. Although intermarriage has blurred ethnic divisions in terms of lineage, the livelihoods, living spaces, social networks, and even the accents of the villagers are still ethnically divided. Traditionally, the local Malay preferred to engage in fishing activities in coastal areas and lived near mangrove swamp forests in the study site. In the 1990s, the Javanese began to cultivate agricultural crops in inland forest areas. Thus, they tended to build their houses about 1 km inland from the coast. A Javanese man who migrated from Java in the 1980s explained, "We [Javanese] are peasants (*petani*), and we opened the land for agriculture in R Village. The local Malay are fishermen and only began imitating us and gradually engaging in agriculture in the 2000s."

Other ethnic groups living in R Village include the Kampar Malay people, who voluntarily migrated (*merantau*) from Pelalawan District to Siak District; the Minangkabau, who migrated from West Sumatra; the Batak from North Sumatra; the Bugis from Sulawesi; Chinese from other areas; the Sasak from Lombok; and the Suku Asli (see the footnote 6). The Malay live in the coastal areas of the study site, while the Kampar Malay people migrated from the Kampar River Basin in Pelalawan District to S Village in the 1980s and settled in the inland forest. Therefore, the local Malay consider the Kampar Malay to be migrants. The Kampar Malay fish in rivers and gather non-timber forest products in the inland forest, and they tend to build their houses inland near the rivers. These differences in traditional livelihood activities led to the formation of separate residential spaces for each ethnic group.

11.2.3 History of the Study Site: Changing Livelihoods and Ecology

Even before Indonesian independence, people known as Rawa and the local Malay lived in the area of S Village and southeast of S Village (in P Village). In recent years, the Rawa have started to call themselves "Suku Asli Anak Rawa" (hereafter "Suku Asli").[6]

Because the local Malay people fished in the rivers and coastal areas for a living, they established a settlement in S Village.[7] Suku Asli did not have permanent houses, but rather practiced semi-nomadic ways of life, in which they moved from place to place every few years and depended on hunting animals, gathering forest products, and cultivating swidden fields in the hinterland forest. The Indonesian government's 1984 resettlement policy "settled" Suku Asli in permanent houses in the coastal areas of P Village. According to the policy, each household received a house, food, and 2 ha of agriculture land, and were encouraged to change their traditional way of life.

Since the 1980s, Javanese people from Java and other areas of Indonesia and Kampar Malay people from the middle and upper reaches of the Kampar River in Pelalawan District migrated to S Village to take advantage of logging opportunities. In addition, non-local Malay have voluntarily migrated to S Village for marriage.

During the 1980s, logging companies were granted concessions to operate in KH lands. According to Decision No. 173 of the Ministry of Forestry of 1986,[8] a part of the coastal area inhabited by the local people of the study site is designated as non-KH. Local people in R Village report that until the early 2000s, the main livelihood of local people was illegal logging, followed by fishing.[9] During the 1990s, Javanese people living in S Village (about five households) began to engage in agricultural activity in inland area on non-KH lands, while logging activity continued. The inland area is more suitable for agriculture than the swamp mixed with seawater soil of the coastal area. Particularly, since some non-peat soil is distributed along the river, the Javanese who were the first to start cultivating

---

[6]"Suku Asli" translates as "indigenous people" in English. In this chapter, I adhere to the local definition of indigeneity, in which both the Suku Asli and local Malay refer to themselves as "*Suku yang asli di kampung sini* (meaning "an ethnic group which is indigenous to this kampung"). To avoid confusion, "Suku Asli" is written in Indonesian. For a review of the historical changes of the names Rawa and Suku Asli, see Osawa (2016).

[7]Interview with Mr. B., June 14, 2021.

[8]Forest map governance agreement (Decision of the Ministry of Forestry No. 173/1986).

[9]According to the Forestry Law (No.41/1999), clearing forest areas without obtaining permission from the local government is defined as illegal logging.

farmland tended to settle inland rather than along the coastline of S Village. As a result, the percentage of migrant Javanese was higher in R Village than in S Village at the time of this study.

In 1996, an acacia plantation company (Company A) received industrial forest plantation rights (*Izin Usaha pemanfaatan Hasil Hutan Kayu Hutan Tanaman Industri*, IUPHHK-HTI)[10] for an area of 299,975 ha in Riau Province, including almost 8000 ha of production forest[11] in today's R Village.

As Company A constructed large drainage ditches to establish plantations, the peatland became dry and flammable. Drying upstream also affected the downstream areas where local people live. The drainage of peatlands causes an irreversible lowering of the surface (subsidence) from peat shrinkage and biological oxidation, with the latter resulting in a loss of carbon stock (Hooijer et al. 2011). Interviews with local people reveal that fires have occurred often on local people's agricultural land since the 2000s.[12] Logging continued in the areas surrounding R Village, except in Company A's plantation site.

During the 2000s, the logging company stopped operating, labor opportunities decreased, and many local people previously engaged in logging in KH forests were forced to shift livelihoods. Some continued logging on their own, selling to other companies. [13] In 2005, the Presidential Instruction on the Eradication of Illegal Logging[14] required the Ministry of Forestry and local governments to crack down on illegal logging in KH zones (Onda et al. 2014). As a result, almost no one in R Village has engaged in logging since the late 2000s. The loggers and the women who ran dining houses catering to them have since shifted their livelihoods to agriculture and agricultural and non-agricultural day labor[15] (hereafter, day labor), and livelihood opportunities have decreased further.[16] Since 2010, local Malay and Kampar Malay, whose main livelihood was fishing, have also shifted their livelihoods, to agriculture and day labor, due to the decline in fish catches. Since the development of acacia plantations, driftwood from the drains have made rivers shallow, worsened the river environment, and reduced the variety and quantity of fish. Fishing in inland lakes with wooden boats (*kapal pompon*) has also become impossible. In addition, the development of infrastructure following the independence of R Village made it easier for people to access inland areas, accelerating the agricultural cultivation

---

[10] Decree of the Minister of Forestry No.743/1996.

[11] Designation of forest areas in Riau Province (Decision of the Ministry of Forestry No.173/1986) and Forestry Law No.41/1999.

[12] Interview with Mr. S., July 3, 2019.

[13] Logs were exported to neighboring islands (Interview with Mr. S., July 3, 2019).

[14] The eradication of illegal logging in KH areas and throughout the territory of the Republic of Indonesia (Presidential Instruction No. 41/2005).

[15] In R Village, agricultural day labor includes cultivating, planting, fertilizing, and harvesting in oil palm plantations as non-agricultural day labor includes construction work and loading and unloading at the port.

[16] Interview with Mr. S., July 3, 2019.

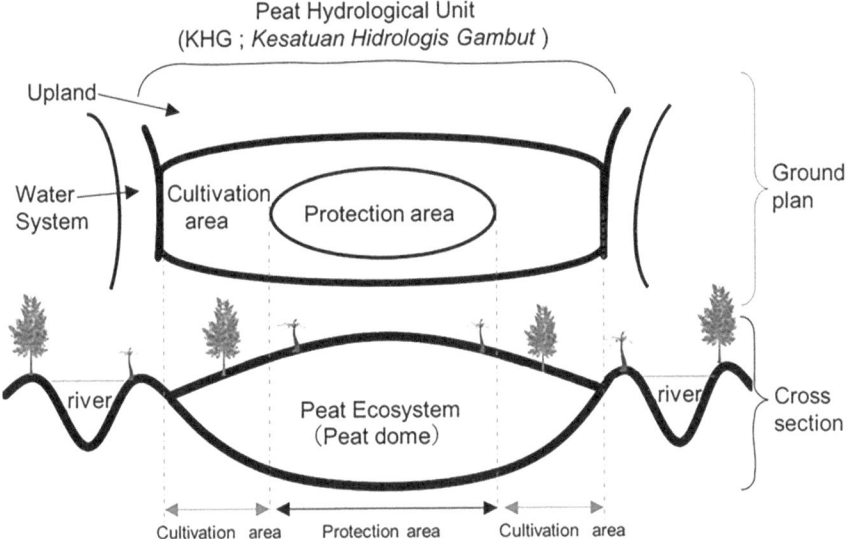

Fig. 11.4 A peat hydrological unit

there. Today, the main livelihood activities in R Village are day labor, agriculture, and fishing.

11.2.4   Frameworks for the Management of Peatlands

R Village is located in a "peat hydrological unit" (*kesatuan hidrologis gambut*, KHG). Government Regulation No. 71 of 2014 for the Protection and Management of Peat Ecosystems defined KHGs as "peat ecosystems located between two rivers, between a river and a sea, and/or in a swamp," as depicted in Fig. 11.4. The concept of the KHG formed the basis of subsequent peatland conservation activities, and with the establishment of the BRG in 2016, KHG-led peatland governance commenced (Januar et al. 2021). The spatial planning of peatlands was carried out via KHG maps that demarcated protected and cultivated areas, as determined by the relevant ministries. The construction of drainage facilities on KHGs affects the whole area, revealing the need for the coordination of all actors involved in peatland conservation activities in individual KHGs. As such, based on existing regulations and policies, the BRG emphasizes cooperation among all actors, including international NGOs, plantation companies, local governments, and local people living in a KHG.

The local regulations applicable to R Village reflect the regional and national peatland laws and policies. In 2018, Siak District promulgated the Green Siak District policy (*Siak Kabupaten Hijau*) to target sustainable resource management and economic growth.[17] This policy also emphasized the collaboration of local governments, companies, local NGOs, and local people to fight peatland fires, and following its promulgation, the NGOs forum (*Sedagho Siak*) and a coalition of seven international/national companies was formed. In addition, a 2019 Riau Provincial Regulation requires companies to implement "3Rs-based peatland conservation" on their business sites and the surrounding areas. According to the regulation, companies and local governments must also organize and support the MPA, or firefighting group, and local people must participate in peatland conservation activities through organizations such as MPAs and NGOs.

11.2.5 Data Collection and Analysis

This study was conducted in two steps. First, I identify and analyze the key implementers and relevant stakeholders of peatland conservation activities in R Village. This involved a series of in-depth interviews with key players in the village administration, the village MPA, and community groups. Second, to clarify the participation and equity of peatland Livelihood Improvement Programs (LIPs) in the study site, a total of 63 randomly selected households (22% of total households) were surveyed. The households represent a total of 250 people, 129 men and 121 women, with an average of four people per household (two single-person households headed by women who had lost their husbands were included in those surveyed). The survey was conducted from April to August 2019. Questionnaires requested basic information about household members, land ownership, and employment. Information on land ownership, income, and ethnicity was then cross-referenced with LIP participation. Of the 124 heads and spouses of the 63 surveyed households, 65 self-identified as Malay (64 people [98.5%] are local Malay and 1 person [1.5 person] is Malay from West Kalimantan province).

11.3 Findings and Discussion

11.3.1 Peatland Fire Fighting and Conservation Activities in R Village

Peatland fires often occurred during the 2000s in Indonesia, and during the 2010s, they became more serious in R Village. Until 2012, the Forest Fire Brigade of the Forestry Department in Siak (*Manggala Agni Daerah Operasi Siak*) was the only entity focused on fighting fires at the village level. In that year (before state and local regulations required the organization of an MPA), a group of five volunteers set up a

---

[17] Green Siak District (Regulation of Regent No. 22/2018).

group to fight fires in R Village (*relawan pemadam dari masyarakat*). This group became the MPA in 2013. As of the time of this study, the MPA had not received any village budget funds, because MPAs are regulated as volunteer (*sukarela*) organizations for the prevention of forest and peat fires.[18] A 2016 decision of the village head also states that MPA members do not receive any honorarium or wages because the MPA is a volunteer organization.[19]

Mr. S., the head of the R Village MPA since its formation in 2013, is a second-generation Javanese whose family migrated to S Village from Java in the 1980s. Until the early 2000s, Mr. S. was engaged in logging, but following the tightening of regulations, he retired from logging and began to engage in agriculture and day labor. Beginning in 2008, Mr. S. and his brother (the head of R Village from 2012–2017) began pushing for R Village to split from S Village. In 2010, R Village became independent and Mr. S.'s brother won the first village head election in 2012 and Mr. S. became the village secretary. Both remained in these positions until 2017. During this time, the non-Javanese villagers were dissatisfied with the Javanese dominance of the village administration and a rumor spread that the MPA was receiving large amounts of village budget to fight fires.

In the 2017 village head election, Mr. S.'s brother lost, and Mr. H. (an ethnic Kampar Malay) became the village head. Mr. S. stepped down from the village secretary and focused on peatland conservation activities as the head of the MPA. From 2013 to 2014, an MPA training project was implemented by the Firefighting Department of the Forestry Department and 15 new local people joined the MPA. The project provided firefighting training and firefighting equipment to each village MPA but did not provide regular financial support for them. Thus, the MPA must collect financial and material resources for peatland conservation through its own network (e.g., from NGOs and acacia plantation companies). In other words, the financial status of the MPA is based on its network with external actors and thus, the financial condition of the MPA is different in each village. Also, the number of MPA members is different in each village. In case of R village, the total number of MPA members is 24 people.[20]

Village task forces (*tim siaga bencana masyarakat* or *satgas relawan*) were also established for each village (*kampung*) in Siak District in 2014 by regulation of the district head. In 2015, village task force in R Village established and received budget from the village fund (*Anggaran Pendapatan Belanja Kampung*, APBKam[21]). In 2019, Rp17,300,000 was allocated from the APBKam. The five members of the task force are responsible for all disaster responses in R Village. Due to the different

---

[18] According to Regulation of the Ministry of Forestry No.12/2009 and Regulation of Directorate General of Natural Resources and Ecosystem Conservation No.2/2014.

[19] Decision of village head in R Village No.140/2016.

[20] Decision of village head in R Village No.140/2016.

[21] Organize the community disaster prevention team (Decision of village head No.17/2019).

funding sources of the village task force and the MPA, it is difficult for the two to collaborate on fire prevention activities.[22]

The MPA carries out peatland conservation activities in collaboration with a variety of actors both inside and outside the village. The MPA receives activity fees and equipment for firefighting from external actors. Each member of the MPA receives around Rp100,000 per day from external actors to patrol and implement peatland conservation programs. The MPA activity fee is almost same as the daily wage of a day laborer in the study site and therefore provides an essential economic incentive for the peatland conservation activity of the MPA. For the MPA's firefighting efforts, the head of the MPA (Mr. S.) was nominated by the Ministry of Environment and Forestry for the *Kalpataru* Award for environmental conservation in 2019.[23] According to the Ministry of Environment and Forestry Regulation No.32/2016, "each concession holder is obliged to support the MPAs." Thus, the acacia plantation companies in R Village support the MPA. The Green Siak policy of 2018, which aims to improve the livelihoods of the local people and to protect the environment in Siak District, also articulates the need for companies, NGOs, and local people to collaborate in peatland conservation. In R Village, the companies implement community development programs based on the 3Rs using their own budgets.

The MPA has cooperated with Company A, which has implemented village development programs as part of its CSR activities, since 2016. In 2016, Company A introduced a village fire prevention program (*Desa Makmur Peduli Api*, DMPA) in 14 villages. Under the program, each MPA conducts fire prevention activities jointly with Company A and attends monthly meetings at the company office. As mentioned above, companies are required to carry out 3Rs activity in villages adjacent to their plantations. The DMPA program introduced both fire prevention measures and Livelihood Improvement Programs, or LIPs. The village head and the secretariat (at the time, Mr. S. and his brother) participated in LIP meetings. In addition to Company A, Company R, which was also granted rights for an industrial forest plantation covering an area of 350,165 ha, has been implementing community development program in Riau Province to help alleviate poverty and improve quality of life through economic development, health, education, and social infrastructure programs.[24]

The MPA in Village R has also cooperated with local NGOs in Riau Province since 2015. Environmental NGOs in Riau Province have been introducing projects

---

[22] Interview with a member of the task force in R Village, July 27, 2019.

[23] Mr. S. also has received several awards, including the Mangrove Nature Tourism Activist Award from the Siak District government in 2017, an award from the Siak Angle Community and SMI Chapter Siak in 2017, the Tourism Awareness Award from the Riau Province Tourism and Creative Economy Office in 2017, and an Award from PT BOB BSP Pertamina Hulu in 2017, according to KLHK (2019).

[24] Company R was granted the rights according to Decree of Minister of Forestry No.130/Kpts-II/1993. Its acacia plantation is in not R Village, but in P Village; however, approximately 8000 ha of its total area falls within R Village.

related to peatland conservation at the village level since the large-scale peatland fires of 2015. The BRG also coordinates with NGOs in implementing the 3Rs.[25] The NGOs need to carry out and demonstrate the success of the programs within the budget and timeframes set by their donors. Thus, the selection of a liaison in each village is critical for the NGOs.

Mr. S. has acted as Village R's liaison with NGOs since it separated from S Village in 2010, devoting himself to building active cooperation as the village secretary (2010–2017) and the head of the MPA (2012–). Even after stepping down from village administration, he continued in this role, acting also as an intermediary for the LIPs. Both companies and NGOs reported that, "If we ask Mr. S. to help to introduce the program to R Village, he mediates well and ensures a quick implementation." Mr. S. also tends to choose the households who he believes have the best ability to carry out the programs to participate in them.

Following the establishment of the BRG, village cooperation with external actors became more active and diverse. These connections with external actors are a form of local peoples' participation in peatland conservation through the MPA, as proscribed in the Riau Province Regulation of 2019, which aimed to achieve successful implementation of the 3Rs in villages through such cooperation. Although the MPA is active, there remains a lack of socialization and participation among members of the village, especially the local Malay. Thus, there is still room for further "integration of local knowledge and more inclusive community-based fire groups" (Thoha et al. 2018).

11.3.2 Livelihood Improvement Programs (LIPs)

Four groups were implementing LIP programs at the time of the study. These included three groups of chili farmers and one group of honey collectors. Interviews with the heads of each of the chili farmer groups revealed that there were five farmers groups in R Village before 2015, but three of those groups dissolved due to a lack of members and unity, and the former members shifted their livelihoods to non-agricultural activities, especially day labor.

The head of the first chili farmers group (Group A) is a Javanese man who migrated from Java around 1990. He engaged in logging until 2004, but began farming instead due to the declining forest resources and the enclosure of KH areas in the 2000s. When he began vegetable farming in 2009, he found it difficult to sell his product due to incomplete road construction. The independence of R Village in 2010 led to infrastructure investment and roads were constructed. Thus, brokers began to visit the village and he could sell vegetables. As a result, he started chili

---

[25]BRG coordinates with NGOs, particularly in establishing Peat Care Villages (*Desa Peduli Gambut*). In addition, the list of BRGM board members includes the names of executives from Yayasan Mitra Insani, an environmental NGO in Riau.

farming in 2014. In 2015, Company R implemented a LIP to support chili farming through Mr. S. The 10-member Group A became the implementer of the LIP. All the members own non-peatland and cultivate their own land and 80% of the members were Javanese in 2021.

The head of the second chili farmers group (Group B) is of Javanese and Malay heritage and migrated to R Village in 2000. He started to farm vegetables in 2010. In 2012, Company A introduced a CSR program through Mr. S., and Group B was formed with 19 members. The program was intended to reduce the amount of abandoned land in the area and develop the village community by improving the livelihoods of local people. Company A provided opportunities for local people to participate in chili cultivation workshops and distributed funds to pay for initial costs and fertilizer. In 2017, the village development program was grouped under the peatland conservation program and became a LIP implemented as part of the DMPA. But in 2017, only five members who owned non-peatland were farming chili. Others did not cultivate chili, instead earning a living in the non-agricultural sector. The reasons for this were that some did not have non-peatland, or they wanted to earn income daily instead of having to rely on the unstable income of each harvest season. In 2021, 68% of Group B members were Javanese.

The head of the third chili farmers group (Group C) is of Javanese heritage. Group C was formed in 2018, when the LIP was introduced by an NGO through Mr. S. and 25 local people joined the program. The head of the group participated in a chili cultivation training program sponsored by the NGO and received support for the initial costs of cultivation. Because this support was only for the first year, the members had to pay for their own expense after that. In 2019, there were only three members in this group. According to additional interviews conducted in 2021, Group C has been dissolved and the group head has already changed his occupation from farmer to day laborer.

The fourth group (Group D) is of honey collectors. The head of Group D, Mr. N., is a Kampar Malay who migrated from P District of Riau Province in 1988 and the core members of the group are his family members. The group had ten members at the time of its formation, but this number had grown to 26 by 2019. No group members are Javanese. Mr. N. is a brother of the current village head (2017–). He was a river fisherman until 2009; however, due to the declining size of his hauls from 2009 to 2017, he started to collect honey on a self-employed basis in the hinterland forest, including in the plantation area of Company A. Honey is collected by combing the *sialang* trees (see p. 54 in Chap. 3) after using smokes (*asap*), to make the bees docile. This method was a concern for Company A because it could potentially cause peatland fires. In 2017, Company A introduced its LIP to Mr. N.'s group of honey collectors through Mr. S. Through a mediation process facilitated by Mr. S., Company A gave the group permission—and a license—to collect honey in its acacia plantation area. Since then, only members of Group D have been able to collect honey inside the plantation area of Company A. All members of Group D enter the acacia plantation area together once per month.

11.3.2.1 Limited Participation in LIPs

Interviews with the four heads of the LIP groups revealed limited participation in LIPs. This is partially due to long-standing livelihood patterns according to ethnic background. As noted earlier, the local Malay tend to live in coastal areas and engage in fishing, Javanese tend to live inland for agriculture, and the Kampar Malay tend to live in inland forest areas, fishing in rivers and gathering non-timber forest products. Based on fieldwork in a neighboring village, Osawa (2016) explains that "each [ethnic group] did not encroach on the different landscapes of the others, as the economic basis of their lives was established in a particular space." Such designated living spaces were observed in the study site. Some villagers also noted that although R villagers do not allow stereotypes based on ethnicity, each person tends to have a profession (*profesi*) according to their ethnicity. One noted, "Even if honey collection is appealing, if they don't think it's their profession, they won't do it."

As the traditional livelihoods of each ethnic group become less viable, the rigid boundaries of living spaces and livelihoods are fading to some degree. However, owning and/or buying land is another obstacle to LIP participation. Locals in R Village report that as population density increased due to the migration of Javanese and Kampar Malay in the 1980s, land prices also increased.[26] All members of the chili farmer groups have non-peatland, which is suitable for chili farming. According to the heads of the groups, the price of non-peatland land is twice that of peat land. Therefore, while rich local people can buy non-peatland and join the chili farmers group, those with lower incomes are not motivated to join the LIP. However, even if one can afford to buy land, it is not always desirable to do so. For example, although landless local Malay fishers could grow profitable chili crops and improve their livelihood if they buy non-peatland, they tend to shift their livelihood to day labor rather than agriculture. One local Malay explained, "It is too difficult for us to buy land. Even if we buy land, we prefer to get paid immediately rather than wait for the harvest season." Households that own only peatland—and face the risk of peatland fire—also tend not to gain access to the chili LIP groups.

In the case of Group D, members collect honey from seven *sialang* trees that are owned by the group head (Mr. N.) and his family members. Mr. N. and his family members (who are all Kampar Malay) receive two thirds of the profits, while the workers, mainly neighbors of Mr. N., receive the remaining one third. Thus, the local Malay people, who engage in fishing and face diminishing livelihood returns due to degradation of the river environment, tend not to access the benefits of either the chili or the honey LIPs.

Similar to the findings of other studies (see Silvianingsih et al. 2020; Nurlia et al. 2021; Januar et al. 2021), this study finds that the chili farming LIPs in R Village do not pay sufficient attention to off-farm livelihoods, particularly fishing activities (as also found by Thornton et al. 2020). Thus, local Malay, who have historically not

---

[26] Interviews with Mr. S. and Mr. A. in 2021.

cleared peatlands for agriculture (Furukawa 1992) and now face deteriorating fishing conditions, tend to miss out on the benefits of LIPs. The study of R Village therefore suggests that a lack of ample attention to these groups may entrench or even accelerate existing economic disparities. The study also echoes Budiman et al. (2020), who conclude that it is important for programs to consider the inland fisheries.

11.3.2.2 Income, Land Ownership, and Ethnic Disparities among LIP Participants and Non-Participants

As mentioned previously, since the 2000s, the local people of R Village have been changing their livelihood activities from logging and fishing to day labor and agriculture. The survey revealed that agriculture accounted for less than one third of annual non-agricultural incomes. Of the 63 households surveyed, 34 were land-owners and 29 were landless. Five households rented land from relatives or neighbors for free, bringing the total number of land-owning households to 39. A total of nine households possessed land from which nothing had been harvested. Eight of these nine households were cultivating oil palm trees that were less than five years old and thus not yet ready to harvest. Six of those eight had experienced crop burning due to fires. Such destruction of crops due to fire is a common economic burden for oil palm farmers for about five years until the crops reach a harvestable age (as it generates additional replanting costs). This helps explain why, like landless households, those who own only peatland also earn the bulk of their income from the non-agriculture sector.

The relationship between annual income and the amount of non-peatland owned indicates that high-income earners (excluding shopkeepers) tend to own non-peatland land and participate in the LIPs (see Fig. 11.5). The households participating in LIPs had relatively higher incomes than other households (although one household had an annual income of only Rp20,000,000 despite participating in an LIP, because the household had only been involved in the chili farmers group for less than three months at the time of the survey (in June 2019) and had yet to receive any profits). The annual income of households that do not own non-peatland is less than Rp40,000,000. They face the risk of losing their livelihood due to low productivities caused by peatland fire and prohibitive replanting costs. They make a living as day laborers and are the lowest income earners in R Village. Therefore, the households which face the risk of livelihood deterioration and are at the bottom of the income ladder do not access the benefits of LIPs and conversely, participants in the LIPs are high income earners.

To understand the characteristics of households based on differences in annual income, I examined the main source of income and the ethnicity of the head of household and spouse. In the analysis, 63 households were categorized by a median (27.1) annual income. The upper group ($n = 31$) was defined as above the median, while the bottom group ($n = 32$) was defined as below the median (including the median). The main sources of income for those in the upper group were agriculture

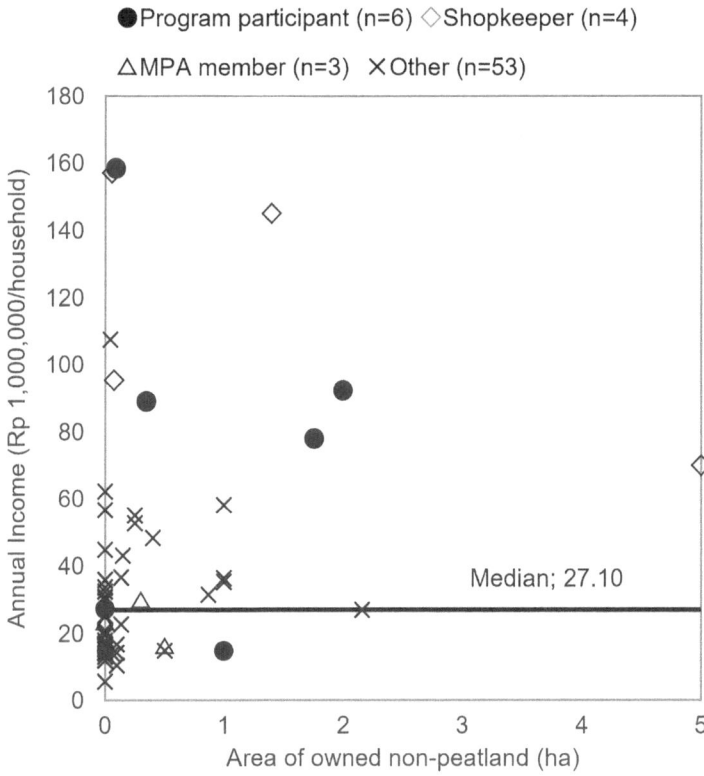

Fig. 11.5 Relationship between annual income and the amount of non-peatland owned ($n = 63$)

(28%), self-employment, such as running a motorcycle repair or other shops (28%), village government officials or company employees (25%), and day labor (16%). Those in the bottom group were day laborers (48%), self-employed in hunter-gathering, fishing, logging, and construction (45%), and farming (10%). Those earning below median incomes were strongly dependent on day labor.

Next, households were analyzed based on whether they were indigenous people (local Malay or Suku Asli) or migrant (Kampar Malay, Javanese, and other migrants including non-local Malay). This variable was introduced because actor analysis revealed that the local Malay were not benefitting from peatland conservation governance, which is indeed indicated by the results depicted in Fig. 11.6. The households in which both the husband and wife are local Malays or Suk Asli are classified as "local Malay/Suku Asli households." Households in which the husband or wife is local Malay or Suku Asli and the wife or husband is a migrant are classified

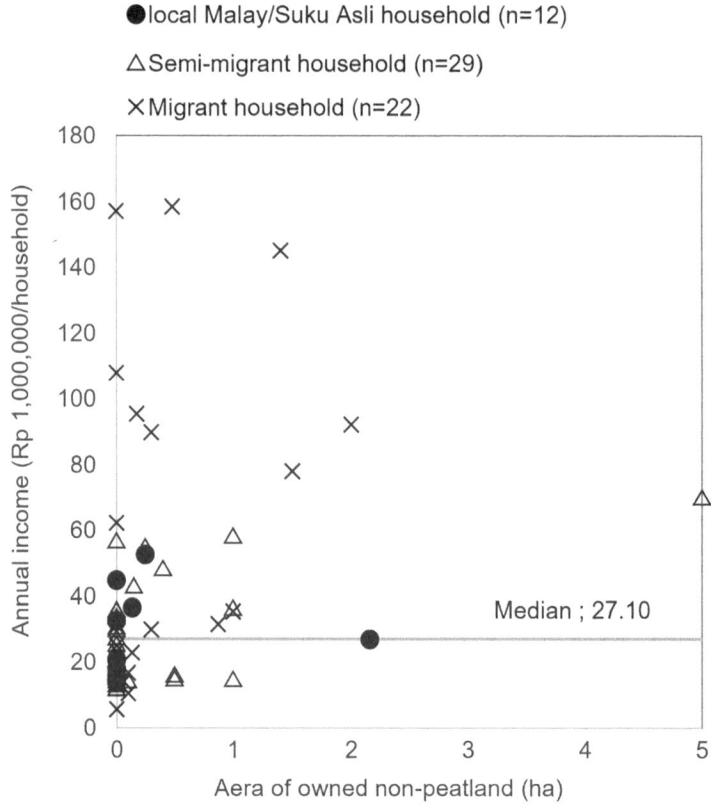

● local Malay/Suku Asli household (n=12)

△ Semi-migrant household (n=29)

✕ Migrant household (n=22)

Fig. 11.6 Ethnic distribution according to annual income and amount of non-peatland owned ($n = 63$)

Table 11.1 Descriptive statistics of annual income ($n=63$)

Descriptive statistics ($n = 63$)	
Mean	38.5
Median	27.1
Standard deviation	34.5
Standard error or mean	4.3

as "semi-migrant households."[27] All other households are defined as "migrant households." After cross-referencing income level and indigenous, local Malay/ Suku Asli households and semi-migrant households made up 71% of the household whose income is below the median. Conversely, the migrant households made up 90% of the ten highest earning households (Table 11.1).

---

[27] Women who were born outside the surveyed area were not included in the definition of "semi-migrant households" because these women tend to have attained higher education levels and have higher income occupations.

Table 11.2 Average production area and annual sales volume from crops (Rp1,000,000/ha)

Crop name	Total plot area (ha)	Production area (Annual average, ha)	Sales volume (Annual average, Rp1,000,000)
Chili	1.35	0.27	180.4
Oil palm (in peatland)	13.84	1.73	6.3
Oil palm (in non-peatland)	9.28	0.58	20.3
Others	10.89	0.99	19.1
Total	35.2	0.88	36.8

Table 11.3 Average annual income of non-agricultural sectors

Non-agricultural sector	Number of laborers	Annual income (Rp)/laborer
Honey collection	3	53.7
Day labor	28	14.5
Public service labor, etc.	22	21.8
Independent business (excluding honey collection)	31	25.2
Total	84	21.7

From the above analysis of the survey results, we have determined that (1) high-earning households that own non-peatland participate in the existing LIPs in R Village, (2) low-earning households primarily rely on day labor, fishing, and logging, and (3) the majority of local Malay/and semi-migrant households have incomes below the median. Given these factors, what happens when an LIP program is implemented? Let us remember that chili farming is suitable in non-peatland. Although profitability per hectare fluctuates with climate, on average, chili can generate 5–30 times more revenue per ha than that of other crops (see Table 11.2). Similarly, honey collection demonstrates tremendous income potential, with annual revenues measuring 2–4 times higher than those gained from other non-agricultural sectors (see Table 11.3). Notably, the annual earnings of the LIP participants from honey collection[28] amount to more than three times those from day labor, which has become the main source of income for local Malay since inland plantation developments have precipitated declines in fish catches. This disparity in revenue generation necessarily means that those participating in the LIPs can generate significantly more income than those who do not participate, potentially accelerating existing economic disparities in the village.

As described in the introduction, the government's zero-burning policy resulted in uneven impacts and economic burdens depending on land ownership, livelihoods,

---

[28] The head of the Group P receives one third of the total sales revenues, three long-term members receive another third, and the remaining workers receive the remaining one third.

and ethnicity. According to interviews with land-owning households in R Village, on average, Rp5,000,000/ha is needed for land preparation using only labor under the zero-burning policy. Land-owning households own an average 1.24 ha, excluding abandoned land. Therefore, the average cost of land preparation per household is Rp6,200,000. The average monthly income of landowner households is Rp2,600,000/household. The cost of land preparation is an economic burden on land-owning households. The PLTB technique (managing land without burning), which reduces negative impacts and contributes to all aspects of the 3Rs (Gunawan et al. 2020), is important for land-owning households to achieve the 3Rs of peatland restoration. However, the 24 landless households in R Village do not receive any benefit from the PLTB.

As we can see from these findings, not only is there an inequity in the negative impacts of peatland policies, but also in the benefits of programs like the LIPs. The LIP programs in R Village benefits certain villagers and not others, particularly those who are landless.

If we exclude the nine households involved in the MPA and LIPs, we find that the Gini coefficient of the annual income of the remaining 54 households was 0.40. On the other hand, the Gini coefficient of all 63 households' annual income is 0.42. Therefore, peatland conservation programs have the potential to exacerbate existing economic disparities. This in turn affects the level of acceptance or rejection of the 3Rs. Thus, mitigation measures to enhance equity should be considered when implementing the LIP programs and other peatland policies.

11.4   Conclusion

This chapter first identified how—and by whom—peatland conservation is implemented in a multi-ethnic village that has both peatland and non-peatland soils. It found that the local firefighting group, the MPA, has been implementing active peatland conservation measures together with acacia plantation companies and local NGOs. The activity fees from external actors are essential for MPA members to maintain their livelihoods and motivation for peatland conservation. Although the MPA is active, there remains a lack of socialization and participation among all members of the village. Employing a household survey, the chapter then examined participation in LIPs. It found that low-earning households that face threats from peatland fires and deteriorating environmental conditions tend not to participate in LIPs, indicating that the programs do not pay adequate attention to non-farm livelihoods. It also found that, given who does participate in the programs and the benefits that can be attained from such programs, the LIPs have the potential to exacerbate existing disparities. Thus, the distribution of benefits from 3Rs programs as well as the implementation and intermediation of the MPA have accelerated existing economic disparities. The presence of income inequality may well destabilize the social and political situation in such villages, and consequently hinder economic growth.

This chapter reveals several challenges to sustainable peatland governance that need to be addressed. First, there is a need to re-examine the institutional character-istics of the MPAs as volunteer organizations, which are the key to peatland management at the village level. Second, it is recommended that households facing threats to their livelihood should be selected for participation in LIPs to mitigate economic disparities in peat communities. Third, even though ethnic boundaries are fading, understanding of ethnic differences, particularly vis-à-vis livelihood choices, should be considered when introducing LIPs.

Implementation of programs based on a more comprehensive approach can produce more equitable access and benefits for local people. For example, granting rights to collect non-timber forest products (NTFPs) in forest areas (KH) to all local people, not just members of an LIP, could be an opportunity to improve the liveli-hoods of villagers regardless of their land ownership status. This study used a field survey to examine a peat community that features a mixture of peatland and non-peatland farmers. However, there is a strong need to conduct additional empir-ical studies in peat communities that solely depend on agriculture in peatland.

References

BPS (Badan Pusat Statistik) Kabupaten Siak (2017) Kecamatan Sungai Apit dalam angka 2017. BPS Kabupaten Siak, Siak

BPS Kabupaten Siak (2019) Kecamatan Sungai Apit dalam angka 2019. BPS Kabupaten Siak, Siak

Budiman I, Bastoni, Sari EN et al (2020) Progress of paludiculture projects in supporting peatland ecosystem restoration in Indonesia. Glob Ecol Conserv 23:art e01084. https://doi.org/10.1016/j.gecco.2020.e01084

Daeli W, Carmenta R, Monroe MC et al (2021) Where policy and culture collide: perceptions and responses of swidden farmers to the burn ban in West Kalimantan. Indonesia Human Ecology 49(2):159–170. https://doi.org/10.1007/s10745-021-00227-y

Furukawa H (1992) Indoneshia no teishicchi (coastal wetlands of Indonesia). Keiso shobo, Tokyo

Gunawan H, Afriyanti D, Humam IA et al (2020) Pengelolaan lahan gambut tanpa bakar: upaya alternatif restorasi pada lahan gambut basah. J Nat Resour Environ Manage 10(4):668–678. https://doi.org/10.29244/jpsl.10.4.668-678

Hooijer A, Page S, Jauhiainen J et al (2011) Subsidence and carbon loss in drained tropical peatlands: reducing uncertainty and implications for CO_2 emission reduction options. Biogeosci Discuss 8:9311–9356. https://doi.org/10.5194/bgd-8-9311-2011

Januar R, Sari ENN, Putra S (2021) Dynamics of local governance: the case of peatland restoration in Central Kalimantan, Indonesia. Land Use Policy 102:art 105270. https://doi.org/10.1016/j.landusepol.2020.105270

KLHK (Kementerian Lingkungan Hidup dan Kehutanan) (2019) Penghargaan Kalpataru 2019: perintis, pengabdi, penyelamat dan pembina lingkungan. KLHK, Jakarta. https://klhkkalpataru.wordpress.com/pdf/. Accessed 27 Oct 2022

Knieling J (2020) Smallholder perceptions of sustainability criteria related to forest and peatland. Dissertation, Georg-August-Universität Göttingen

Koizumi Y, Nagata J (2018) Indoneshia Riau-shu jumin no shusseichi minzoku haikei to sangyo betsu shugyo kozo: 2000 nen, 2010 nen jinko sensasu kohyo Deta no bunseki wo chushin ni (population by birthplace and ethnicity and employment structure by industry in Riau Province, Indonesia: an analysis of the raw data of the 2000 and 2010 population censuses). Jpn J Southeast Asian Stud 56(1):3–32. https://doi.org/10.20495/tak.56.1_3

Murniati, Suharti S (2018) Towards zero burning peatland preparation: incentive scheme and stakeholders role. Biodiversitas 19(4):1396–1405. https://doi.org/10.13057/biodiv/d190428

Nurlia A, Rahmat M, Waluyo EA et al (2021) Gender role in farmers' livelihood strategies at peatland area of fire-prone in Ogan Komering Ilir regency South Sumatra Province. In: Sriwijaya international conference on earth science and environmental issue, Palembang, October 2020. IOP conference series: earth and environmental science, vol 810. IOP Publishing, Bristol., art 012028. https://doi.org/10.1088/1755-1315/810/1/012028

Okamoto K (2013) Indoneshia-koku no shinrin tochi Kasai mondai no genjo to taisaku no hoko ni tsuite (present situation and direction of the forest and land fire control in Indonesia). Jpn J Int For For 87:14–19. https://doi.org/10.32205/jjjiff.87.0_14

Onda N, Ota M, Shiga K (2014) Iho bassai ni taisuru shinrin keisatsu no yakuwari to sono kadai: Indoneshia, Gunumparun kokuritsu koen wo jirei toshite (the role of the forest police and problems faced countering illegal logging in Indonesia: a case study of Gunung Palung National Park in West Kalimantan). For Econ 67(8):1–18. https://doi.org/10.19013/rinrin.67.8_1

Osawa T (2016) At the edge of mangrove forest: the Suku Asli and the quest for indigeneity, ethnicity and development. Dissertation,. University of Edinburgh

Silvianingsih YA, Hairiah K, Suprayogo D et al (2020) Agroforests, swiddening and livelihoods between restored peat domes and river: effects of the 2015 fire ban in Central Kalimantan (Indonesia). Int For Rev 22(3):382–396. https://doi.org/10.1505/146554820830405645

Siscawati M, Banjade MR, Liswanti N et al (2017) Overview of forest tenure reforms in Indonesia. In: Working Paper 223. CIFOR, Bogor. 10.17528/cifor/006402

Thoha AS, Saharjo BH, Boer R et al (2018) Strengthening community participation in reducing GHG emission from forest and peatland fire. In: Conference on agriculture, environment, and food security, Medan, November 2017. IOP conference series: earth and environmental science, vol 122. IOP Publishing, Bristol., art 12076. https://doi.org/10.1088/1755-1315/122/1/012076

Thornton SA, Setiana E, Yoyo K et al (2020) Towards biocultural approaches to peatland conservation: the case for fish and livelihoods in Indonesia. Environ Sci Pol 114(1):341–351. https://doi.org/10.1016/j.envsci.2020.08.018

Ward C, Stringer LC, Warren-Thomas E et al (2021) Smallholder perceptions of land restoration activities: rewetting tropical peatland oil palm areas in Sumatra, Indonesia. Reg environ. Change 21:art 1. https://doi.org/10.1007/s10113-020-01737-z

Watts JD, Tacconi L, Hapsari N et al (2019) Incentivizing compliance: evaluating the effectiveness of targeted village incentives for reducing burning in Indonesia. For Policy Econ 108:art 101956. https://doi.org/10.1016/j.forpol.2019.101956

World Bank (2016) The cost of fire: an economic analysis of Indonesia's 2015 fire crisis. Indonesia sustainable landscapes knowledge note, no 1. World Bank Group, Washington, D.C. http://documents.worldbank.org/curated/en/776101467990969768/The-cost-of-fire-an-economic-analysis-of-Indonesia-s-2015-fire-crisis. Accessed 9 Aug 2021

Chapter 12
Integrated Spatial Ecosystem Services Valuation Approach with Community Participation in a Social Forestry Scheme

Dheny Sampurno

Abstract In 2016, the Indonesian government established seven village forests in East Tebing Tinggi Sub-district, Riau Province. These social forestry schemes grant rights to the local society to manage communal land for ecological and livelihood benefits. To do this, they need to identify and value the products and services of the natural resource assets. This study conducts a rapid spatial assessment for an ecosystem service valuation with the participation of local representatives, demonstrating that the integration of a spatial approach and local participation is scientifically implementable for the village forest authority. Using the peat ecosystem services approach, the study estimates that seven village forests contain approximately 36.2 million tons of carbon stocks from the peat soil and peat forest biomass in the form of regulating services. Supporting services are evident in the government's regulation of ecological conditions based on its designation of peat ecosystem function. The agro-ecosystem of sago plantations for food production offers provisioning services. Local residents and governments support the potential of ecotourism to enhance socio-cultural value via cultural services. All these services demonstrate how the environmental returns for both local livelihoods and a sustainable ecosystem are possible to achieve at the local level. However, support from governments and organizations is required to ensure that local communities can continue to hold the communal land right.

Keywords Tropical peat ecosystem · Ecosystem services · Social forestry · Peat ecosystem valuation · Community participation

D. Sampurno (✉)
Geospatial Information Agency, Cibinong, Indonesia
e-mail: dheny.trie@big.go.id

© The Author(s) 2023
M. Okamoto et al. (eds.), *Local Governance of Peatland Restoration in Riau, Indonesia*, Global Environmental Studies,
https://doi.org/10.1007/978-981-99-0902-5_12

261

12.1 Introduction

Tropical peatlands have sustained the lives of millions of people in Indonesia. While Indonesians have not yet utilized organic peat soil as fuel, they have used various peatland resources by harvesting forest products, harnessing fisheries, and engaging in agriculture. How peatland is used is intertwined with ill-defined land ownership certification and conflicting land permits among locals, private interests, and government entities (Lye et al. 2003). Therefore, the issues of land ownership and natural resource management in peat environments have been critical among various stakeholders.

In 2015, the Indonesian government divided 24.2 million hectares of tropical peatland into 865 Hydrological Peat Units (or *Kesatuan Hidrologis Gambut*, KHG) (Menteri Lingkungan Hidup dan Kehutanan Republik Indonesia 2017). As mentioned in Chaps. 2 and 11, based on a 2017 regulation,[1] the Ministry of Environment and Forestry (MoEF) leads a peatland protection and management program. One of the requirements to protect and manage these KHGs is geospatial data, or maps. Since 2015, geospatial data of the KHGs has been provided in the form of a rough scale map and a more detailed resolution peatland map.[2] This KHG map contains information about the peat ecosystem and defines the purpose of peat management according to the ecosystem functions of either conservation "*fungsi lindung*" or cultivation "*fungsi budidaya*." Based on the ecosystem function classification, the peat ecosystem provides numerous benefits, including carbon sequestration for ecological sustainability (conservation function) and agriculture and plantation production (cultivation function) for economic benefits (Osaki and Tsuji 2016).

Occasionally, local peat society, in the role of appropriator, bears the accusation of forest or peat degradation, although in reality, they desire the environmental returns from the natural processes of peat forests and peatland agriculture production to support their livelihoods (Angelsen et al. 2014). The concept of ecosystem services is introduced as a way to link the production and consumption roles of ecosystems and societies (Schleyer et al. 2017). Hypothetically, the flow of sustainable ecosystem services is strengthened by the active participation of local society in peat ecosystem protection and management on communal land.

The ability to identify the peat's natural and agricultural assets allows us to place a value on the ecosystem services of a particular peatland area. As such, ecosystem services valuation serves as a tool to achieve environmental sustainability, social justice, and economic viability in the long term (Craig et al. 2002). Such valuation captures the ecological, social, and economic values of an ecosystem (Potschin et al. 2016). The use of spatial data, together with the involvement of local representatives in the assessment of ecosystem services, are key supports to the valuation. This study examines communal land that the Indonesian Government has already designated as

---

[1] Indonesian Government Regulation Number 57, 2017.

[2] The first KHG Map was launched in 2015 at a scale of 1:250,000. This was followed by a more detailed 1:50,000 scale map in subsequent years.

Village Forest together with members of the local forest village authority (*Lembaga Pengelola Hutan Desa*, LPHD), who have been assigned to manage the village forest. A village forest is a type of social forestry in a designated area within a forest concession. The LPHDs were created by the MoEF and their members are both non-village administrators and those appointed by the local society. Based on a 2016 government regulation,[3] the LPHD holds an important position in village forest management; it must be able to work horizontally with village level institutions and vertically with the MoEF.

Most LPHD members are aware of peat forest conservation and peatland management and have already cooperated with the government for the peat restoration program. These experiences and awareness could be an advantage to them in improving village forest management. As it complies with Indonesia's peat and forest regulations, peat management based on ecosystem services valuation within social forestry areas has turned out to be a possible way for local environmentalists to hold their own bargaining power and work at the same level with the local authorities.

The aim of this study is to observe the valuation of peat ecosystem assets in seven village forests by combining the results of spatial analysis and field data provided by LPHD members. This valuation is based on the ecological condition of the peatland and all its remaining resources. The social forestry scheme itself is a means for the community to legally secure these resources and benefit from shared products and services without causing peat degradation, which complies with the ecosystem services program introduced by the government. In addition, the appointed community representatives, the LPHD members, have participated in fieldwork and discussions in the valuation process for the better social forestry management. Failing to combine the spatial analysis with the participation of LPHD members would result in an unbalanced peat ecosystem management. While spatial analysis provides thorough datapoints and reduces the need for a terrestrial survey, the participation of LPHD members in this study creates a synergistic benefit. Their local knowledge provides more accurate information on the forest, and in return participants gain information and confidence to manage their village forest.

Due to limitations in time and the level of scientific knowledge of the LPHD members, it is preferable that the results of the valuation are quantifiable values that are easier for locals to understand. As mentioned in Chap. 9 (Prasetyawan), measuring Willingness to Pay (WTP) can tell us the willingness of a community to conserve an environment that they depend on for income. In this study, the participation of the local community, represented by LPHD members, in the ecosystem service valuation process is a step in measuring such willingness. The LPHD members are chosen for participation because the focus of the study is on social forestry in communal land areas. By transferring peat ecosystem knowledge to the local peat society through geospatial data to develop spatial awareness that can be applied to decision making in village forest management, this study contributes to retention of the village forest right.

---

[3] Ministry of Environment and Forestry Regulation Number 83, 2016 about Social Forestry.

12.2   The Study Area

Unlike Pelalawan District, which is in mainland Sumatra, Kepulauan Meranti District is an archipelago administration area off eastern Sumatra Island, close to Malaysia and Singapore. The study area comprises seven village forests with 138,000 hectares of coastal tropical peatlands on Tebing Tinggi Island, which is one of the islands of Kepulauan Meranti District (see Fig. 12.1). These villages are home to a mixed ethnic population of 8418 people (Melayu, Java, Batak, and Suku Asli (or indigenous people), see also Chap. 11, Kasori); the majority are sago plantation smallholders (BPS 2017). A village forest is one type of social forestry right to communal land that is designated by the government through the national agrarian reform program. Recently, this program was provided with accurate maps derived from the Indicative Map of Social Forestry Allocation (*Peta Indikatif Alokasi Perhutanan Sosial*, PIAPS).

In terms of the local history of social forestry establishment, local activists and NGOs struggled to evict the *Lestari Unggul Makmur* (LUM) concession from East Tebing Tinggi Sub-District beginning in 2009. The approximately 10,000 hectares of acacia plantations in the concession area had degraded the peat ecosystem, causing the water table to lower and culminating in the massive peat fire in 2014. After the massive fire, the struggle intensified and finally bore fruit. The village community in East Tebing Tinggi Sub-District obtained the LUM's HTI (*Hutan Tanaman Industri*, Industrial Forest) concession area and converted the area into seven village forests.

Village forests in a tropical peat ecosystem involve multiple layers of land management: village administration, state forest designation and peat ecosystem

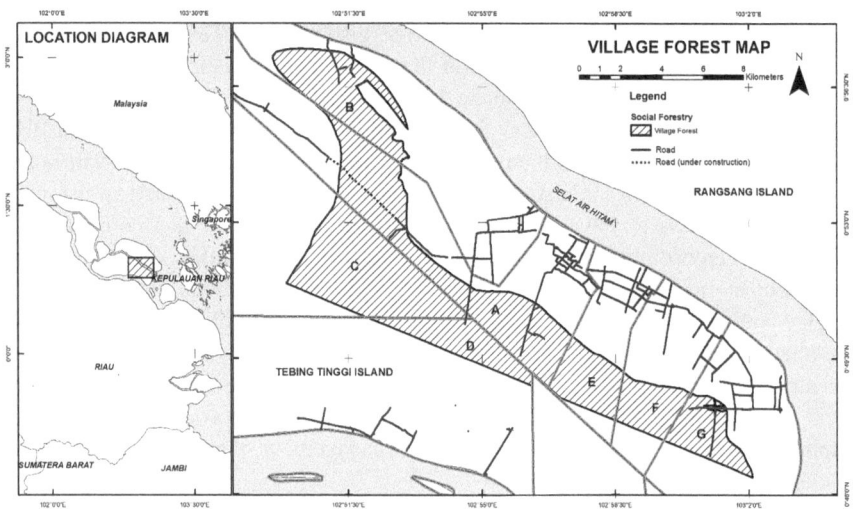

Fig. 12.1 Map of the seven village forests studied (A=Sungai Tohor, B=Sungai Tohor Barat, C=Lukun, D=Kepau Baru, E=Nipah Sendanu, F=Sendanu Darul Ikhsan, and G=Tanjung Sari)

Table 12.1 Basic features of the seven village forests

Village administration	Code (Map)	Area (Hectares)	State forest designation	Peat ecosystem function
Sungai Tohor	A	2940	HP	Conservation
Sungai Tohor Barat	B	1482	HP, HPK	Conservation
Lukun	C	2446	HP	Conservation
Kepau Baru	D	844	HP	Conservation
Nipah Sendanu	E	838	HP	Conservation, cultivation
Sendanu Darul Ikhsan (SDI)	F	650	HP, HPK	Conservation
Tanjung sari	G	760	HP, HPK	Conservation

function. Visualizing the village forests according to these multiple layers provides initial geospatial information about their status. Due to the separation of Kepulauan Meranti District from Bengkalis District, some of these villages are new village administrations. Therefore, the village forests are distributed unevenly across each of the seven villages. Sungai Tohor has the largest area of village forest, with 2940 hectares, while Sendanu Darul Ikhsan has the smallest area, with 650 hectares (Table 12.1).

According to regulations enacted by MoEF to manage forests, each forest is designated a state forest status. Although all seven village forests of this study have been designated production forests[4] (*Hutan Produksi*, HP), three have also been designated as conversion forests[5] (*Hutan Produksi yang dapat dikonversi*, HPK). Basically, these are the production forest zones where any forest production by government or private entities is permissible. In addition to state forest status, each forest area is also designated as Peat Ecosystem Function (*Fungsi Ekosistem Gambut*, PEF). All seven village forests, with the exception of a small area of the Nipah Sendanua forest, are designated as a conservation function zone (due to the peat dome) and play critical roles to provide water to the surrounding agricultural land. Cultivation is either forbidden or limited in conservation PEF areas, while the government allows any type of cultivation in cultivation PEF areas.

When the seven villages assumed the right of communal land in the form of village forests, they had to face the challenges of forest management given the existing landscape. The peat landscape is a part of the geophysical structure of the peat ecosystem, which holds many of the important elements of peat soil formation (Burkhard and Maes 2017). The existing land use and cover conditions represent the tropical peat landscape in the village forest. The condition of forest density depends on its utilization. The vast majority of the village forests are secondary forests, except for the Lukun village forest, which is almost entirely undisturbed forest.

---

[4]Production forests are state forest land allocated mainly for timber and non-timber forest production.

[5]Conversion forest is a production forest reserved for development outside of forest activity.

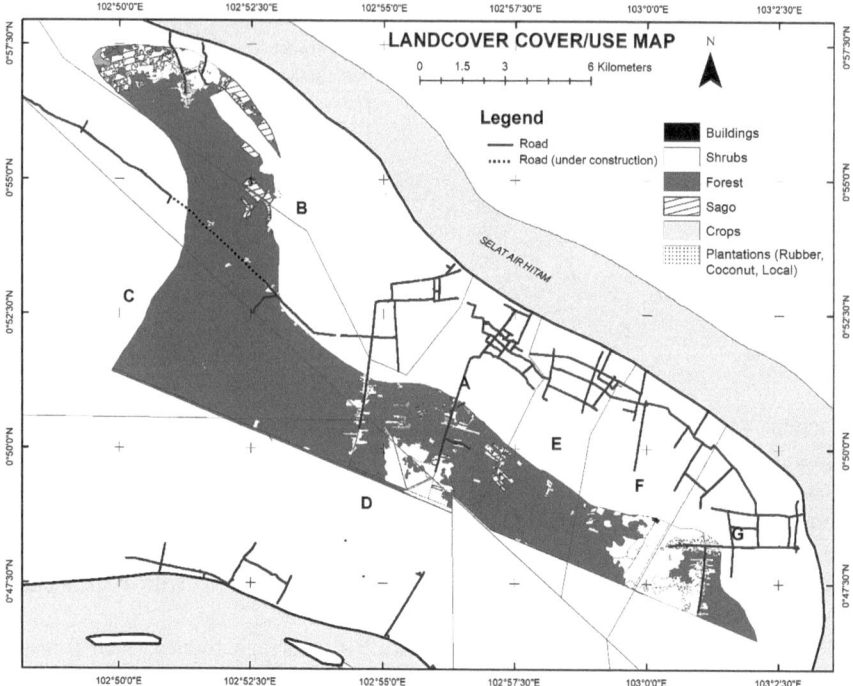

Fig. 12.2 Land use/cover map of the seven village forests

Remote sensing captures the presence of dense tree canopy cover through the high density of the tree trunks per area and the high soil wetness index (Vickers and Palmer 2000). This existing forest condition combined with other land use and cover datapoints can provide holistic landscape information and help identify the spatial distribution of the remaining forest that needs to be conserved (Fig. 12.2).

In 2017, approximately 74% of the village forests remained forested (Fig. 12.2) even though the natural forest ecosystem has not totally recovered from the timber extraction in the Logging Forest Concession era (*Hak Pengusahaan Hutan*, HPH)[6] and sporadic deforestation continues as a form of agricultural expansion (Sampurno 2019). Lukun has the most forest cover, with 2441 hectares of forest, or 83% of the total area, while Tanjung Sari village has the least, with 96 hectares of forest cover (Fig. 12.3).

Figure 12.3 depicts the different types of land use in the seven village forests. Although forest cover is dominant in each village forest, sago cultivation represents

---

[6] Started in the 1970s, the Logging Forest Concession (*Hak Pengusahaan Hutan*, HPH) accounted for more than 50 percent of Riau's regionally generated revenue (*Pendapatan Asli Daerah*) (Hidayat 2016). However, most HPH were converted to HTI (*Hutan Tanaman Industri* – Industrial Forest Plantations), which seldom produce equal benefits and affect the wellbeing of nearby local remote communities.

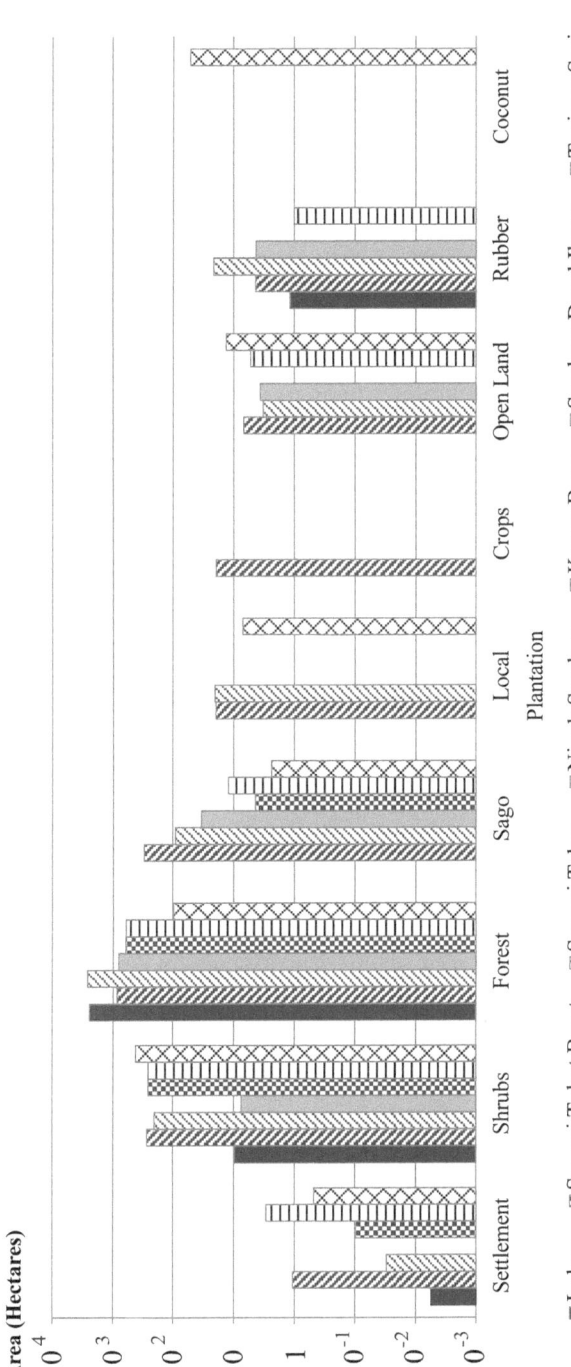

Fig. 12.3 Land-use distribution in the seven village forests

a significant proportion, among other agricultural production, in Sungai Tohor Barat, Sungai Tohor, Nipah Sendanu, and Sendanu Darul Ikhsan. The efforts to improve rural livelihoods affect tropical peat ecosystem resource extraction (Angelsen et al. 2014). In the northwestern parts of the Sungai Tohor Barat village forest, locals extract mangrove trees and sell them to wood briquette factories. The mangrove ecosystem may continue to degrade and cause sea abrasion, but these villagers must earn an income by extracting mangroves, because for the most part, they have very little production land. Still, the economic benefit from natural resource extraction is insufficient for household needs, while sustainability is at stake. Another land use management issue is occurring in Sungai Tohor village forest, in the area where the ex-HTI canal was built. Previously, locals planted 6–7-year-old rubber and sago trees on a group of local agricultural plots. In 2014, the opened canals caused excessive drainage and resulted in massive peat fires, damaging most of these plots. Those who did not have capital abandoned the damaged plots, while others gradually planted sago once the effects of El Niño diminished in 2016. These small-scale (less than two hectares) plantation systems depend on the smallholders' capital and a stable peat ecosystem.

12.3 Methodology

12.3.1 Ecosystem Service Valuation Approach

In this study we adopt the concept of ecosystem services from the Millennium Ecosystem Assessment (MEA), which categorizes four types of services: supporting, provisioning, regulating, and cultural (Hassan et al. 2005; Hester and Harrison 2010). For the most part, beneficiaries only recognize the final services in a cascade of ecosystem services as quantifiable products (Burkhard and Maes 2017; Turner et al. 2008). This study, however, applies an omnidirectional ecosystem services scheme and analyzes a whole island as a unit of the ecosystem so that we can value not only the specific lowland peat ecosystem but also the surroundings that earn ecological benefits of the energy flow from the peat dome. Defining the supporting, provisioning, regulating, and cultural services may vary depending on the valuation approach.

According to the general concept of ecosystem services, each service is related to other services. In this case study, the supporting services generate other ecosystem services based on the peat characteristics, the flow of energy of material, and ecological conditions (Maltby 2009). To value regulation services, the calculation of potential carbon stocks targeted measurement of both tropical peat forest and the peat soil (Chave et al. 2005; Lal et al. 2013; Rudiyanto et al. 2018). To value tropical peat forest carbon stocks, the spatial assessment utilized high-accuracy LiDAR data and aerial photos that were suitable for integration with direct ground measurements (de Jong and van der Meer 2004). In this valuation of regulating services, the participation of LPHD members is needed not only to identify the trees, but also to measure and roughly estimate the carbon stock using a simple technique. The valuation of provisioning services targeted sago production as a potential food production, estimating values of existing smallholder sago plantations (Pratama et al. 2018; The Society of Sago Palm Studies 2015). However, the sago production estimation considered not only sago's economic value to local community livelihoods, but also its ecological value as an aquatic plant. While both sago and tropical forests support the ecological sustainability of peatlands, human activity may contribute to the socio-cultural value of a peat ecosystem. Peatland eco-tourism is chosen to represent socio-cultural value and becomes an indicator of cultural services, which aim to enrich biodiversity restoration, improve aesthetic value, and educate about tropical peat ecosystems (Raymond et al. 2009; Schmidt et al. 2017).

During the valuation process, LPHD members joined the direct ground measurements and discussions as the service receivers and providers. By participating in this valuation process, the LPHD members gradually came to understand the potency of each service not only given its economic value, but also its contribution to improving standards of living in peatland areas. As mentioned above, community involvement in ecosystem services positions community members as appropriators, evaluators, and decision-makers in village forest management.

12.3.2   Peat Ecosystem Services Valuation Framework

In the development of a peat ecosystem services valuation framework, we began with geospatial and primary field data as the basic parameters of the spatial analysis. These two types of data are combined for analysis in the ecosystem services valuation approach. The high-resolution remote-sensing data derived from LiDAR and aerial photographs visualize the surface characteristics of peat, while the soil map provides below ground (BLG) data for estimating the peat carbon stocks. Ground measurements by LPHD members supports the acquisition of primary field data (Table 12.2).

12.3.2.1   Carbon Stock Estimation

The above-ground (ABG) carbon stock assessment of forest cover used both LiDAR and aerial photo data[7] to create models of forest volume. The remote-sensing data derived from airborne sensors was advantageous in that it offered better accuracy and spatial resolution.[8] The estimation of ABG biomass also used 22 valid, direct

---

[7] Aerial photos were recorded on January 2017 and provided by the Peatland Restoration Agency (*Badan Restorasi Gambut*, BRG).

[8] The spatial resolution for LiDAR and aerial photos was less than 10 cm, which was able to present information on a scale of 1:2500.

Table 12.2 Measuring the indicators of ecosystem services (Burkhard and Maes 2017)

Ecosystem service		Spatial indicator	Measurement		
			Direct	Indirect	Model
Regulating	Carbon stocks	ABG[a] forest carbon stocks	Plotting	Regression with remote sensing data	Carbon stocks
		BLG[b] carbon stocks	Boring	GIS interpolation with LiDAR	Carbon stocks
Provisioning	Food production	Sago plantation production	Plot measurements and statistics	Remote sensing of sago plantation condition	Sago plantation production
Cultural	Biodiversity enrichment, improved aesthetic, and education	Socio-cultural environment	Participatory scenario planning	Infrastructure and remote sensing data spatial interconnection	Ecotourism
Supporting	Peat function	Peat characteristics	Government regulation (reviewed)	Remote sensing of landscape composition (reviewed)	PEF

[a] Above ground biomass
[b] Below ground biomass

plot measurements of forest areas that provided diameter of breast height (DBH) in centimeters and tree height in meters (BSN 2011; Darusman et al. 2009). This study uses an allometric equation for the measurement of the peat forest, based on the tree species found in the village forest (Brown 1997; Chave et al. 2005; Manuri et al. 2014). The plot areas for ground measurement reflected the accuracy of the estimation of the tree heights measured with LiDAR data (Mutwiri et al. 2017).

To acquire primary field data, the LPHD members were involved in ground measurements of the biomass estimation of the plots. In this step, measurements were conducted based on the Indonesian standard carbon-counting method, which is useful for the locals to learn so they may carry out future assessments independently.

To estimate below ground (BLG) carbon stocks, defining the accuracy of final data was dependent on the level of information targeted to be produced. For local forest management, we use an interpolation model of the peat depth to represent spatial information of necro mass (Eidsvik et al. 2015) in this study. The analysis of BLG peat soil carbon stocks considered the depth of peat soil and its characteristics together with elevation data and peat characteristics (KLHK 2015).

Here it is useful to note that because community representatives participated in a survey conducted by MoEF in 2015, they already had information about the general peat depth. The characteristics of the peat on Tebing Tinggi Island are similar to those of peat on Sumatra's eastern coastal islands. Previous studies of the peat carbon content on these islands were also referenced for this study (see Dommain et al. 2011; Chadirin et al. 2015; Giesen 2015; Rixen et al. 2016; Murdiyarso et al.

2017; Rudiyanto et al. 2018). Instead of using laboratory data, the analysis of peat soil biomass used an algorithm and GIS (geographic information system) model to create a more practical estimation of the vast area of peat soil (Warren et al. 2012). For the estimation of peat carbon stocks, this study obtained recent data on the dry density and carbon content using measurements from a previous study (Haidar 2013).

Estimations of the ABG and BLG biomasses constitute both the value of carbon stock—which is substantial for the flow of energy and soil nutrients—as well as the value of conservation to reducing greenhouse gas emissions. These methods use scientific measurements and models that offer quantifiable values that are clearly understandable for the community.

12.3.2.2 Food Production Estimation

Aerial photos identify existing sago trees and quantify the estimation of sago production in the village forests. The quantity and quality of sago trees depend on the plantation system and peat habitat, which are only supported by sufficient water due to a lack of nutrients in the peat soil (Stanton 1980; The Society of Sago Palm Studies 2015). The blocking canals maintain the water table and keep the peatland bogged. The paludiculture system is well known to the locals as a better plantation system than monoculture.

Indirect measurement with LiDAR uses a three-dimensional distribution of point clouds to identify the canopy cover size and height of the sago trees. In addition to direct measurement of sago plantation production, sago plantation statistics and previous research were also used as a reference (Pratama et al. 2018), with verification via fieldwork of ten plots in two villages (Sungai Tohor and Sungai Tohor Barat). Verification from the indirect measurement confirmed the age of the sago trees. The participation of LPHD members, who had been well educated on identification and measurement, accelerated the verification process. In the future, they will be able to estimate the production of sago trees by high-resolution spatial data. The result of the integration of direct and indirect measurement methods resulted in a food production model based on the number of sago trees (in various conditions), which was then converted into an economic value based on the production. This spatial assessment of the valuation of sago production represents the initial value of food production as a provisioning service.

12.3.2.3 Examination of the Socio-Cultural Environment

The term *value* refers to the economic benefits, the extensive ecosystem and biodiversity benefits, as well as cultural, artistic, inspirational, educational, spiritual, and aesthetic benefits, or value(s) (Burkhard and Maes 2017; Schmidt et al. 2017). In the valuation of ecosystem assets, accounting for non-monetary units is seldom done, and it is even neglected. The valuation of cultural services uses participatory community scenario planning through interviews and discussions about the ecotourism model. The assessment for this study involved interviews and discussions with 20 local peat environmentalists,[9] members of the non-governmental organization (NGO) *Wahana Lingkungan Hidup* (WALHI), and village authorities. Unlike the valuation of regulating and provisioning services, in the valuation of cultural services, the local community takes a greater part in the discussion; in this case, the community representatives from LPHD led the discussion with various ideas. This study suggests that the ecotourism model represents the socio-cultural value based on the strategic interconnection between attraction, infrastructure, and local participation. This cultural service is selected due to the development of peat eco-tourism by the local community.

12.4 Implementation of a Peat Ecosystem Services Approach

12.4.1 Ecosystem Services Valuation

The perception of timber collection as a source of profitable income remains for some locals. Gradually, shifting the paradigm of forest utilization toward conservation has paralleled the peat restoration actions and grassroots encouragement by environmental NGOs. However, the concept of ecosystem services delivering ecological and economic benefits has not yet been introduced to the local activists and residents. Governments and NGOs have introduced non-timber forest production (NTFP) as a means to fulfill economic household needs with the concomitant aim of conserving the forest for ecological benefit due to the vital role of the peat ecosystem. The local economy for sago production is growing on Tebing Tinggi Island due to market and ecosystem suitability (The Society of Sago Palm Studies 2015). Since 2015, local wisdom about the peat ecosystem has grown to consider the equity of peat sustainability and land utilization for sago production (particularly with the well-known paludiculture system). The preservation of forest and sago cultivation as food production has the potential to strike a balance between ecological and economic benefits.

While the peat ecosystem provides regulating and provisioning services, supporting services are determined by the spatial distribution of conserved and utilized peatland (Turner et al. 2008; Hester and Harrison 2010; Lal et al. 2013). Conservation of the natural peat dome benefits areas beneath it, providing sufficient groundwater for agriculture, household, and peat fire prevention uses. Although generally local residents only consider economic benefits, natural peat preservation has become a priority for peat ecosystem management. Government Regulation No. 71/2014 supports sustainability by categorizing peat areas into one of two

---

[9]In Sungai Tohor: 6 people, Sungai Tohor Barat: 3 people, Lukun: 2 people, Nipah Sendanu: 2 people, SDI: 2 people, Tanjung Sari: 3 people, and Kepau Baru: 2 people.

functions: conservation (protection) or cultivation. The implication of this government regulation for the village forests is that it helps to determine the ecological benefits of most locations in the peat dome.

The locals in the seven villages have several strategies at their disposal for managing the village forest. According to the policy for forest utilization in production, villages can apply for a forest utilization permit (*Ijin Usaha Pemanfaatan Kawasan*, IUPK), a forest timber product permit (*Ijin Pemanfaatan Hasil Hutan Kayu*, IPHHK), a non-timber forest product permit (*Ijin Usaha Pemanfaatan Hasil Hutan Bukan Kayu*, IUPHHBK), or an ecosystem services permit (*Ijin Usaha Pemanfaatan Jasa Lingkungan*, IUPJL).

Due to a lack of bureaucracy knowledge, LPHD members were not aware that they, as decision makers, must follow the proper social forestry management process. Part of this process is to identify the natural peat assets and develop an appropriate program to protect and manage the forest for the community's livelihood. In managing the forest, they face the following challenges: (1) they must establish the village forest program according to proper procedures and government regulations, (2) they must protect the remaining forest and avoid any deforestation by illegal logging or land grabbing, and (3) they must monitor and report on the status of the peat village forest. In this study, we focused on the planning process, which consists of assessing the natural assets and creating a communal land utilization plan. Although we have not yet considered the budget necessary to sustain the village forest right, we hope the method of ecosystem services valuation we introduced can support an assessment of assets and that the transferring knowledge about this method to the LPHD members can improve understanding and help them with planning and monitoring in the future. The government might revoke the social forestry right without a good planning and monitoring system by LPHD at any time.

12.4.2 Carbon Stock Estimation

Valuing ecosystem services by using spatial data and community participation not only enables visualization of the resources under management, but also develops the understanding of the process of village forest management. To value regulating services, ABG and BLG biomasses were estimated according to the model depicted in Fig. 12.4a. First, the points of return from LiDAR remote sensing estimate tree height conditions for the majority of the forest area. Next, a regression model uses each plot[10] measurement within an area of 400 m^2 and applies it to the entire canopy.

The distribution of forest biomass and carbon stock related to tree height is shown in Fig. 12.4b, c. Areas with 80% canopy density dominate the village forests. The natural condition of the peat ecosystem with a dense canopy of tropical wetland forest is the best condition for preserving water in the peat dome, with a humidity

---

[10] The ABG was measured in 22 verified plots in December 2018, with allometric calculations based on the tree height (m), tree breast width (cm), and wood density (g . cm^{-3}).

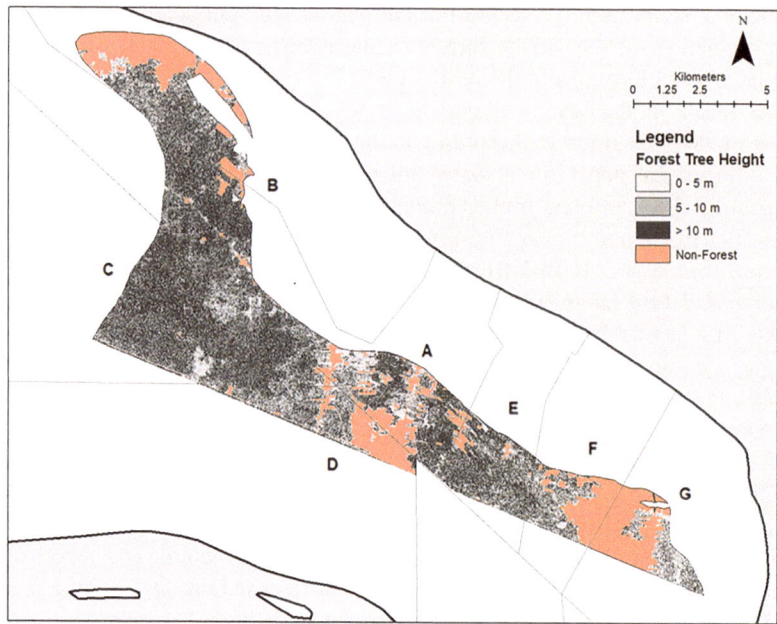

a) Tree height model

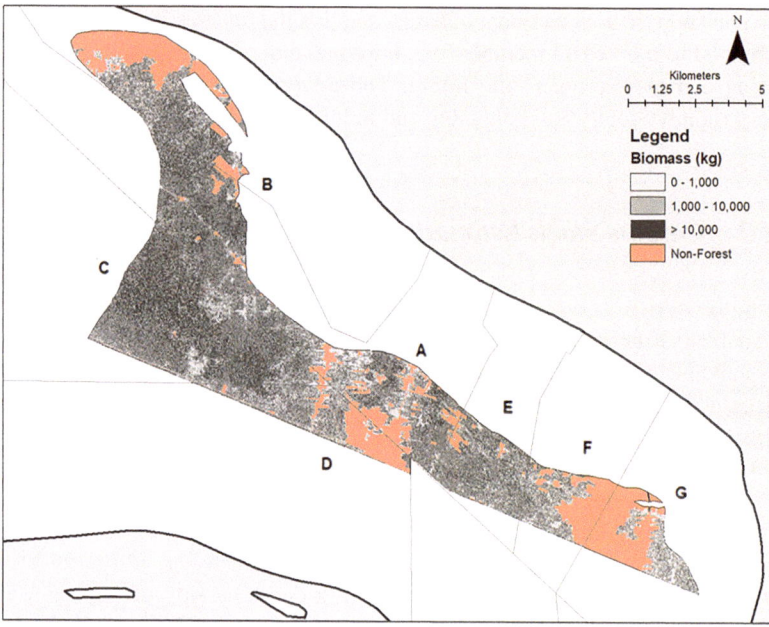

b) Biomass estimation

Fig. 12.4 Estimation of carbon stock

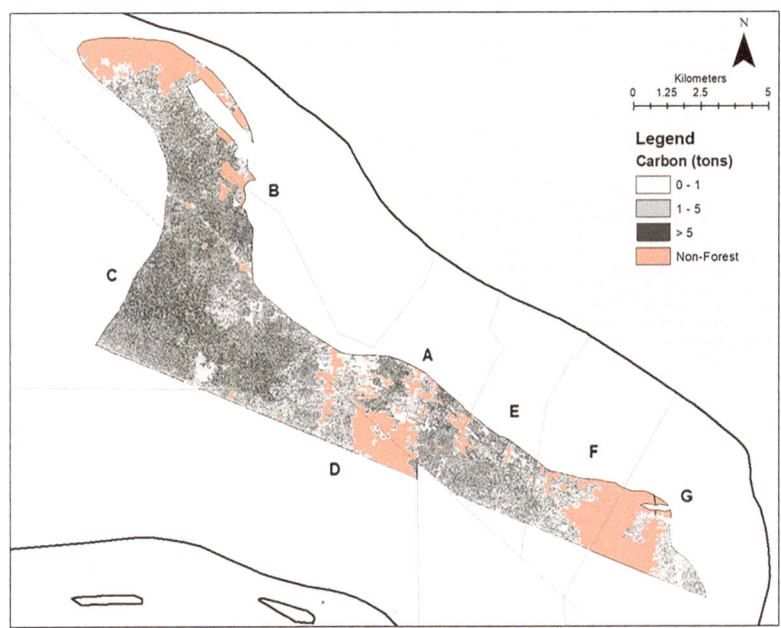

c) Above ground carbon stock

Fig. 12.4 (continued)

level of above 60% to prevent fire. Mangrove forests with less than 70% canopy density are present in the northwest part of Sungai Tohor Barat village forest, close to the shore. These mangroves are unable to reach maturity stage due to their extraction for wood briquette production. The tallest tree heights were found in the Lukun village forest, where the forests are quite dense (Fig. 12.4a). The Nipah Sendanu village forest also contains trees with a height of more than ten meters, but they are less dense than in the Lukun forest. Approximately 2.53×10^5 tons of ABG carbon stock was estimated in the Lukun village forest (Table 12.3), demonstrating its significant ecological value; in contrast, Tanjung Sari village forest had the lowest level of ABG carbon stock (8.08×10^3 tons), with only a small amount of forest cover on the far eastern part. The area of limited forest cover in the Tanjung Sari village forest is the area that has been left unoccupied by locals, while the other parts consist of settlements and small-scale coconut plantations. A similar situation also exists in the Sendanu Darul Ikhsan village forest, where the forested areas are mixed with settlements and small-scale plantations.

In the plot-direct measurement of ABG biomass, community representatives from each LPHD were able to identify tree species and general characteristics. They identified 68 tree species across the village forests, with the majority comprised of *Macaranga hypoleuca mucil* (*Mahang*), *Cratoxylum* (*Gronggang*), and *Parartocarpus* spp. (*Tenggayun*). *Gronggang* is a tree species that can grow to a size of more than 50 cm in diameter by the age of 7–8 years. *Punak* (*Tetramerista glabra miq.*) hardly existed in the village forests as a mature tree, as it has a high

Table 12.3 Estimated values of ecosystem services in seven village forests on Tebing Tinggi Island

Ecosystem service			Village forest						
			Sungai Tohor	Sungai Tohor Barat	Lukun	Kepau Baru	Nipah Sendanu	Sendanu Darul Ikhsan (SDI)	Tanjung sari
Village area (hectares)			2940	1482	2446	844	838	650	760
Regulating service									
Carbon sequestration	Forest area (hectares)		2599	858	2441	607	806	603	96
	ABG								
	Biomass (kg)		3.66×10^8	9.57×10^7	5.38×10^8	6.62×10^7	9.70×10^7	3.36×10^7	1.72×10^7
	Carbon (tons)		1.72×10^5	4.50×10^4	2.53×10^5	3.11×10^4	4.56×10^4	1.58×10^4	8.08×10^3
	BLG								
	Volume (m^3)		1.88×10^8	8.44×10^7	1.84×10^8	5.59×10^7	4.39×10^7	3.69×10^7	3.93×10^7
	Carbon (tons)		1.06×10^7	4.84×10^6	1.01×10^7	3.14×10^6	2.57×10^6	2.13×10^6	2.32×10^6
Provisioning service									
Food production	Existing sago cultivation								
	Area (hectares)		89	301	0	4	34	12	2
	Productivity (chunks/year)		28,621	96,493	0	1404	10,891	3916	763
	Value (dollar)		80,622	271,812	0	3956	30,678	11,031	2150
Supporting service									
	Spatial zonation		PEF	PEF	PEF	PEF	PEF, CEF	PEF	PEF
	PEF area (hectares)		2940	1482	2446	844	816	650	760
	CEF area (hectares)		0	0	0	0	22	0	0

Cultural service							
Education / recreational	—	—	—	—	—	—	Educational park
Ecotourism	Peat and mangrove	Peat vegetation	—	Peat vegetation	—	Peat vegetation	Educational park
Enrich biodiversity	Peat vegetation	Peat vegetation	Peat vegetation	Peat vegetation	Peat vegetation	Peat vegetation	Peat vegetation

Note: PEF signifies conservation (or protected) and CEF signifies cultivation ecosystem function

Source of data: Based on fieldwork conducted in December 2018

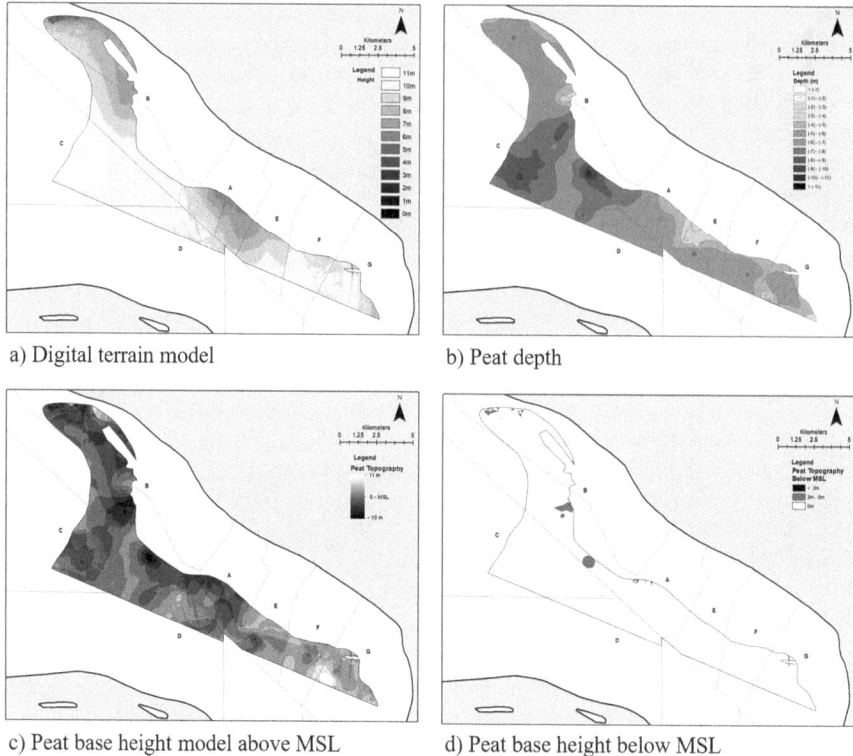

a) Digital terrain model   b) Peat depth

c) Peat base height model above MSL   d) Peat base height below MSL

Fig. 12.5 Peat depth models in the village forests

timber value and price. Regardless of the distribution of tree species, this study estimated biomass using allometric measurements and tree height modeling. The tree height spatial data represents the distribution of forest density and will effectively help to create a program for reforestation or recolonization with a dispersal system.

The condition of the peat ecosystem depends on the composition of organic material in soil formation. The deepest peat soil on Tebing Tinggi Island was measured at 12 m, while in the Sungai Tohor village forest it was about 11 m. The LiDAR data provided a high-resolution digital terrain model (DTM) and was able to describe the micro-topography of the peat ecosystem. In traditional peat depth modeling, the peat surface is often displayed as a flat topography. The microrelief is often neglected due to lowland peat ecosystem conditions. Integration of the interpolated peat depth[11] and DTM allows us to determine the peat base height. According to the spatial distribution, 98.77% of the total forest area of the seven villages exhibited peat soil above the mean sea level (MSL) (Fig. 12.5), while the highest point on the peat surface was 11 m above MSL.

---

[11] The peat depth spatial model was created by the Krigging interpolation method (Rudiyanto et al. 2015) using 1386 points of the peat depth measurements completed by MoEF in 2015.

The carbon sequestration model estimates the volume of organic material in the peat by calculating peat thickness and maturity. The village forests in Sungai Tohor and Lukun have more than 1.01×10^7 tons of peat carbon stock in their necro masses (Table 12.3). Given the high porosity of peat soil and the slope of the peat surface, maintaining the water table, preserving water, and preventing carbon emissions during the long dry season become severe challenges. The strategy for maintaining the balance of ecological and economic benefits involves balancing the composition of peat forest in the peat dome and the sago plantations in the buffer zone. Another strategy is to construct canal dams to prevent water from flowing out to sea, and canal dams have been built via government funding, NGO aid, and community self-budgeting.

Due to the remaining forest condition, the seven villages support a small amount of ABG carbon stock in their forested areas. Tanjung Sari village exhibited the least amount of ABG and BLG carbon storage (Table 12.3). The average landscape-level ABG carbon stock was about 71 t C / ha in the studied village forests, whereas van Breugel et al. (2011) estimated carbon stocks to be 61 t C / ha in the Panama Canal Watershed, using similar allometric biomass models in a similar secondary forest. Although the forests seemed dense from the top, because they are mainly secondary forests with an average of 13.23 cm of DBH, they produce a small amount of carbon stock at the landscape level, while the BLG necro mass of the peat soil showed potential carbon stocks with a volume of 634 M m^3. It is important to note here that these calculations of carbon stocks are not only the result of spatial analysis, but also the contributions of LPHD members to the fieldwork, and thus result in improved awareness of the potential benefits from such stocks.

To sustain the regulating services, local livelihoods must be supported with incentives for peat conservation and reforestation (Sathaye et al. 2001). The uncertainty of the carbon markets in Indonesia, especially through local governments with the REDD+ (reducing emissions from deforestation and forest degradation) scheme, has become a challenge for carbon creditors and carbon investors in terms of application to forest conservation (Djaenudin et al. 2016). Discussion of carbon stock values with community representatives and social forestry agents has motivated to start conservation schemes, with incentives from forestry government institutions and village fundraising (Sungai Tohor village has begun an ecotourism program).

12.4.3 Sago Production Estimation

The provisioning services represented by food production in the village forests do not provide the same services value as carbon storage. The composition of land use and local accessibility to the village forest differentiates the value of the sago plantation production. Unlike the six other village forests, sago plantations hardly exist in Lukun, as the forest is in a remote location more than 17 kilometers from the Lukun settlements. In contrast, vast sago plantations exist in Sungai Tohor Barat,

Sungai Tohor, Nipah Sendanu, and Sendanu Darul Ikhsan, and most sago starch mills are scattered among these villages. Sungai Tohor Barat has 301.54 hectares of sago plantations, which are mostly owned by a landlord who also owns a sago starch mill on the shore. The 13 sago starch mills owned by local Sungai Tohor villagers also boost the sago production in the neighboring villages, including in the six village forests. Discussion with the local community and LPHDs about sago production as part of the ecosystem services program allowed by the government increased motivation to manage the production zones of the village forests. According to the social forestry regulation, the village forests are divided into conservation and cultivation zones. However, the community already understood that according to the peat ecosystem characteristics of the village forests, peat preservation is a crucial part of the sago plantation system.

In the planning process of each village forest, LPHDs have their own strategy to maximize the space for the community needs. For example, Sungai Tohor divides its village forest into areas for forest conservation and for food production, where the food production areas are designated for sago plantations. They proposed that some of the sago areas be designated for shared management while the status of the remainder is still being negotiated with the owners, especially those not from the village.

12.4.4   Cultural Services Identification

Placing value on non-monetary services shifts the motivation of the local community from economic benefits (for example from timber collection) to preserving the natural peat ecosystem. The most apparent result of this shift was the noticeable local interest in ecotourism. From the local perspective, peat ecotourism is a system that balances the peat biodiversity and water conservation with the sago (or any other agriculture) production, together with improvements of living standards.

The Sungai Tohor villagers, pioneering peat conservation environmentalists, have already established a 14.57-hectare site for ecotourism and research. In 2018, 30–40 visiting domestic and international researchers, NGO staff, members of the media, and tourists visited this site.

The Sungai Tohor peat forest ecotourism and sago education site is about 5 km from the local harbor, while the mangrove ecotourism attraction is 2 km closer (Fig. 12.6). Unlike the peat forest, the mangrove ecotourism area has attracted more than 100 visitors per month. The village government collects entrance fee revenue and uses it for infrastructure development, such as the road access to the mangrove site and maintenance of tourism facilities. Ecotourism has not yet been implemented in the other villages due to a lack of accessibility and the composition of land use; for example, in Tanjung Sari and Sungai Tohor Barat villages, most land is used for plantations and settlements.

The benefits of peat ecotourism include education via peat forest monitoring, improved living conditions, sago agro-ecosystem education, sago production,

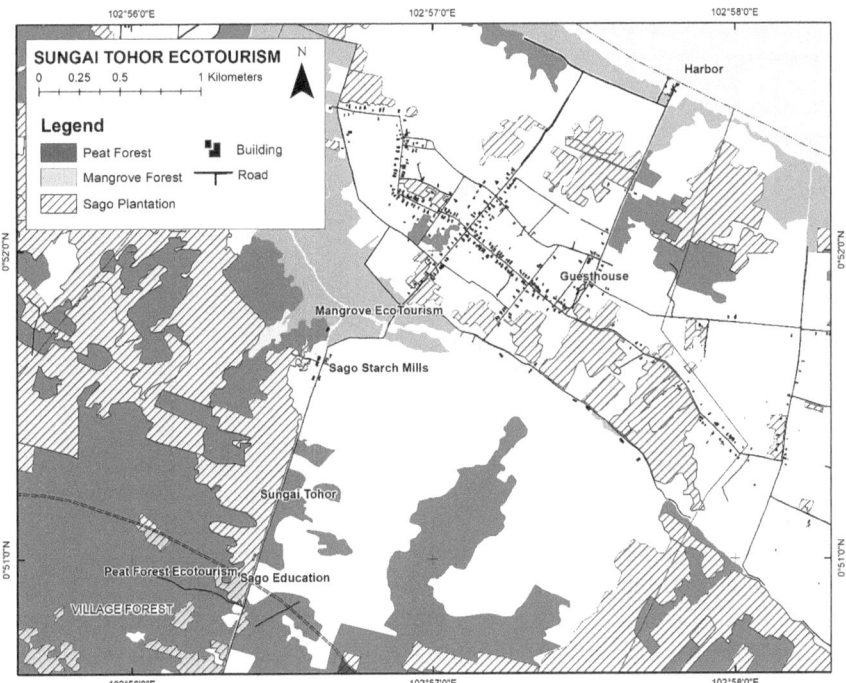

Fig. 12.6 Sungai Tohor ecotourism site

community involvement in peat conservation, endemic tree species planting, protection of canal water levels, and protection of the natural peat forest. Peat ecotourism also offers aesthetic value by allowing a mixed landscape of peat forest, sago plantations, sustainable sago production systems, and sago home industries. Although the ecotourism program of Sungai Tohor Village has not yet been formally established, the fame of the village as a peat research site and the eagerness of local environmentalist community have already demonstrated the benefits of peat ecotourism. The composition of the peat landscape, from the peat dome forest to the shore (Fig. 12.7), creates an ideal sustainable peat forest ecosystem, highlighted by a sequenced transfer of energy to sago food production.

This sequence begins with the peat forest in the peat dome, which intersects with the village forest (Fig. 12.7). This section consists of walking along the peat track routes and planting trees. At lower elevations, sago is harvested from the sago plantations. Local activities around the processing of the harvested sago take place in the sago starch mills and various sago industries—including home industries. Distribution and sales of sago is then conducted in the markets and at the shipping port. This geolocation introduces tropical peat village ecotourism, attracts visitors, and improves the locals' standard of living with the enjoyment of the forest scenery, clear air, and abundant water for swimming and fishing. The sustainable sago

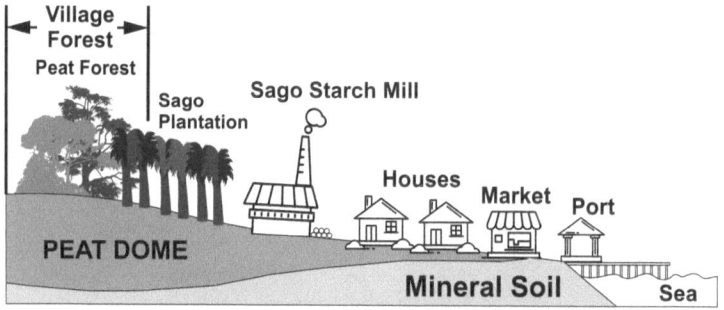

Fig. 12.7 Composition of a tropical peat ecotourism model

production system together with the forest scenery and clean air provide aesthetic value, which encourages visitors to support the ecotourism program's revenue generation.

Fresh air, birdsongs, forest cover, and abundant water provide aesthetic value to the location. The improved aesthetic value together with the sustainable semi-traditional sago production systems in turn create socio-cultural value. It is impossible to quantify these values in the spatial assessment, but for future valuations of cultural services, the spatial information, together with the monitoring of village forest management, will be able to supplement a qualitative approach. Biodiversity enrichment through tree planting and peat education in the form of research activity and ecotourism have also become part of cultural services in the broader scope of forest management.

12.5 Conclusion

The peat ecosystem service valuation conducted for this study was determined by spatial data utilization, community participation, and the coverage area of valuation. However, the purpose of this study was also to support the LPHD members in their village forest management by providing the services data. Based on the results of this approach, we agree with the prioritization of ecological values to conserve the remaining forest to ensure the sustainability of peatlands. However, trading carbon may not be a feasible mechanism to conserve the peat forest at the local village level. On the contrary, the persistent efforts of local environmentalists such as the LPHD members to conserve and enrich the natural peat biodiversity has attracted not only the Indonesian government, but also international organizations and researchers to support conservation action.

The participatory role of the community (mainly implemented via the LPHD members) is essential for residents to quickly understand the holistic character of the local natural resources. The LiDAR datapoints and aerial photographs were able to capture the natural resource assets using a simplified scientific method, which is useful for the initial valuation, and the LPHD can further evaluate ecosystem services using drone images. The spatial assessment with active community participation contributed to a more comprehensive local understanding of the tropical peat ecosystem. The integrated spatial assessment involved community participation based on local environmental knowledge. The keys to the success of assessment were its modest techniques, community participation, and quantifiable values.

The village social forestry scheme has accommodated the rights of use regarding state forest land. In the case of the seven villages of this study, the absence of customary land (*Tanah Adat*), non-contested village boundaries, and community control over the village forests are advantages for the village community in their management of the village forests. Conserving the natural condition of the 74% secondary forest cover in the seven village forests has become an ecological priority for the local communities, and ecosystem services can be considered a basis for sustainable management. The contribution of the remaining peat and mangrove forests to above ground carbon stocks varies according to each designated village forest. The tremendous peat forest volume in Lukun Village is quite valuable in this regard, with the other six village peat forests following behind. Preservation of natural forest in peat domes stabilizes the hydrology system, which is necessary for sago plantations.

The local people prefer the eco-friendly economic communal benefit as the basis for estimating the value of ecosystem services. The active community participation in the food production assessment suggested that sago could be considered a shared asset. The ecosystem service approach not only promoted economic benefits for sago plantation shareholders and conservation of the remaining forest, but also broke the perception that ecosystem services should be treated as a hindrance. The local community has realized that the socio-cultural value of cultural services is mainly non-monetary. However, some village communities have started to establish tropical peat ecotourism not only to reap educational, aesthetic, and socio-cultural values, but also to generate support for the local economy. Therefore, a social forestry scheme with economic, socio-cultural, and ecological considerations motivates the community to achieve sustainable peat forest management.

References

Angelsen A, Jagger P, Babigumira R et al (2014) Environmental income and rural livelihoods: a global-comparative analysis. World Dev 64(S1):S12–S28. https://doi.org/10.1016/j.worlddev. 2014.03.006
BPS (Badan Pusat Statistik) (2017) Kabupaten Kepulauan Meranti dalam Angka. BPS, Kepulauan Meranti District
Brown S (1997) Estimating biomass and biomass change of tropical forests: a primer. FAO forestry paper 134. FAO, Rome, p 55
BSN (Badan Standardisasi Nasional) (2011) Pengukuran dan penghitungan cadangan karbon: pengukuran lapangan untuk penaksiran cadangan karbon hutan. SNI 7724 BSN, Jakarta

Burkhard B, Maes J (eds) (2017) Mapping ecosystem services. Pensoft Publishers, Sofia, p 374. https://doi.org/10.3897/ab.e12837

Chadirin Y, Saptomo SK, Setiawan BI et al (2015) CO_2 emission from bare peat land using continues measurement. Adv Environ Biol 9(24):180–183

Chave J, Andalo C, Brown S et al (2005) Tree allometry and improved estimation of carbon stocks and balance in tropical forests. Oecologia 145(1):87–99. https://doi.org/10.1007/s00442-005-0100-x

Craig WJ, Harris TM, Weiner D (eds) (2002) Community participation and geographic information systems. Taylor and Francis, London

Darusman T, Mulyana A, Budiono R (2009) Pengukuran biomassa permukaan dan ketebalan gambut di hutan gambut DAS Mentaya dan DAS Katingan. https://www.researchgate.net/publication/311539591_Pengukuran_Biomassa_Permukaan_dan_Ketebalan_Gambut_di_Hutan_Gambut_DAS_Mentaya_dan_DAS_Katingan. Accessed 27 Oct 2022

de Jong SM, van der Meer FD (eds) (2004) Remote sensing image analysis: including the spatial domain, Remote sensing and digital image processing, vol, vol 5. Springer, Dordrecht. https://doi.org/10.1007/978-1-4020-2560-0

Djaenudin D, Lugina M, Ramawati R et al (2016) Perkembangan implementasi pasar karbon hutan di Indonesia. J Analisis Kebijakan Kehutanan 13(3):159–172. https://doi.org/10.20886/jakk.2016.13.3.159-172

Dommain R, Couwenberg J, Joosten H (2011) Development and carbon sequestration of tropical peat domes in south-East Asia: links to post-glacial sea-level changes and Holocene climate variability. Quat Sci Rev 30(7–8):999–1010. https://doi.org/10.1016/j.quascirev.2011.01.018

Eidsvik J, Mukerji T, Bhattacharjya D (2015) Value of information in the earth sciences: integrating spatial modeling and decision analysis. Cambridge University Press, Cambridge

Giesen W (2015) Utilising non-timber forest products to conserve Indonesia's peat swamp forests and reduce carbon emissions. J Indones Nat Hist 3(2):17–26

Haidar M (2013) Terrestrial carbon loss from degraded peatland in Bengkalis Island, Indonesia. Dissertation,. Yamaguchi University

Hassan R, Scholes R, Ash N (eds) (2005) Ecosystems and human Well-being: current state and trends, Millennium ecosystem assessment series, vol 1. Island Press, Washington

Hester RE, Harrison RM (2010) Ecosystem services. Issues in envrionental science and technology, vol 30. The Royal Society of Chemistry, Campbridge

Hidayat H (2016) Forest resources management in Indonesia (1968–2004): a political ecology approach. Springer, Singapore. https://doi.org/10.1007/978-981-287-745-1

KLHK (Kementerian Lingkungan Hidup dan Kehutanan) (2015) Laporan inventarisasi karakteristik ekosistem gambut di KHG Pulau Tebing Tinggi, Kabupaten Kepulauan Meranti. KLHK Republik Indonesia, Provinsi Riau

Lal R, Lorenz K, Hüttl RF et al (eds) (2013) Ecosystem services and carbon sequestration in the biosphere. Springer, Dordrecht. https://doi.org/10.1007/978-94-007-6455-2

Lye TP, de Jong W, Abe K (2003) The political ecology of tropical forests in Southeast Asia: historical perspectives. In: Kyoto area studies on Asia, vol 6. Kyoto University Press\Trans Pacific Press, Kyoto\Melbourne

Maltby E (ed) (2009) Functional assessment of wetlands: toward evaluation of ecosystem services. Woodhead Publishing, Cambridge

Manuri S, Brack C, Nugroho NP et al (2014) Tree biomass equations for tropical peat swamp forest ecosystems in Indonesia. For Ecol Manag 334:241–253. https://doi.org/10.1016/j.foreco.2014.08.031

Menteri Lingkungan Hidup dan Kehutanan Republik Indonesia (2017) NOMOR SK.129/MENLHK/SETJEN/PKL.0/2/2017 – Penetapan Peta Kesatuan Hidrologis Gambut Nasional. Kementerian Lingkungan Hidup dan Kehutanan Republik Indonesia

Murdiyarso D, Hergoualc'h K, Sasmito S et al (2017) Permanent research plots in Bengkalis, Riau: carbon dynamics and water regimes of re-wetted peatlands. In: The science behind peatlands. Center for International Forestry Research, Bogor

Mutwiri FK, Odera PA, Kinyanjui MJ (2017) Estimation of tree height and forest biomass using airborne LiDAR data: a case study of Londiani Forest block in the Mau complex, Kenya. Open J For 7(2):255–269. https://doi.org/10.4236/ojf.2017.72016

Osaki M, Tsuji N (eds) (2016) Tropical peatland ecosystems. Springer, Tokyo. https://doi.org/10.1007/978-4-431-55681-7

Potschin M, Haines-Young R, Fish R et al (eds) (2016) Routledge handbook of ecosystem services. Routledge, Oxon and NY

Pratama GR, Hardjomidjojo H, Iskandar A et al (2018) Analisis rantai nilai agroindustri sagu di Kabupaten Kepulauan Meranti. J Teknologi Industri Pertanian 28(2):199–209. https://doi.org/10.24961/j.tek.ind.pert.2017.27.1.1

Raymond CM, Bryan BA, MacDonald DH et al (2009) Mapping community values for natural capital and ecosystem services. Ecol Econ 68(5):1301–1315. https://doi.org/10.1016/j.ecolecon.2008.12.006

Rixen T, Baum A, Wit F et al (2016) Carbon leaching from tropical peat soils and consequences for carbon balances. Front. Earth Sci 4:art 74. https://doi.org/10.3389/feart.2016.00074

Rudiyanto, Setiawana BI, Ariefa C et al (2015) Estimating distribution of carbon stock in tropical peatland using a combination of an empirical peat depth model and GIS. Procedia Environ Sci 24:152–157. https://doi.org/10.1016/j.proenv.2015.03.020

Rudiyanto, Minasny B, Setiawan BI et al (2018) Open digital mapping as a cost-effective method for mapping peat thickness and assessing the carbon stock of tropical peatlands. Geoderma 313: 25–40. https://doi.org/10.1016/j.geoderma.2017.10.018

Sampurno D (2019) Spatiotemporal analysis in monitoring landscape dynamic patterns in tropical peat ecosystem (study in Tebing Tinggi Island, Riau, Indonesia). J Environ Sci Sustain Dev 2(1):75–96. https://doi.org/10.7454/jessd.v2i1.33

Sathaye JA, Makundi WR, Andrasko K et al (2001) Carbon mitigation potential and costs of forestry options in Brazil, China, India, Indonesia, Mexico, The Philippines and Tanzania. Mitig Adapt Strateg Glob Chang 6(3–4):185–211. https://doi.org/10.1023/A:1013398002336

Schleyer C, Lux A, Mehring M et al (2017) Ecosystem services as a boundary concept: arguments from social ecology. Sustainability 9(7) art 1107. https://doi.org/10.3390/su9071107

Schmidt K, Walz A, Martín-López B et al (2017) Testing socio-cultural valuation methods of ecosystem services to explain land use preferences. Ecosyst Serv 26(Part A):270–288. https://doi.org/10.1016/j.ecoser.2017.07.001

Stanton WR (1980) SAGO: the equatorial swamp as a natural resource. In: Flach M (ed) Proceedings of the 2nd international sago symposium, Kuala Lumpur, September 1979, World crops: production, utilization, description, vol 1. Martinus Nijhoff, The Hague. https://doi.org/10.1007/978-94-009-8928-3

The Society of Sago Palm Studies (ed) (2015) The sago palm: the food and environmental challenges of the 21st century. Kyoto University Press\Trans Pacific Press, Kyoto\Melbourne

Turner RK, Georgiou S, Fisher B (2008) Valuing ecosystem services: the case of multi-functional wetlands. Earthscan, London and VA

van Breugel M, Ransijn J, Craven D et al (2011) Estimating carbon stock in secondary forests: decisions and uncertainties associated with allometric biomass models. For Ecol Manag 262(8): 1648–1657. https://doi.org/10.1016/j.foreco.2011.07.018

Vickers AD, Palmer SCF (2000) The influence of canopy cover and other factors upon the regeneration of scots pine and its associated ground flora within Glen Tanar National Nature Reserve. Forestry 73(1):37–49. https://doi.org/10.1093/forestry/73.1.37

Warren MW, Kauffman JB, Murdiyarso D et al (2012) A cost-efficient method to assess carbon stocks in tropical peat soil. Biogeosciences 9(11):4477–4485. https://doi.org/10.5194/bg-9-4477-2012

Chapter 13
Conclusion

Takamasa Osawa, Wahyu Prasetyawan, Akhwan Binawan, and Masaaki Okamoto

Abstract The issues, stakeholders, and solutions related to the peatland problem cannot be easily defined or easily addressed. The problem can thus be seen as one of today's "wicked problems," which have no true or false solutions, but rather better or worse ones. In addressing this kind of problem, academic researchers and NGO staff can play an important role to identify social and ecological issues and their inter-relationships, and to facilitate communication among local residents and other stakeholders. These activities should be done as continuous and flexible collaboration with local communities to find the better solutions to the peatland problem—or to realize a better future for society—together with those involved in the problem. This is the transdisciplinary approach that the authors seek.

Keywords Complexity of the peatland problem · Transdisciplinary approach · Collaboration · Commitment

13.1 The Peatland Problem: Defying Definitive Formulation

This volume addresses a range of issues that emerge from the peatland degradation and fires occurring in Indonesia. It specifically looks at the local governance of peatland, using Rantau Baru Village in Riau Province as its main case study.

T. Osawa (✉)
Institute of Liberal Arts and Science, Kanazawa University, Kanazawa, Ishikawa, Japan
e-mail: tosawa@staff.kanazawa-u.ac.jp

W. Prasetyawan
Syarif Hidayatullah Islamic State University, Tangerang Selatan, Banten, Indonesia

A. Binawan
Perkumpulan Ara Sati Hakiki, Jl. Kayu Putih, Perumahan Athaya IV, Pekanbaru, Riau, Indonesia

M. Okamoto
Center for Southeast Asian Studies, Kyoto University, Kyoto, Japan

© The Author(s) 2023 287
M. Okamoto et al. (eds.), *Local Governance of Peatland Restoration in Riau, Indonesia*, Global Environmental Studies,
https://doi.org/10.1007/978-981-99-0902-5_13

Researchers from a range of academic and activist backgrounds participated in the investigation, allowing them to address the issues from multiple perspectives.

First analyzing the mapping of peatlands, the volume immediately identifies the dysfunction of government policies in both the formulation and implementation of spatial planning (Chap. 2). The volume then turns to Rantau Baru, investigating the link between land use and villagers' aspirations for the future (Chap. 3). The next two chapters examine fishing in the Kampar River, as it is the main source of livelihood in the village. Chapter 4 points to the decrease of fish resources in recent years in village fishing grounds, while Chap. 5 highlights the limits of traditional fishing methods and the potential of fishing tourism for village development. The following five chapters shed light on the social issues that emerged in Rantau Baru during the process of implementing government policies to prevent peatland fires and improve livelihoods. These social dimensions include the idealized promotion of local wisdom (Chap. 6), unequal gender participation in village spatial planning (Chap. 7), the asymmetry of political power in decision-making about village budget expenditures (Chap. 8), the impact of low incomes and limited access to education on environmental protection measures implemented in cash-poor areas (Chap. 9), and the consequences of land grabbing (Chap. 10). Chapters 11 and 12 examine government policies aimed at addressing peatland degradation and fires in two additional villages in Riau. Chapter 11 highlights how government programs have accelerated economic disparities, while Chap. 12 examines the advantages of promoting ecosystem services valuation to achieve sustainable peatland use. Several chapters provide concrete suggestions to resolve or mitigate the issues raised. These suggestions include establishing a freshwater protected area (Chap. 4), promoting fishing tourism from urban areas (Chap. 5), and expanding customary common lands with the support of the district government (Chap. 10). The authors in this volume have concretely engaged with the social and ecological problems affecting peatlands, committed to affected communities, and collaborated with the various actors involved—that is, we have carried out the transdisciplinary approach we believed is necessary to address peatland degradation in its real complexity.

Although we summarize various issues emerging around peatlands as "the peatland problem," the chapters confirm that the problem also involves a wide range of inter-related environmental and social challenges. For example, while increasing global carbon emissions (Chap. 1) and land grabbing (Chap. 10) are each distinct problems of their own, they are also caused by the exploitation of the peat environment. Responses to such problems are also inextricably linked, as the response to any one issue may create new challenges. For example, as demonstrated in Chaps. 8 and 12, policies that support the prevention of peatland fires exacerbated the inequality of political and economic power in local communities. While the peatland problem begins with the drainage, drying, and burning of peatlands, these actions in turn cause a cascade of additional, and distinct, social and environmental problems that is not completely predictable. Therefore, the "peatland problem" cannot be definitively formulated in any single dimension.

Similarly, the agencies of the stakeholders involved in the peatland problem cannot be addressed in a formulaic way. Stakeholders' attempts to respond to (or ignore) peat degradation and fire are complex and sometimes contradictory. These responses therefore cannot be simplified into a single motivation or trajectory. For example, while the central government has adopted a moratorium policy on peatland development, it has not yet taken steps to ban peatland development in a strong and effective manner (Chap. 2); while oil palm companies provide a certain amount of budget to prevent peatland fires, they also promote the expansion of oil palm plantations that may cause or increase fire events (Chaps. 8 and 10); and, while villagers are threatened by frequent peatland fires and strongly concerned about the future of their community, they also sell peatland to outsiders, plant oil palm, and drain peatlands themselves (Chaps. 3 and 10). The peatland problem is the product of a complex interplay of diverse stakeholders' knowledge, interests, and aspirations. It is therefore impossible to assign stakeholders' attitudes to a reductive dichotomy of development versus conservation.

In this situation, we cannot identify a single formulated or fixed solution to the peatland problem. Trade-offs between environmental conservation and economic activities are complex phenomena that cannot be solved with a simple to-do list. Banning peatland use entirely, for example, may deprive local people of their economic development potential and lead to poverty, even in areas where oil palm can be grown without causing fire. Or it may create new, unpredictable patterns of peatland exploitation that leads to increased fire incidents and ecosystem degradation.

In short, the issues, stakeholders, and solutions related to the peatland problem cannot be easily defined or easily addressed. Peatland degradation is one of today's "wicked problems", with "no definitive formulation, no stopping rule, and no test for a solution" (Ludwig 2001, p. 759; see also Rittel and Webber 1973). Such problems have no true or false solutions, but rather have "better or worse" or "satisfying or unsatisfying" ones (Rittel and Webber 1973, pp. 162–163). Fundamentally, there is no single solution to the peatland problem, because with each response, new challenges arise, including the breakdown of ecosystems and conflicts between stakeholders.

13.2 A Transdisciplinary Approach: Continuously Directed toward the Future

How, then, given its local, national, regional, and global significance, should the peatland problem be addressed? One possible way is to promote understandings of how different stakeholders think about the peatland problem and to increase mutual understanding among them (Ludwig 2001). As noted above, the motivations and orientations of stakeholders are not one-dimensional or held in isolation. Each actor has unique intentions, interests, and desires formed in a complex social context. Promoting mutual understanding of these intentions, interests, and desires positively impact communication, negotiation, and cooperation among different stakeholders, which can contribute to "better" solutions to the peatland problem.

It is also necessary to pay attention to the specific interface where the concrete social and environmental issues of the peatland problem emerge—that is, the local communities and the environment around them. Each location has its own unique social and environmental characteristics, to which no one-size-fits-all solution can be applied. Even within a location, a variety of social and environmental issues result from peatland degradation and fires, and one issue may cause several additional and inter-related issues to emerge. These must be addressed in both concrete and long-term ways. Given that long-term and detailed efforts are required, the main agents of peatland restoration activities are the people of local communities. However, central and local governments, companies, and urban residents are also principal drivers of the peatland problem; local people do not have sufficient capital or labor to address the problem alone (Chap. 6). Therefore, all actors must be involved in solving the peatland problem, and appropriate support should be provided by non-locals who are also implicated in the problem.

Academic researchers and NGO staff play an important role in the pursuit of "better" solutions by: (1) identifying social and ecological issues and their inter-relationships; and (2) facilitating communication among local residents and other stakeholders. These efforts require flexible, continuous, and tailor-made approaches to address the challenges that emerge in the community. Researchers and NGOs may not always provide innovative, ultimate, or universal solutions. However, given that the peatland problem involves so many stakeholders and at the same time affects such a broad array of social and environmental conditions, it is necessary to maintain continuous and patient responses, staying close to and following up on issues as they emerge in the local community and surrounding environment.

With this outlook in mind, the researchers involved in this investigation have begun to implement suggestions to mitigate or solve the social and ecosystem challenges in Rantau Baru. Here, recent negotiations with the district government around customary land rights and the conservation of floodplains and peat swamps along the Kampar River are of particular importance.

First, to strengthen customary land rights, we are working to reinforce mutual understanding between the villagers and district government officials under the initiative of the local NGO, Hakiki. From 2020 to 2022, we held several meetings with senior officials of the Pelalawan District government, in which we presented maps of Rantau Baru customary territory (Chap. 10) and requested protection of rights to customary territory.[1] In addition, in 2021, we produced a short film on honey collection from *sialang* trees (see Chap. 3), which demonstrated the village custom of maintaining the *sialang* forest as common lands. We also organized a workshop to screen the film in the village for government officials. We plan to hold

---

[1] This action was supported by the Association of Malay Adat (*Lembaga Adat Melayu Riau*), which requested the district head (*bupati*) of Pelalawan to protect *sialang* forests from deforestation by industries (Chap. 3).

similar workshops for additional stakeholders, including urban residents and oil palm/acacia companies. Although these meetings and workshops will not immediately change the situation in Rantau Baru, the maps will be an important reference when the district government designates the *adat* area (*wilayah adat*: see Chap. 10), and demonstration of customary management of *sialang* forests appearing in the film will support negotiations aimed at strengthening the customary rights of the villagers.

Second, in October 2022, we began a new project to establish protected areas in the floodplain and swamps of Rantau Baru. As mentioned in Chap. 3, during the last decade, oil palm plantations have been expanding onto the floodplain and peat swamps near the Kampar River. However, most of the palm trees planted in these areas were destroyed by seasonal floods and are not providing sufficient returns to the villagers (see Chap. 3). Wetlands are drained for oil palm plantations, destroying the submerged forests and peat swamps that provide important spawning grounds and habitat for small fishes at the bottom of the river's food chain (see Chap. 4). Protecting these wetlands will increase fish resources in the future. As part of our project, ecologists and social scientists will assess the social and ecological processes associated with these protection activities. Hakiki members are also working to spread awareness of the potential of protection areas. The villagers have begun to embrace the aims of the protection activities and actively engage in discussing the details of the protection scheme.

These activities do not provide ultimate solutions to the various issues that have emerged due to peatland degradation and fire in the village. It is also quite possible that our attempts will give rise to new issues. However, what characterizes our activities is our continuous and flexible engagement with the social and environmental issues in the village, including any new issues that may emerge. In this process, Rantau Baru villagers, NGO activists, university researchers, local government officials, and other stakeholders are closely linked and work together in a collaboration. This is done to find better solutions to the peatland problem —or, to realize a better future for the society—together with those involved in the problem. This ongoing and dynamic commitment to the village community is the transdisciplinary approach we believe is necessary to mitigate and resolve the highly complex and mixed-interest problems of peatland degradation and fire.

References

Ludwig D (2001) The era of management is over. Ecosystems 4(8):758–764. https://doi.org/10.1007/s10021-001-0044-x

Rittel HWG, Webber MM (1973) Dilemmas in a general theory planning. Policy Sci 4(2):155–169. https://doi.org/10.1007/BF01405730

Column 1: Using Small, Unmanned Aircraft (Drones) to Create Maps

Kazuo Watanabe, Akhwan Binawan, and Masaaki Okamoto
Center for Southeast Asian Studies, Kyoto University, Kyoto, Japan
Perkumpulan Ara Sati Hakiki, Jl. Kayu Putih, Perumahan Athaya IV, Pekanbaru, Riau, Indonesia

Around 2014, the use of small, unmanned aircraft called drones began to spread throughout the world. The most common uses for drones include video and landscape photography. It is also possible to create ground images of large areas by overlaying multiple photos taken by drones. Although it is possible to do the same with high-resolution satellite imagery, such as that used to create the village map described in Chap. 2, it is difficult to obtain clear images of the entire area of interest with satellite imagery due to the influence of clouds and the high cost, therefore the use of drones to create maps has been spreading rapidly recently. In our research in M Village, we attempted to create a base map for a part of the RW1 hamlet. To do this, local NGOs flew drones over the area to obtain cloud-free, timely, and higher resolution images than satellite images.

The area surrounded by the red frame in Fig. 1 was set in advance by the flight plan creation software, and a drone was flown over the area with DJI's Phantom 3 Professional with automatic navigation. The actual flight took about 4 h (not including preparation and waiting time due to sudden changes in weather conditions) to cover an area of 412 hectares, including lectures to local NGO staff so that they can continue to operate the drone in the future, and acquired 3500 images. The acquired photos were processed by SfM (Structure from Motion) using Agisoft's Metashape (formerly PhotoScan) to create a single orthomosaic image. The ground resolution of this image is 5 cm, or ten times finer than the WorldView satellite image.

Figure 2 compares a satellite image and a drone image, and demonstrates how clear the image acquired by the drone is. In addition, since the drone flies at a low altitude of about 100 meters above the ground, no cloud cover obscures the ground,

© The Author(s) 2023   293
M. Okamoto et al. (eds.), *Local Governance of Peatland Restoration in Riau, Indonesia*, Global Environmental Studies,
https://doi.org/10.1007/978-981-99-0902-5

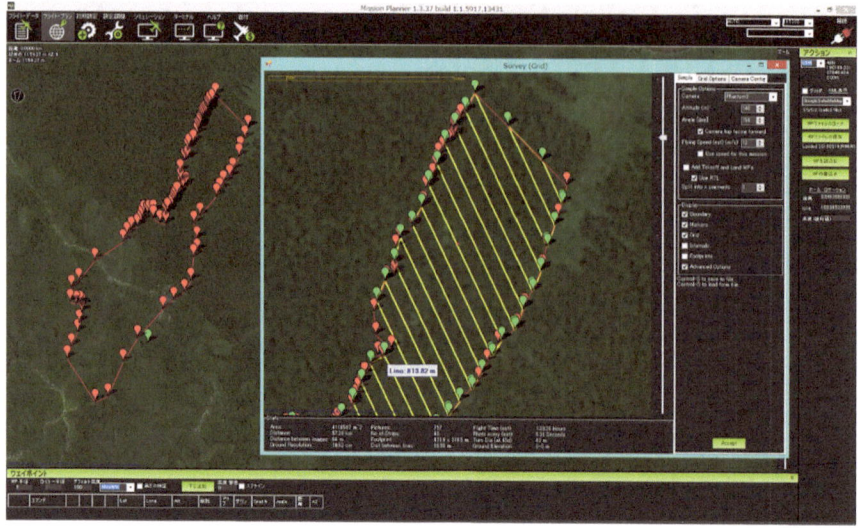

Fig. 1 Automatic navigation route of drone above Village M (Photo taken by Watanabe, February 2016)

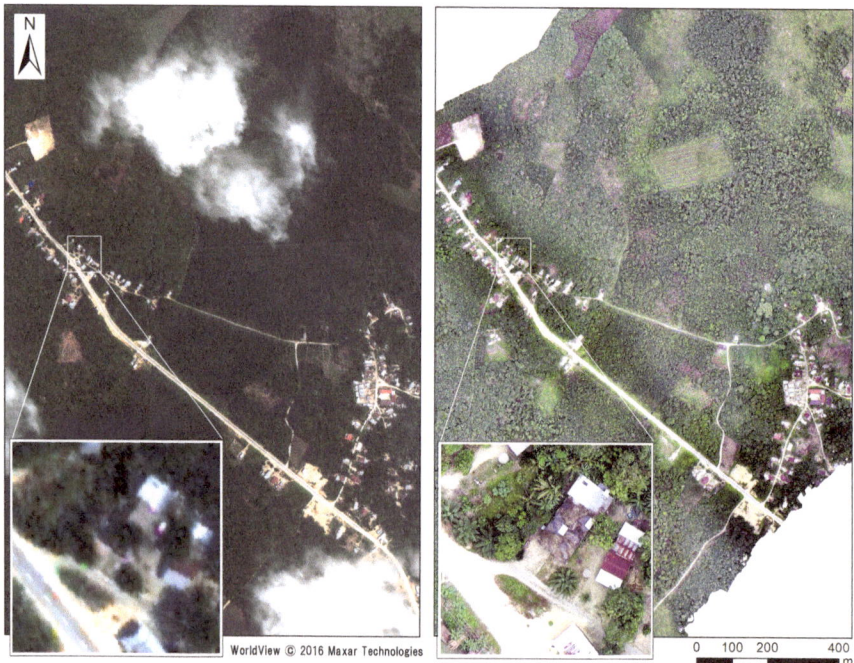

Fig. 2 WorldView satellite image (Left: photo provided by © 2016 Maxar Technologies) and the drone image (right) of Village M (Photo taken by Watanabe, February 2016)

as is the case with manned aircraft and satellite images. By using drones, we can acquire images of the necessary locations by ourselves whenever we need them, making it very convenient when updating maps.

However, mapping with drones also has its limitations. The most important of these is the range of images that can be taken. During our study, the actual flight time took approximately 4 h, but the battery lasted only less than 20 minutes, forcing us to raise and lower the drone about 20 times to change the battery. In addition, as we only brought six batteries to the site, so even though we recharged them during the flight, they could not recharge fast enough, resulting in a very long standby time. Therefore, it actually took two full days to capture only one part of one RW. Since it would take about 1 month to acquire data for the entire M Village using this drone, improvements such as increasing the number of drones or using a type of drone with a longer flight time capacity are necessary. In addition, as the number of images taken becomes enormous, high-specification workstations with the ability to create orthomosaic images will also be necessary. It has become clear that there are also hardware issues that must be resolved when using drones to create wide-area village maps.

Column 2: Questionnaire Survey in Rantau Baru

Wahyu Prasetyawan and Takamasa Osawa
Syarif Hidayatullah Islamic State University, Tangerang Selatan, Banten, Indonesia
Institute of Liberal Arts and Science, Kanazawa University, Kanazawa, Ishikawa, Japan

The main research site of Chaps. 3–10 is the village of Rantau Baru, where our multidisciplinary research team conducted various research activities, including participant observation and interviews, online interviews (see Column 3), participatory mapping (Binawan and Osawa, Chap. 10), and a household questionnaire survey. Table 1 outlines the survey, which is frequently referred to in the chapters; the complete set of questions is available in Appendix 1.

After preparing the questions regarding the six main topics noted in Table 1, we conducted two pre-tests to confirm the relevance and adjust the content and wording of the questions. The pre-tests revealed that the overly scientific and complicated language of some questions were a barrier and that older respondents found it difficult to understand the official Indonesian language used in the questions. Based on this feedback, we revised the questionnaire several times before finalizing a total of 123 questions.

The survey targeted all 164 houses that had registered residents living in the village on January 27, 2020. The "head of household" (*kepala keluarga*) is a role addressed in the Indonesian family register. Usually, the husband of a married couple living in the same house takes this role, while an ex-wife or widow comes to take it in the case of divorce or bereavement. In some cases, more than one head of household lived in a house. Considering the gender balance of the interviewees, we selected one respondent from each house, either the head of household or their spouse. One NGO member and six post-graduate students from Riau University who are well versed in the local dialect of Malay stayed in the village and conducted the survey through face-to-face interviews. They were able to gather responses from 152 of the 164 houses in Rantau Baru, or 94% of the total.

© The Author(s) 2023
M. Okamoto et al. (eds.), *Local Governance of Peatland Restoration in Riau, Indonesia*, Global Environmental Studies,
https://doi.org/10.1007/978-981-99-0902-5

Table 1 Parameters of the Questionnaire Survey in Rantau Baru Village

Data collection period	From January 27 to February 2, 2020
Data collection site	Rantau Baru Village, Pangkalan Kerinci subdistrict, Pelalawan District, Riau
Total population of the village	715 people (in 2018)
Survey population	Head of the household (*kepala keluarga*), or their spouse, of all 164 houses in the village
Respondents	152 households (94% of the total houses): 116 households from the subvillages of Danau Sepunjung and Malako Kocik; and 48 households from the subvillage of sei Pebadaran
Gender of the respondents	Male 74; female 78 (controlled to achieve an approximately even split across all interviews)
Number of questions	123 questions (see Appendix 1)
Main topics of the questions	(1) Basic information of the household (2) Engagement in peatland restoration (3) Engagement in social activities (4) Engagement in fishing activities (5) Family land use (6) Economic situation
Duration of survey interview	30–45 mins
Enumerators	6 postgraduates from Riau university and 1 person from the local NGO Hakiki foundation (4 males and 3 females). All of the enumerators could speak the dialect of Malay used in the Kampar River basin.
Pre-tests	(1) from august 31 to September 2, 2019 (2) from November 17 to November 19, 2019

Column 3: Fieldwork during the Covid-19 Pandemic

Kurniawati Hastuti Dewi, Akhwan Binawan, and Takamasa Osawa
Research Center for Politics, National Research and Innovation Agency (BRIN),
Jakarta Selatan, Indonesia
Perkumpulan Ara Sati Hakiki, Jl. Kayu Putih, Perumahan Athaya IV, Pekanbaru,
Riau, Indonesia
Institute of Liberal Arts and Science, Kanazawa University, Kanazawa, Ishikawa,
Japan

The coronavirus (COVID-19) pandemic that began in 2019 had become a significant global issue by 2020. As such, COVID-19 affected our research activities in Rantau Baru. The disease spread to Indonesia at the beginning of March 2020. By April 17, large-scale social restrictions referred to as the *Pembatasan Sosial Berskala Besar* (PSBB) were applied in the Riau provincial capital, Pekanbaru. By 12 May, the Ministry of Health (Kementerian Kesehatan) agreed to apply the PSBB in several other districts within Riau, including Pelalawan (Johanes 2020; Sehatnegeriku 2020). The PSBB imposed strong restrictions on crossing district and province borders. These restrictions were then imposed intermittently throughout 2020 and 2021.

After finishing the quantitative research questionnaire in February 2020, we intended to carry out qualitative research through fieldwork within the village. Due to the PSBB, however, it became impossible to visit Rantau Baru and communicate with the villagers. Therefore, we decided to conduct online meetings and interviews instead. Akhwan Binawan living in Pekanbaru visited Rantau Baru when the PSBB was temporally lifted (Fig. 3). As such, he was able to arrange the meetings and interviews.

The interviews, which involved 15 participants (five female, ten male), were conducted in November and December 2020. The selection procedure for the villagers was based on purposive sampling in that they were chosen to join the online interviews according to the purpose of the interview questions. As previously noted, targeted participants included village officials, traditional leaders, female

M. Okamoto et al. (eds.), *Local Governance of Peatland Restoration in Riau,
Indonesia*, Global Environmental Studies,
https://doi.org/10.1007/978-981-99-0902-5

Fig. 3 After long days with no school due to Covid-19, the children of Rantau Baru came back to their elementary school. All children had their body temperatures checked before entering the classroom. (Photo taken by Binawan, September, 2021)

village leaders, leaders in women's farming groups such as the *Kelompok Wanita Tani* (KWT) and members of *Pemberdayaan Kesejahteraan Keluarga* (PKK). Binawan arranged the schedule so that a maximum of two respondents were interviewed per meeting or day. Before the online interviews, he met the respondents to ensure that no changes to the online interview schedule were needed and then arranged similar schedules for the following participants. On the day of each interview, Binawan met the respondents in their homes and brought them to nearest restaurant/coffee shop. Reservations for each location were made in advance to ensure a stable internet connection and a productive online interview. The duration of each interview was approximately 1 h per person. In these interviews, each researcher was located in either Japan or Jakarta and asked participants different questions according to their research focus.

Through the interviews, we obtained a variety of interesting data including villager experiences, detailed economic circumstances, opinions on peatland use, and historical anecdotes as shown in relevant chapters. In order to understand the respondents' opinions and experiences of the online interviews, we designed a simple feedback questionnaire. Each of the 15 participants completed the questionnaire after their interview. As shown in Table 2 below, the questionnaire consisted of nine questions.

Table 2 Impression of the online interviews

		Not problematic/as usual	Problematic/ stressful
Q1.	Nervousness before interview	10	5
Q2.	Stress during online interview compared with face-to-face one	12	3
Q3.	Length of online interview	11	4
Q4.	Interview with plural interviewers	12	3
Q5.	Interviewer's expressions on the monitor	13	2
Q6.	Listening through loudspeaker	13	2
Q7.	Putting into words in online interview	14	1
Q8.	Additional online interview later	15	0
Q9.	Researcher's visit at the village	11	4

Interview results indicate that the majority of participants did not feel nervous before their interview. In addition, they did not experience more stress during online interviews compared to face-to-face interviews. In terms of the duration, the majority of participants did not feel this was a problem. Participants did not express any issues with having multiple interviewers either. The majority of participants did not indicate any problems with the interviewer's expressions (of words, attitude, and emotions on the monitor) or with listening to the audio through a loudspeaker. In terms of answering questions, most participants did not have any difficulty with explaining their thoughts. All participants were happy to attend an additional online interview, and most were willing to accept the researcher's visit from outside the district despite the pandemic.[1] This summary suggests that the online interviews went smoothly. This may be because the researchers had visited the village before (December 2018, August 2019, November 2019, and January and February 2020) and so the villagers knew the researchers beforehand.

Online interviews were not the ideal method to gather data due to difficulties such as lack of internet access alongside limited communication and interaction with respondents. In addition, as the researchers had not participated in the village life for a long time, it was often difficult to propose situational and timely questions to the participants. Nevertheless, based on our research, we noted some merits of online interviews. In terms of cost and time, online interviews were cheaper and more efficient compared to face-to-face interviews. In addition, online interviews enabled multiple researchers to propose different questions according to their focus. In turn, this allowed us to better understand issues with multidisciplinary perspectives. In sum, according to our research, online interviews can be a viable alternative to face-to-face interviews. When combined with other research methods such as fieldwork, online interviews contribute to a valid social science methodology in relation to the COVID-19 pandemic.

---

[1] Researcher's visits from outside the district during the pandemic were offered in the question on the following terms: "if the researchers are isolated for two weeks in a hotel in Pekanbaru before coming to the village."

References

Johanes (2020) Hari keempat PSBB di Pelalawan, begini kondisi aktivitas masyarakat dan tempat usaha yang beroperasi. Tribun Pekanbaru, 18 May. https://pekanbaru.tribunnews.com/2020/0 5/18/hari-keempat-psbb-di-pelalawan-begini-kondisi-aktivitas-masyarakat-dan-tempat-usaha-yang-beroperasi. Accessed 15 Aug 2020

Sehat Negeriku (2020) Disetujui menkes, 5 wilayah di Riau siap-siap terapkan PSBB. Sehat Negeriku, 13 May. https://sehatnegeriku.kemkes.go.id/baca/rilis-media/20200513/3733876/disetujui-menkes-5-wilayah-riau-siap-siap-terapkan-psbb/. Accessed 15 Aug 2020

Appendix 1: Questionnaire

House Code :

Interviewer Code :

Questionnaire Number :
(number interviewed)

A. Adress:	RT: _____ RW: _____ Subvillage: 1.Sepunjung 2. Malako Kecil 3. Sei Pebadaran
B. Gender	1. Man 2. Woman
C. Name of the selected respondent	_____
D. Relationship with the family head	1. The family head 2. Husband/wife of the family head 3. Others _____(no interviews allowed)
E. Name of the spouse of the selected respondent	 _____ 8. None
F. Number of families living in this house	

Questionnaire Control	Nama	Date	Signature
G. Interviewer			
H. Coordinator			

I. **Start time of the interview:**...........................

© The Author(s) 2023
M. Okamoto et al. (eds.), *Local Governance of Peatland Restoration in Riau, Indonesia*, Global Environmental Studies,
https://doi.org/10.1007/978-981-99-0902-5

1. Basic Information

1. **Gender** of the respondent:

   1. Male   2. Female

2. Age, date, and place of birth

A. Age	
	.... years old
B. Date, month, and year of birth	
	..................................; ....
C. Place of birth (District/City/Village/Town)	1. Rantau Baru Village
	2. Kiyap Jaya Village
	3. Other villages in the Pelalawan District
	Describe the name of the village ....
	4. Other village outside the Pelalawan District
	Describe the name of the District/City ....

3. What is your current marital status?

   1. Not married
   2. Married
   3. Divorce
   4. Widow/Widower

4. Age, date, and place of birth Husband/Wife Mother/Father

A. Age	
	.... years old
B. Date, month, and year of birth	
	.................................; ....
C. Place of birth (District/City/Village/Town)	1. Rantau Baru Village
	2. Kiyap Jaya Village
	3. Other villages in the Pelalawan District
	Describe the name of the village ....
	4. Other village outside the Pelalawan District
	Describe the name of the District/City ....

5. What **religion** do you follow?

    1. Islam   2. Catholic   3. Protestant

    4. Others, please specify: ……………………………

6. What is your **ethnicity** and your spouse's ethnicity?

    1. Malay   2. Javanese   3. Batak
    4. Minang   5. Chinese/Chinese Indonesian
    6. Nias   7. Others, please specify: …………………

A. Myself	……………………………
B. My husband/wife	……………………………

7. **SHOW THE CARD A.** If you and your husband/wife are following **the Adat Melayu Petalangan**, what is the sub-ethnicity of you and your husband/wife?

    1. Meliling   2. Melayu Datok Mudo   3. Melayu Datok Tuo
    4. Coastal Malay   5. Others, Please specify: ………   8. Don't know

A. Myself	……………………………
B. My husband/wife	……………………………

8. What is the main **language** you speak at home?

    1. Indonesian   2. Local language (Malay)   3. Javanese
    4. Batak   5. Minang   6. Others, please specify: …………………

9. **SHOW THE CARD B.** Do you strongly disagree **(STS),** disagree **(TS),** neutral **(N),** agree **(S),** or strongly agree **(SS)** with the statements below?

Statement	STS	TS	N	S	SS	Don't know
Men use Indonesian more often than women	1	2	3	4	5	8

10. How many years have you **lived** in this village?

    Write: ……………………………………………… years

11. How many **children** (alive) do you have now?

    …………… people

12.   How old is each child now? What is gender?

1.   .... years (L / P)   6.   .... years (L / P)

2.   .... years (L / P)   7.   .... years (L / P)

3.   .... years (L / P)   8.   .... years (L / P)

4.   .... years (L / P)   9.   .... years (L / P)

5.   .... years (L / P)   10. .... years (L / P)

13. How many people are living in this house?

A. Total/all (including myself)	.... people
B. My own children among cohabitants	.... people
C. My mother/father	.... people
D. My mother/father-in-law	.... people

14.   What is **the main job** and **side job** (if any) of yourself and your husband/wife?

	1. Main job	2. Side job
A. Myself	....	....
B. My husband/wife	....	....

15.   **SHOW THE CARD C.** Which job are you most proud of? (Choose any one)

1. Teacher 2. Doctor   3. Civil servant   4. Fisher   5. Farmer   6. Private employee
7. Lecture 8. Politician   9. Soldier 10. Police   11. Others, please specify: ....

2. Peatland Preservation

16.   Can you tell the difference between peatland and mineral soil?

1. Yes
2. No⇒ **skip to Question 23**

17. **SHOW THE CARD D.** Where did you get to know the word "peat" for the first time? (Choose any one)

   1. World of mouth
   2. From my parents
   3. From my husband
   4. From the extension workers
   5. From an oil palm plantation company
   6. From the village office
   7. From a farmer's group
   8. Mass media
   9. Others, please specify: ....
   10. Don't know

18. Do you have peatland?

   1. Yes
   2. No ⇒ **skip to Question 23**

19. If you own peatland, how have you used it in the last year?

   1. Farming/Gardening
   2. Fish roasting
   3. No used

20. What crops are planted on the peatland that you have?

....

21. Below are the various types of peatland utilization activities for farming/gardening that you do.

In your experience, which activities are usually carried out by women only, men only, or by both women and men?

Activities	Female	Male	Women and Male	Don't know
A. Clear the peatland	1	2	3	8
B. Apply fertilizer	1	2	3	8
C. Harvest the production	1	2	3	8
D. Sell the production	1	2	3	8

22. **SHOW THE CARD E.** Where did you learn how to cultivate peatland? (Choose more than one)

   1. From the ancestors (parents or grandparents)
   2. Husband/wife
   3. Government information
   4. NGO information
   5. Farmers' group
   6. Women farmers' group (KWT)
   7. Oil palm plantation company
   8. Mass media
   9. Others, please specify: ....
   10. Don't know

23. **SHOW THE CARD F** . How often do you participate in training or socialization on peatland conservation?

Activities	Once a month or more	Once in six months	Once a year	Every two year	Less than every two years	Not at all
	1 month 1×	6 month 1×	1 year 1×	2 year 1×	<2 years 1×	
Peatland ecosystem socialization/training	5	4	3	2	1	0

24. **SHOW THE CARD B.** Do you strongly disagree **(SD)**, disagree **(D)**, neutral **(N)**, agree **(A)**, or strongly agree **(SA)** with the statements below?

Statement	SD	D	N	A	SA	Don't know
A. One of the efforts to prevent peatland fires is to clear new land without burning.	1	2	3	4	5	8
B. To prevent forest and land fires, canal blockings and boreholes should be constructed in this village.	1	2	3	4	5	8
C. I care about the preservation of peatland in this village	1	2	3	4	5	8

25. **SHOW THE CARD G.** In the past year, how often have you received information about the construction and management of canal blocks and boreholes in this village?

Activities	Once a month or more	Once in two months	Once in six months	Once a year	Not at all
	1 month 1×	2 months 1×	6 month 1×	1 year 1×	
Received info on canal blocking and boreholes in this village	4	3	2	1	0

26. What is your assessment of the stage of the peat and its environment in Rantau Baru? Is it good enough, deteriorating or damaged?

1. Good enough
2. Deteriorating
3. Damaged

27. State the reason why you choose the answer above

....

28. **SHOW THE CARD H.** If there is a peatland management and conservation program implemented by the community in Rantau Baru and you can enjoy the benefits, and all the villagers share the costs together.

So, how much are you willing to pay for the program per month?

1. Less than Rp.9,999
2. Between Rp.10,000 to Rp.19,900
3. Between Rp.20,000 to Rp.29,900
4. Between Rp.30,000 to Rp.39,900
5. Between Rp.40,000 to Rp.49,900
6. More than Rp.50,000
7. Now willing

29. **SHOW THE CARD I.** How often do you receive information about the activities of the Fire Prevention Community Group (Masyarakat Peduli Api, MPA)?

Activities	Once a month or more	Once in six months	Once a year	Every two years	Less than every two years	Not at all
	1 month 1×	6 Months 1×	1 Year 1×	2 Years 1×	<2 Years 1×	
Receive information about the activities of MPA	5	4	3	2	1	0

30. SHOW THE CARD J. How often have you met with the leader or members of MPA in the past year?

	Every day	Once a week	Once a month	Once in six months	Less than once every six months	Never at all
	1 day 1×	1 Week 1×	1 Month 1×	6 Months 1×	< 6Months 1×	
MPA Chairman/ Member	5	4	3	2	1	0

31. **(For women only)**
SHOW THE CARD J . How often do you get involved in the activities of the following organizations/groups?

Organization/group	Every day	Once a week	Once a month	Once in six months	Less than once every six months	Not at all
	1 day 1×	1 Week 1×	1 Month 1×	6 Months 1×	<6 Months 1×	
A. PKK (Family Welfare Development)	5	4	3	2	1	0
B. NOT (Women Farmers Group)	5	4	3	2	1	0

3. Community Participation

32.   SHOW THE CARD K. In the last six years, how often have you attended the development planning meetings or musrenbang at each of the following levels? The event is held once a year.

Activities	Every year	Four or five times in six years	Every two years	One or two times in six years	Never
	6 years 6×	6 years 4,5×	6 years 3×	6 years 1,2×	6 years 0×
A. Musyawarah di tingkat dusun (Rembug RW/Dusun)	5	4	3	2	1
B. Musyawarah di tingkat desa (Musrenbang desa)	5	4	3	2	1

**33.   **The village office posts information on the use of village funds in several places. Have you seen (read) it in the past year?

1. Yes
2. No

34.   SHOW THE CARD J. How often have you met the following people in the past year?

Activities	Every day	Once a week	Once a month	Once in six months	Less than once every six months	Never at all
	1 day 1×	1 week 1×	1 month 1×	6 months 1×	<6 Months 1×	
A. Head of neighborhood organization or hamlet	5	4	3	2	1	0
B. Village secretary, village officials in general affairs and in finance (village officials)	5	4	3	2	1	0
C. Village head	5	4	3	2	1	0
D. Member of village consultative body (BPD)	5	4	3	2	1	0

4.   Fisher Activities

35.   In the past year, did you do any fishing (for sale or for your own consumption)?

1.   Yes
2.   No ⇒ **skip to Question 45**

36.   **SHOW THE CARD L.** Where do you fish on a daily basis? (you can choose more than one)

1. Kampar River
2. Canal created by companies
3. Karang Lake
4. Suwak Teluk Bederas
5. Boko-Boko River
6. Sepunjung Lake
7. Others, mention: ………………………….

37.   Circle the place where you catch fish.

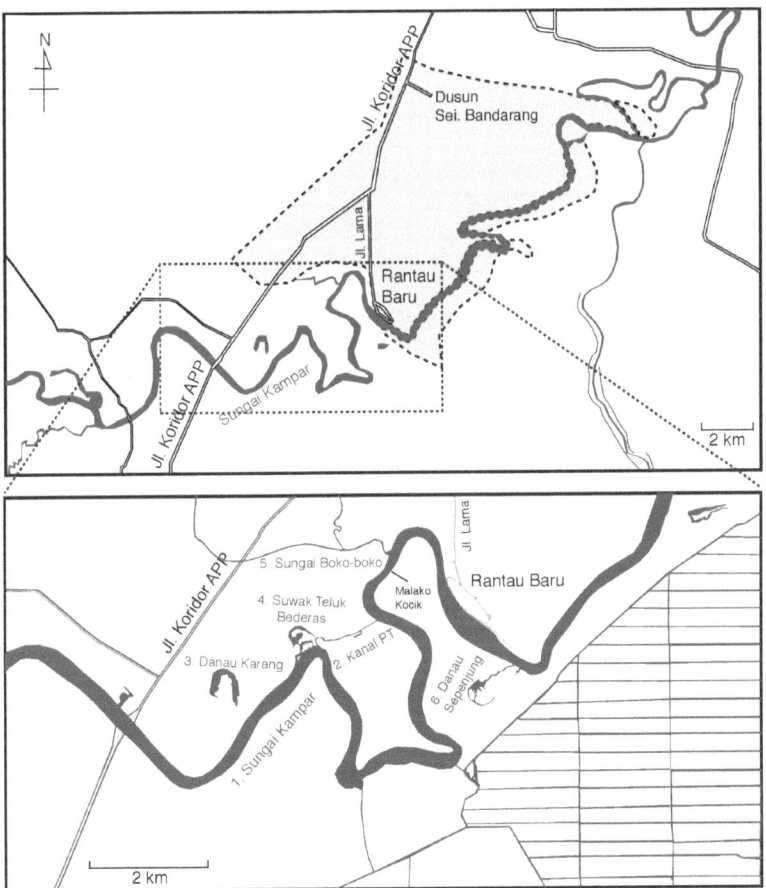

38.   SHOW THE CARD M. What equipment do you use to catch fish?

     1. Hand net
     2. Square net
     3. Cast net
     4. Gill net
     5. Trap shorter than 1 m (Small trap, < 1 m)
     6. Trap longer than 1 m (Large trap, > 1 m)
     7. Fixed fish net (Trap larger than 10 m)
     8. Long line fishing
     9. Others, please specify: ....

39.   Is the amount of fish caught more or less compared to 10 years ago?

     1. Far More than 10 years ago
     2. More than 10 years ago
     3. Same
     4. Less than 10 years ago
     5. Far less than 10 years ago

40.   Please write down the types of fish that you have caught in the last 1 year.

---

41.   Are there any fish species that were caught in the past, but no longer found in the last 5 years? If so, please list the name of the fish species. (may be more than one).

---

42.   Are there any fish species that did not exist before but are now found? If so, please write down the name of the fish species. (may be more than one).

---

43.   What is the approximate kilogram of fish that can be caught per day during the rainy season (September February) and the dry season (April-August)? Name the **top three types of fish** caught.

Fish	1. Rainy Season September-February		2. Dry Season April-Augustus	
	a. Tertinggi	**b.**Terendah	**a.**Tertinggi	**b.**Terendah
A.	kg	kg	kg	kg
B.	kg	kg	kg	kg
C.	kg	kg	kg	kg

44.   The following is a profile of the daily activities of men and women in fishing families. Which activities are usually done by women only, men only, or by both women and men ?

Activities	Female	Male	Female and Male	Don't know
A. Catch fish in rivers/lakes	1	2	3	8
B. Process fish	1	2	3	8
C. Sell fish	1	2	3	8

45. What is your assessment of the fish catch in the rivers and lakes in Rantau Baru? Is it good enough or is it declining?

   1. Good enough
   2. Start decreasing
   3. Decreasing

46. State the reason why you chose the answer above.

---

47. **SHOW THE CARD N.** If there is a river and lake conservation program implemented by the community in Rantau Baru for the sustainability of fish catches and you can enjoy the results and all the village members bear the cost together.

   So, how much are you willing to pay for the program per month?

   1. Less than Rp.9,999
   2. Between Rp.10,000 hingga Rp.19,900
   3. Between Rp.20,000 hingga Rp.29,900
   4. Between Rp.30,000 hingga Rp.39,900
   5. Between Rp.40,000 hingga Rp.49,900
   6. More than Rp.50,000
   7. Not willing

48. Do you own your own rowing boat (sampan) or motorized boat (pompon)? If you own, how many?

   1. Small Flat Bottom Rowing Boat   (A......................buah)
   2. Small Flat Bottom Motorized Boat   (B......................buah)
   3. Large Wooden Motorized Boat   (C......................buah)
   4. Does not belong ⇒ **skip to Question 50**

49. In total, how many rowing boats and motorized boats do you own?

   ....Boats

50. Did you participate in river auction activities this year?

   1. Yes
   2. No ⇒ **skip to Question 53**

51. Did you participate in the auction as a group or individually?

   1. Group
   2. Individual

52. Did you win the auction this year? If yes, please state the name of the river/river tributary that you won.

   Write: ....

53. Do you process smoked/salted fish? If yes, where are the products sold?

   1. Market
   2. Collector/agent
   3. Household/consumption ⇒ **skip to Question 55**
   4. Not processing fish ⇒ **skip to Question 55**

54.   In the past three months, on average, how much is the price of processed fish per kilogram? Name the top three types of processed fish sold.

Type of Fish	Price/kg
A. Fish	Rp. / Kg
B. Fish	Rp. / Kg
C. Fish	Rp. / Kg

5. Land Tenure

55.   **SHOW THE CARD O.** Have you or your family ever received land (including land that has been sold) or do you or your family currently own land 1) around Old Road, 2) around New Road, 3) next to Malako Kucik in Sei. Kampar, 4) around your house, 5) other places/villages? (may select more than one)

  1. Around old road ⇒ **skip to Question 56**
  2. Around new road ⇒ **skip to Question 67**
  3. Next to Malako Kucik in Sei. Kampar ⇒ **skip to Qustion 78**
  4. Around the house ⇒ **skip to Question 89**
  5. Other places/villages, please specify: ....
                                              ⇒ **skip to Question 100**
  6. Never received/owned land ⇒ **skip to Question 111**

<1. Old road>

56.   How much land is that? (total around the old road)

  Describe: .... ha

57.   When was the land acquired? (you can name more than one)

  Describe: ....

58.   **SHOW THE CARD P.** Whose name is recorded as the owner of the land on the land certificate? (you can choose more than one) (if it has been sold,what is the name on the land certificate that youonce had)

  1. Myself
  2. My Husband/Wife
  3. Biological father
  4. Biological mother
  5. Father-in-law
  6. Mother-in-law
  7. Son
  8. Daughter
  9. Grandson
  10. Granddaughter
  11. Others, please specify: ....
  12. Have Never had the certificate
  88. Don't know

59.   How was the land acquired?

1. Grant
2. Inheritance
3. Purchase
4. Pawn
5. Others, please specify: ....

60.   What type of soil is it?

1. Peat
2. Mineral soil
3. Mix -more peat
4. Mix -more mineral soil

61.   Has the land ever been burned? If so, around what year? You can mention more than one if it burned several times.

1. Yes. Describe: year ....,...............................,.........................,..............
2. Never

62.   Has the land been sold? If so, around what year was it sold? How many hectares were sold at that time?

1.   Yes. Describe: year **A**................**(B.**......... ha)
       ⇒ **If all the land has been sold, skip to Question 67**
2.   No

63.   What is the land used for? (You can answer more than one)

1. Palm
2. Shrubs
3. Forest
4. Don't use it for anything
5. Others, please specify: ....

64.   If you are to sell the land to a villager with the same ethnicity, how would the sale price be?

1. Low price
2. Regular price
3. Expensive price
4. Unwilling to sell, even if the selling price is expensive

65.   If you are to sell the land to a person from this village with a differentethnicity, how would the sale price be?

1. Low price
2. Regular price
3. Expensive price
4. Unwilling to sell, even if the selling price is expensive

66.   If you are to sell the land to someone from outside the village, how would the sale price be ?

1. Low price
2. Regular price
3. Expensive price
4. Unwilling to sell, even if the selling price is expensive

<2. New Road>

67.   How much land is that? (total around the new road)

Describe: .... ha

68.   When did you get the land? (you can mention more than one)

Descibe: year ....

69.   **SHOW THE CARD P.** What name is recorded as the owner of the land on the land certificate? (may choose more than one) (if it has been sold, what is the name on the land certificate that you have owned)

1. Myself
2. My husband/Wife
3. Biological father
4. Biological mother
5. Father-in-law
6. Mother-in-law
7. Son
8. Daughter
9. Grandson
10. Granddaughter
11. Others, please specify: ....
12. Never had that letter
88. Don't know

70.   How was the land acquired?

1. Grant
2. Inheritance
3. Purchase
4. Pawn
5. Others, please specify: ....

71.   What type of soil is it?

1. Peat
2. Mineral soil
3. Mix -more peat
4. Mix -more mineral soil

72.   Has the land ever been burned? If so, around what year? You can mention more than one if it burned several times.

1. Yes. Describe: year ....,...........................,...........................,...............
2. No role

73.   Has the land been sold? If so, around what year was it sold? How many hectares were sold at that time?

1. Yes. Describe: year A.....................(B.....................ha) ⇒ **If all the land is sold, skip to**
                                                                        Question 78
2. No

74. What is the land used for? (You can answer more than one)
   1. Palm
   2. Shrubs
   3. Forest
   4. No use for anything
   5. Others, please specify: ....

75. If you are to sell the land to a villager with the same ethnicity, how would the sale price be?

   1. Low price
   2. Regular price
   3. High price
   4. Unwilling to sell, even if the selling price is expensive

76. If you are to sell the land to a person from this village with a differentethnicity, how would the sale price be?

   1. Low price
   2. Regular price
   3. High price
   4. Unwilling to sell, even if the selling price is expensive

77. If you sold the land to someone from outside the village, how would the sale price be?

   1. Low price
   2. Regular price
   3. High price
   4. Unwilling to sell, even if the selling price is expensive

<3. Next to Malako Kucik in Sei. Kampar >
78. How much land is that? (total land next to Malako Kucik)

   Describe: .... ha

79. When did you get the land? (you may mention more than one)

   Describe: year ....

80. **SHOW THE CARD Q.** Whose name is recorded as the owner of the land on the land certificate? (You may choose more than one) (If it has been sold, what   is the name on the land certificate that you   have owned)

   1. Myself
   2. My husband/Wife
   3. Biological father
   4. Biological mother
   5. Father-in-law
   6. Mother-in-law
   7. Son
   8. Daughter
   9. Grandson
   10. Granddaughter
   11. Others, please specify: ....
   12. Never had that letter
   88. Don't know

81.   How was the land acquired?

1. Grant
2. Inheritance
3. Purchase
4. Pawn
5. Others, please specify: ....

82.   What type of soil is it?

1. Peat
2. Mineral soil
3. Mix -more peat
4. Mix -more mineral soil

83.   Has the land ever been burned? If so, around what year? You can mention more than one if it burned several times.

1. Yes. Describe: year......................,...........................,.........................,...............
2. No role

84.   Has the land been sold? If so, around what year was it sold? How many hectares were sold at that time?

1.   Yes. Describe: year **A**......................(**B**......................ha)
   ⇒ **When all the land is sold, skip to Question 89**
2. No

85.   What is the land used for? (You can answer more than one)

1. Palm
2. Shrubs
3. Forest
4. No use for anything
5. Others, please specify: ....

86.   If you are to sell the land to a villager with the same ethnicity, how would the sale price be?

1. Low price
2. Regular price
3. Expensive price
4. Unwilling to sell, even if the selling price is expensive

87.   If you are to sell the land to a person from this village with a differentethnicity, how would the sale price be?

1. Low price
2. Regular price
3. Expensive price
4. Unwilling to sell, even if the selling price is expensive

88.   If you are to sell the land to someone from outside the village, how would the sale price be?

1. Low price
2. Regular price
3. Expensive price
4. Unwilling to sell, even if the selling price is expensive

<4. Around the House>

89.　How big is the land?

　　Describe: ………………….m × ……………….m   = ………………….m²

90.　When did you get the land? (You may mention more than one)

　　Desribe: year …………………………………………………

91.　**SHOW THE CARD P.** Whose name is recorded as the owner of the land on the land certificate? (You may choose more than one) (If it has been sold, what   is the name on the land certificate that you have owned)

　　1. Myself
　　2. My husband/Wife
　　3. Biological father
　　4. Biological mother
　　5. Father-in-law
　　6. Mother-in-law
　　7. Son
　　8. Daughter
　　9. Grandson
　　10. Granddaughter
　　11. Others, please specify: ………………………
　　12. Never had that letter
　　88. Don't know

92.　How was the land acquired?

　　1. Grant
　　2. Inheritance
　　3. Purchase
　　4. Pawn
　　5. Others, please specify: …………………………………………………………………………

93.　What type of soil is it?

　　1. Peat
　　2. Mineral soil
　　3. Mix -more peat
　　4. Mix -more mineral soil

94.　Has the land ever been burned? If yes, around what year? You could mention more than one if it burned several times.

　　1. Yes. Describe: year……………………,……………………….,…………………,……………
　　2. No role

95.　Has the land been sold? If yes, around what year was the land sold? How many hectares were sold at that time?

　　1. Yes. Describe: year A………………….(**B**………………….ha) ⇒ **When all the land is sold, skip to Question 100**
　　2. No

96.　What is the land used for? (You can answer more than one)

　　1. Palm
　　2. Shrubs
　　3. Forest
　　4. No use for anything
　　5. Others, please specify: …………………………………………………………………………

97. If you were to sell the land to a villager of the same ethnicity, how would the sale price be?

1. Low price
2. Regular price
3. High price
4. Unwilling to sell, even if the selling price is high

98. How would the sale price be if you sell the land to a person from this village with a different ethnicity?

1. Low price
2. Regular price
3. High price
4. Unwilling to sell, even if the selling price is high

99. How would the sale price be if you sell the land to someone from outside the village?

1. Low price
2. Regular price
3. High price
4. Unwilling to sell, even if the selling price is high

<5. Other places/villages>
100. How much land is that? (total)

Describe: .... ha

101. When did you get the land? (You can mention more than one)

Describe: year ....

102. **SHOW THE CARD P.** Whose name is recorded as the owner of the land on the land certificate? (You may choose more than one) (If it has been sold, what   is the name on the land certificate that you have owned)

1. Myself
2. My husband/Wife
3. Biological father
4. Biological mother
5. Father-in-law
6. Mother-in-law
7. Son
8. Daughter
9. Grandson
10. Granddaughter
11. Others, please specify: ....
12. Never had that letter
88. Don't know

103. How was the land acquired?

1. Grant
2. Inheritance
3. Purchase
4. Pawn
5. Others, please specify: ....

104. What type of soil is it?

    1. Peat
    2. Mineral soil
    3. Mixed -more peat
    4. Mixed -more mineral soil

105. Has the land ever been burned? If so, around what year? You c ould mention more than one if it burned several times.

    1. Yes. Describe: year ....,........................,.......................,...............
    2. No role

106. Has the land been sold? If yes, around what year was it sold? How many hectares were sold at that time?

    1. Yes. Describe: year **A**.....................(**B**.....................ha) ⇒ **When all the land is sold, skip to Question 111**
    2. No

107. What is the land used for? (You can answer more than one)

    1. Palm
    2. Shrubs
    3. Forest
    4. Don't use it for anything
    5. Others, please specify: ....

108. If you were to sell the land to a villager of the same ethnicity, how would the sale price be?

    1. Low price
    2. Regular price
    3. High price
    4. Unwilling to sell, even if the selling price is high

109. How would the sale price be if you sell the land to a person from this village with a different ethnicity?

    1. Low price
    2. Regular price
    3. High price
    4. Unwilling to sell, even if the selling price is high

110. How would the sale price be if you sell the land to someone from outside the village ?

    1. Low price
    2. Regular price
    3. High price
    4. Unwilling to sell, even if the selling price is high

---

6. Daily Life

111. Do you have your honey tree (*pohon sialang*)? If yes, how many trees?

    1. Yes **A**.. trees
    2. Yes, but can't count
    3. Does not belong ⇒ **skip to Question 113**

112.   How long have you been managing *sialang* trees?

Describe: .... years

113.   In total, **how many bedrooms** are **there** in your house (any rooms used by anyone as a place to sleep regularly)?

Descrive: .... rooms

114.   Does your house have a bathroom?

    1. Have a bathroom in my house
    2. Use shared restrooms
    3. Use the river
    4. Others, please specify: ....

115.   What is the **ownership status of** your **house**?

    1. Owned/family owned
    2. Contact
    3. Others, please specify: ....

116.   **SHOW THE CARD Q.** Where do you get water for drinking and cooking in this house?

    1. Bottled water
    2. Well/groundwater (manual or machine pump)
    3. Spring source
    4. Rainwater
    5. River water
    6. Borehole well
    7. Others, please specify: ....

A. Drink	....
B. Cook	....

117.   How many motorcycles do you own?

Describe: .... motorcycles

118.   What is the **last formal education** for you and the people below? Including the current school or the dropout school. (Choose only one answer)

    1. Elementary School
    2. Junior High School
    3. High School
    4. Diploma
    5. Undergraduate School
    6. Graduate School
    7. Neve went to school

A. Myself	....
B. My husband/wife	....
C. Biological father	....
D. Biological mother	....

119. At what age did you experience the following?

Event	Age	Not applicable
A. Completed school (including the dropout)	....	88
B. Worked for the First Time	....	88
C. Married for the First Time	....	88 _(never married)_
D. Had the first child	....	88 _(no child)_
E. Separated from my spouse (living divorce, death divorce)	....	88

120. The following are types of activities in daily life.
In your experience, which activities are usually carried out by women only, men only, or by both women and men?

Activities	Female	Male	Women and Male	Don't know
A. Parenting	1	2	3	8
B. Attend RT/RW/Village meeting	1	2	3	8

121. In your experience, who is the most decisive in managing money in the family?

    1. Husband
    2. Wife
    3. Husband and wife
    8. Don't know/ Can't use

122. What is the **average expenditure** in this house for **Electricity** per month?

    Rp ....per month

123. Overall, what is the **average expenditure of** your family in this house per month?

    Rp ....per month

<If the respondent has difficulty answering this question, SHOW THE CARD R.>

Including:	Not including:
• Daily meals • Purchase of washing and cleaning products • Child's tuition • Electricity • Water • Cigarettes • Salary for housemaid • Gasoline • Monthly house rent • Lease of other goods • Other routine expenses	• Payment for Mutual financing group (Arisan) • Investment savings • Entertainment • Watching movies • Recreation • Item installment • Annual house contract • Clothing • Other non-routine expensive

THANK YOU VERY MUCH FOR
COOPERATION THAT YOU PROVIDE

Interview completed at: ....

Appendix 2: Photographs

The confluence of the Kampar and Bokol Bokol rivers. The Bokol Bokol is a tributary that flows from the peatland into the Kampar River near Rantau Baru Village

A fisherman setting a trap in a shallow area of the Kampar River

Rantau Baru Village during the rainy season

The flooded main road of Rantau Baru Village

Smoking fish for sale in Rantau Baru Village

Large fish traps (*Pengilar*) in the making

Community members learning how to use a GPS Logger

Discussion of the village and customary community borders

Setting up a drone for the participatory mapping in Rantau Baru Village

Online interview with Rantau Baru villagers

Belida (*Chitala lopis*)

Pantau/Tabingal (*Rasbora* sp.)

Selais (*Kryptopterus* sp.)

Tapa (*Wallago leeri*)

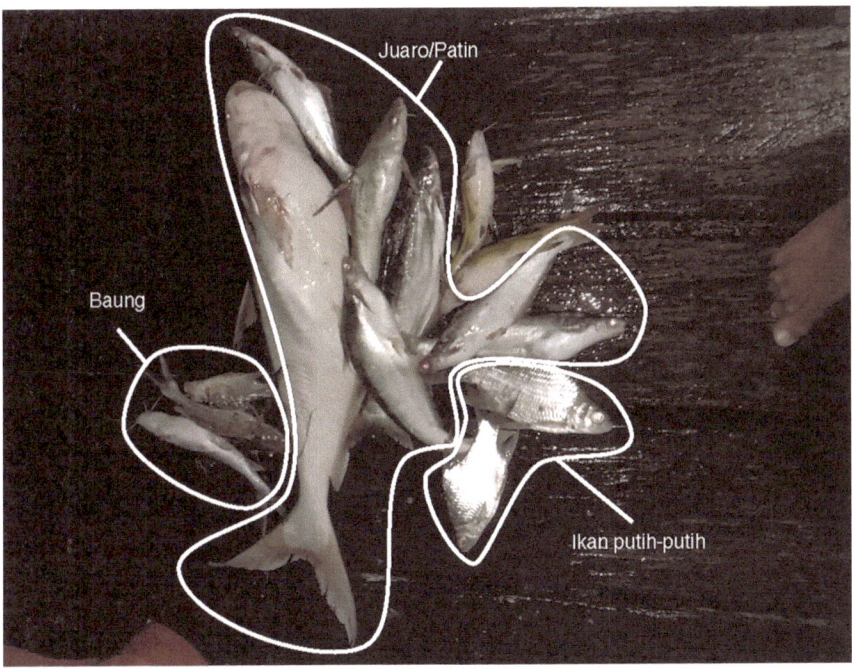

Baung (*Mystus* sp.), Juaro/Patin (*Pangasius* sp.) and Ikan Putih-putih (*Puntioplites waandersi*)